Nano-inspired Biosen
Protein Assay with Cli
Applications

Nano-inspired Biosensors for Protein Assay with Clinical Applications

Edited by

Genxi Li
State Key Laboratory of Pharmaceutical Biotechnology,
Department of Biochemistry, Nanjing University,
Nanjing, P. R. China
Laboratory of Biosensing Technology, School of Life Sciences,
Shanghai University, Shanghai, P. R. China

ELSEVIER

Elsevier
Radarweg 29, PO Box 211, 1000 AE Amsterdam, Netherlands
The Boulevard, Langford Lane, Kidlington, Oxford OX5 1GB, United Kingdom
50 Hampshire Street, 5th Floor, Cambridge, MA 02139, United States

British Library Cataloguing-in-Publication Data
A catalogue record for this book is available from the British Library

Library of Congress Cataloging-in-Publication Data
A catalog record for this book is available from the Library of Congress

ISBN: 978-0-12-815053-5

For Information on all Elsevier publications
visit our website at https://www.elsevier.com/books-and-journals

Publisher: Susan Dennis
Acquisition Editor: Kathryn Morrissey
Editorial Project Manager: Tasha Frank
Production Project Manager: Bharatwaj Varatharajan
Cover Designer: Mark Rogers

Typeset by MPS Limited, Chennai, India

Contents

List of Contributors

Ya Cao Center for Molecular Recognition and Biosensing, School of Life Sciences, Shanghai University, Shanghai, P. R. China

Guifang Chen Center for Molecular Recognition and Biosensing, School of Life Sciences, Shanghai University, Shanghai, P. R. China

Chang Feng State Key Laboratory of Pharmaceutical Biotechnology, Department of Biochemistry, Nanjing University, Nanjing, P. R. China

Tao Gao Center for Molecular Recognition and Biosensing, School of Life Sciences, Shanghai University, Shanghai, P. R. China

Chao Li State Key Laboratory of Pharmaceutical Biotechnology, Department of Biochemistry, Nanjing University, Nanjing, P. R. China

Hai Shi State Key Laboratory of Pharmaceutical Biotechnology, Department of Biochemistry, Nanjing University, Nanjing, P. R. China

Liu Shi State Key Laboratory of Pharmaceutical Biotechnology, Department of Biochemistry, Nanjing University, Nanjing, P. R. China

Lei Wang State Key Laboratory of Pharmaceutical Biotechnology, Department of Biochemistry, Nanjing University, Nanjing, P. R. China

Shuai Wu State Key Laboratory of Pharmaceutical Biotechnology, Department of Biochemistry, Nanjing University, Nanjing, P. R. China

Jingjing Xu Center for Molecular Recognition and Biosensing, School of Life Sciences, Shanghai University, Shanghai, P. R. China

Juan Zhang Center for Molecular Recognition and Biosensing, School of Life Sciences, Shanghai University, Shanghai, P. R. China

Jing Zhao Center for Molecular Recognition and Biosensing, School of Life Sciences, Shanghai University, Shanghai, P.R. China

Ji Zheng State Key Laboratory of Pharmaceutical Biotechnology, Department of Biochemistry, Nanjing University, Nanjing, P. R. China

Nandi Zhou School of Biotechnology and the Key Laboratory of Industrial Biotechnology, Ministry of Education, Jiangnan University, Wuxi, P. R. China

Xiaoli Zhu Center for Molecular Recognition and Biosensing, School of Life Sciences, Shanghai University, Shanghai, P. R. China

Preface

Biosensors are developing so rapidly, even faster than expected. One main reason is the employment of various kinds of nanomaterials and the assemblies of many types of nanostructures for the fabrication of biosensors. Therefore, the interaction between the recognition element and the analyte can be more favorable, the sensing layers can be more easily designed, and the measureable signal can be amplified, etc. While more and more biosensors are proposed, they must also perform with satisfactory sensitivity, selectivity, specificity, and accuracy. In the meantime, the fabricated biosensors have demonstrated their clinical applications. For instance, with the development of proteomic technologies, a lot of disease marker proteins have been discovered for many kinds of diseases, so detection methods are required to be developed towards the assay of the newly discovered disease markers due to the high importance and potential application in the diagnoses of these diseases. Meanwhile, the proposed methods are expected to be simple, rapid, sensitive, and cost-effective, thus they can be feasibly employed for the diagnoses and for the evaluation of the treatment of these diseases. Among the methods, biosensor based techniques may live up to the expectations, and more and more biosensors reported in these years have shown their practical or potential applications in clinical diagnosis and the evaluation of the disease treatment. Remarkable progress has been made over the years on the design and fabrication of sensing systems for disease marker detection with clinical applications. So, it is highly required to write a book that is focused on "nano-inspired biosensors for protein assay with clinical applications." On the one hand, colleagues, especially the young scientists and the graduate students, in the research area of analytical chemistry and biosensors can quickly know the recent advance of protein analysis and the new methodology in designing new kinds of biosensors. On the other hand, for those colleagues from the area of nanotechnology or biosensors, it may provide an opportunity for them to know the possible application of their research results in medical science. Besides, for those readers from the area of diagnosis, this book may provide an opportunity for them to know the frontier of the diagnostic methodology. I hope we have prepared a handbook for the young scientists and the graduate students in the research fields of analytical chemistry, biosensors, biotechnology, and nanotechnology. This book may also provide valuable information for the students majoring in material science, molecular science, biomedical engineering, basic and clinical medicine, etc.

Genxi Li

Acknowledgments

Firstly, I would like to thank George Knott, an Acquisitions Editor from Elsevier, for his invitation to write this book. When George asked me whether I would be interested in writing a book for Elsevier, I gave a positive answer without hesitation. One reason is that I noticed it was highly required to write a book on biosensors. Although there have been several books published on biosensors, it is necessary to present another one that is focused on the employment of various kinds of nanomaterials and the assemblies of many types of nanostructures for the fabrication of biosensors. On the one hand, nano-inspired biosensors are developing so rapidly. On the other, more and more biosensors are proposed towards the assay of disease marker proteins and many sensors have demonstrated their clinical applications. So, an in-time summary is urgently needed. Another reason is that I have written lots of contributions to books about biosensors, such as "Encyclopedia of Sensors," "Nanomaterials for Biosensors," "Biosensors and Molecular Technologies for Cancer Diagnostics," "Electrochemical Analysis of Proteins and Cells," "Engineering in Translational Medicine," etc.; however, these books were not published by Elsevier, although I have published so many research papers with Elsevier. Therefore, the kind invitation by George Knott provides me a good opportunity to publish a book with Elsevier, which is really appreciated. Secondly, I would like to thank Kathryn Morrissey, who is also an Acquisitions Editor from Elsevier, for her kind help and encouragement. Although I have great interest to write a book for Elsevier, I have also realized that it requires extensive work to write a book. Moreover, besides the daily work on research and teaching, I am so busy with so many other obligations. So, I had planned to give up the plan to write this book. Without the kind help from Katy, this book would not have appeared. Thirdly, I would like to thank Tasha Frank, a Senior Editorial Project Manager from Elsevier, for her kind help and support. She is really an excellent EPM. Without her help, everything related to this book would not have gone forward satisfactorily and successfully. Finally, I would like to thank my coworkers. In the last several months, preparation of the manuscript for this book has been the first priority in my labs both in Nanjing and Shanghai. Without their involvement, this great job would not have been completed in due course. Although the names of some coworkers have been listed as coauthors in the chapters, their help should be still appreciated. Certainly, special thanks should go to my coworkers whose names are not listed as coauthors. Please allow me to say my thanks to them, they are Wenxin Chai, Hong Chen, Huinan Chen, Tianshu Chen, Tingjun Chen, Chengjie Duan, Yiwei Han, Yunfei Liu, Jianyang Lu, Dongsheng Mao, Chaoli Mu, Yanxia Wang, Lan Xue, and Yi Yang.

Genxi Li

Introduction

Jing Zhao and Guifang Chen
Center for Molecular Recognition and Biosensing, School of Life Sciences, Shanghai University, Shanghai, P. R. China

I.1 THE PRINCIPLE OF BIOSENSOR

With the tremendous development in biochemistry and molecular biology, a bioassay that provides methods for accurate and efficient measurement of specific analytes in a biological sample has attracted increasing attention. Based on the biosensing principle, a new type of device has been developed that can translate the amount of target molecules into a quantifiable signal to detect analytes for different purposes, such as drug analysis, disease diagnosis, and biomolecule quantification. The device is called a biosensor, which reports the quantifiable, exclusive, and specific signal for biomolecular interactions (Perumal and Hashim, 2014).

In 1962, Clark and Lyons reported the first biosensor by immobilizing a glucose oxidase on the surface of an amperometric oxygen electrode with the assistance of a semipermeable dialysis membrane, which was used for the direct quantification of glucose in a biological sample. According to the definition from IUPAC, a biosensor is a self-contained integrated device based on specific biochemical reactions, which is regulated by isolated biological macromolecules (e.g., enzymes, immunosystems, tissues, organelles, or whole cells) and can detect chemical compounds using different signal outputs (e.g., electrical, thermal, or optical signals). A biosensor is one subtype of chemical sensors, and is composed of two basic elements as a receptor for chemical recognition and a physicochemical transducer for signal reporting (Ali et al., 2017; Mohanty and Kougianos, 2006). A receptor is a recognition system to identify specific molecules based on various intermolecular interactions, thereby ensuring a high degree of selectivity for analyte identification; a transducer, which is also known as a detector or a sensor, transfers interaction information from recognition events to a detectable signal, thereby ensuring a high degree of sensitivity for analyte measurements.

I.1.1 Bioreceptor

The receptor of biosensor, known as a bioreceptor, usually recognizes the target analyte based on biomolecular interactions by making use of various biological molecules (e.g., protein, DNA, cell, tissue) as recognition elements, which is the difference between biosensor and other types of chemical sensors. The biological recognition element is an essential component in the fabrication of a biosensor, which is able to sense both biological and nonbiological targets in a given sample. According to the principle of biomolecular interaction, bioreceptors are generally divided into two categories: biocatalytic recognition elements and bioaffinity recognition elements. Biocatalytic elements rely on specific recognition of a catalyst and substrate that is described as a "lock-and-key" system, while bioaffinity elements make use of the bioaffinity interaction that is naturally observed in the biological system.

I.1.1.1 Enzyme

An enzyme is one typical biocatalytic recognition element, which has been extensively studied and used in biosensor fabrication (Wilson and Hu, 2000). An enzyme is a macromolecular biological catalyst, which catalyzes the conversion of a specific substrate to a new product. An enzyme is quite specific to its substrate for the complementary binding site. The chemoselective, regioselective, and stereospecific interaction with substrates ensures high selectivity of enzyme as a bioreceptor. Specifically, the chemical reaction between an analyte and an enzyme is based on the equation:

$$S + S' \xrightarrow{E} P + P'$$

where S and S' are the substrates of an enzyme, which are also the analytes; while P and P' are the products after enzymatic catalysis. Sensitive and selective monitoring of analyte S is usually realized through the measurement of quantitative changes of another substrate S' or the generation of reaction product P or P', which is also achieved by monitoring the redox state of an active center of the enzyme or direct electron transfer between enzyme and the transducer. A cascade reaction using multiple enzymes promotes extensive use of enzyme-based receptors, in which a product of the first enzymatic reaction is a substrate of another enzyme for different detection purposes (Fu et al., 2012; Wang et al., 2009).

I.1.1.2 Antibody–Antigen

An immunosensor is a good example of a biosensor using a bioaffinity recognition element as a receptor, and it is based on the antigen–antibody interaction (Conroy et al., 2009). An antibody is a Y-shaped immunoglobulin, which is generated from B cells of the adaptive immune system of an organism. Each antibody contains a particular epitope to allow recognition and binding of the antigen with high-affinity and selectivity, which is known as a target analyte. In this sense, identification of the analyte is realized based on its interaction with the antibody using an immobilized biocomplexing agent as a bioreceptor, while antigen–antibody complexion could be monitored by combining with different transducers after reaching a binding equilibrium (Kokkinos et al., 2016; Mauriz et al., 2016; Yin et al., 2010). However, the antibody suffers from several intrinsic limitations in application. For example, preparation of the antibody is very complicated and relatively expensive; an antibody is not stable and is easily disrupted by the external environment and surface modification; an antibody with high immunogenicity does not easily penetrate tissues for in vivo studying.

I.1.1.3 Cell and Tissue

A cell is the fundamental unit of living organisms with dimensions between 1 and 100 μm, which is also recognized as an independent function and structure unit in vivo. A cell is surrounded by a plasma membrane, which contains several organelles and various biological molecules inside, such as electrolytes, proteins, and nucleic acids. Tissue is composed of several cells with similar properties, and thus performs a specific function in vivo. In biology, a tissue is a state of being between a cell and an organ. No matter whether inside the cell or in the tissue, a lot of living activities are taking place constantly. Both enzymes and receptor proteins maintain their native structure and function to the greatest extent. Therefore, cells and tissues as aggregations of different enzymes, antibodies, and receptors are directly used as a natural source of recognition element for bioreceptors (Gui et al., 2017).

I.1.1.4 DNA and Aptamer

DNA is the most popular bioreceptor for the fabrication of a biosensor in the recent decades (Dhiman et al., 2017; Saidur et al., 2017; Ye et al., 2018). DNA hybridization is a natural bioaffinity system in organisms. Two or more complementary nucleic acid strands are intertwined with each other through noncovalent and sequence-specific interactions in a thermodynamic manner. DNA hybridization is highly selective based on hydrogen bonds between A and T or G and C. Probe DNA hybridizes with target DNA, the analyte, according to precise Watson–Crick base pairing. Meanwhile, the introduction of a DNA-assisted signal amplification strategy greatly improves the signal production, acquisition, and amplification, thereby promoting development of biosensors using DNA as a bioaffinity element.

In 1990, the Gold lab and Szostak lab reported the discovery of RNA ligand for specific binding to target molecules, separately (Ellington and Szostak, 1990; Turek and Gold, 1990). The nucleic acid ligand was then named as an aptamer by Szostak, which originated from Latin word "*apto*" meaning "to fit." Two years later, DNA ligand was screened as another type of aptamer, which overcame the shortcomings of unstable RNA (Bock et al., 1992). Since then, aptamer, especially DNA aptamer, replaces antibody as an appealing bioreceptor. Aptamer exhibits the comparable binding affinity and selectivity to antibodies, and also displays several advantages beyond antibodies, including high thermostability, low immunogenicity, easy to undertake large-scale synthesis, and modification. Unlike in vivo preparation of an antibody, an aptamer is screened from a large random DNA or RNA library through in vitro selection process SELEX (systematic evolution of ligands by exponential enrichment). Recently, improved SELEX of cell-SELEX and tissue-SELEX have been developed in order to maintain the native folding status of protein as that in vivo (Mi et al., 2010; Sefah et al., 2010).

A wide variety of molecules have been demonstrated as targets of aptamer from small molecules and proteins to whole cells. Aptamer-based biosensors, known as aptasensors, are realized in different ways according to the structural properties of the binding aptamer (Dhiman et al., 2017; Lan et al., 2017b; Zhou et al., 2014). First, the aptamer is able to bind with a target and thus changes the external conditions (e.g., the mass, steric hindrance, and refractive index) near the transducer surface upon a single binding event. In this model, an aptamer interacts with the target molecule and forms a composite structure similar to an antibody−antigen complex. Second, binding with the target molecule induces conformational changes of an aptamer, and conformational change reduces the distance between the electrochemical tag and electrode surface or fluorophore and quencher, thereby inducing responses of electrochemical or fluorescent signal. Third, the target molecule and complementary DNA compete to bind with an aptamer, and arouse signal changes from conformational switching induced by strand displacement. Compared to conformational changes upon target binding, conformational switching from strand displacement induces much greater changes in distances, thereby increasing signal response and reducing background interference. Fourth, redesign of aptamers is employed to develop biosensors due to their high ability for synthesis and modification. On one hand, a whole aptamer could be divided into two separate parts, while the split aptamers could collaborate to work as a whole aptamer when binding with a target. On the other hand, different aptamers could be combined together through synthesis of a long nucleic acid strand, in which aptamers work to identify different target, simultaneously.

I.1.1.5 Small Molecule Ligand

Binding of a small molecule to a protein is involved in several important physiological processes in living organisms (Scott et al., 2016). The most well-known example is reversible binding of oxygen and hemoglobin, which plays a critical role in oxygen transportation through the circulation system.

Binding of oxygen and hemoglobin stabilizes the mutual conformational adaption for transportation purposes. Besides the example, small molecule–protein interactions are widely found at different stages of life based on conformational accommodation, including enzyme–inhibitor interaction, enzyme–substrate interaction, and ligand–receptor interaction. Small molecule–protein interaction is a stable and high-affinity binding like antibody–antigen, while small molecules benefit from low synthesis cost and increased chemical stability, thereby providing new insight for bioreceptors.

One common use of small molecules is to detect enzymatic activity, in which a small molecule is a native or artificial substrate or inhibitor of a target enzyme (Cao et al., 2010; Su et al., 2016). This detection is based on the native function of enzymes in biochemical processes. The discovery of a small molecule-ligand that binds to the target receptor with high affinity and satisfactory specificity not only provides the opportunity for revealing the function of protein in vivo, but also offers more choice for molecular recognition and drug discovery. In this sense, the native small molecule ligand is becoming a new type of recognition element in biosensing (Kubota and Hamachi, 2015; Low et al., 2008; Pode et al., 2017). For example, native ligand of folate receptor (FR), folate, is one typical small molecule-based recognition element with high binding affinity (Kd$\sim 10^{-9}$ M). Because FR emerges as a potential malignant tumor-selective target, folate is widely used to detect FR-related tumors (Low et al., 2008). Besides limited native ligand, synthetic binders are also extensively investigated with development of supramolecular chemistry and molecular recognition. Generally, small molecule binders are able to specially recognize a particular substructure, surface "hot-spot," or active center of the target protein (Pode et al., 2017). In most studies, the interaction of a small molecule and target protein may induce conformational changes of either the small molecule probe or target protein, thereby arousing signal changes for quantitative measurement.

Besides small molecule chemical compounds, small molecule peptide ligand is another appealing recognition element in the fabrication of biosensors (Liu et al., 2015; Puiu and Bala, 2018; Reverdatto et al., 2015). Both the peptide ligand and antibody are composed of amino acids, but the antibody needs a complex spatial structure to maintain native activity. Compared to whole protein (e.g., antibody), a peptide ligand with biological simplicity has low synthesis cost, high chemical and thermal stability, easy for large-scale production, and modification. Short peptides originating from the in vivo system were firstly used as recognition elements (Pavan and Berti, 2012). For example, RGD peptide that binds to integrin on the cell surface is the most extensively used ligand for cancer detection. Afterward, artificial peptides are selected from a random peptide library using phage-, ribosome-, or mRNA-display strategies, which is quite similar to aptamer screening. Therefore, the peptide ligand from in vitro selection is also called a peptide aptamer. Due to the high stability of the peptide ligand, neither modification nor immobilization has any influence on specificity and binding activity of the peptide ligand, which is superior to antibody. Moreover, a peptide ligand can bind to a specific site of the target and thus regulate the orientation of the immobilized target, which is critical to promote communication of immobilized protein and transducer for signal response.

I.1.1.6 Immobilization of Bio-Receptor

A biosensor translates the interfacial information of biological interaction into a chemical or physical signal, which is used for quantitative measurement of an analyte in a biological sample. Immobilization of the recognition element on the transducer is a critical step to obtain interfacial information upon a biological recognition event. Since the first GOD-based biosensor, bioreceptors have been preferably immobilized on a transducer surface via desired modification. In order to maintain the native biological activities of a bioreceptor, several immobilization techniques have been developed with high biocompatibility, stability, and conductivity, especially for electrochemical biosensors (Bhakta et al., 2015; Gao et al., 2015; Gauchet et al., 2006; Levicky et al., 1998; Li et al., 2014; Pramanik et al., 2012; Trilling et al., 2013; Yang et al., 2012; Zhang et al., 2017). The immobilization methods are divided into two categories: noncovalent immobilization and covalent immobilization. Membrane encapsulation is an example of noncovalent immobilization approach since the development of the first biosensor (Gao et al., 2015). In the model of the first biosensor, GOD was modified on the surface of an electric transducer (the electrode) by trapping it in membrane materials. The membrane material thus became a hot spot in the development of different generations of biosensors, which should be highly stable, environmentally friendly, and suitable for signal conductivity. Biological macromolecules and biocompatible polymers were firstly used for membrane coating, including fish sperm DNA, polyethylene glycol (PEG), poly diallyldimethylammonium chloride (PDDA), Nafion, and chitosan (Trilling et al., 2013). Later, with the development of nanotechniques, nanomaterials become a new-generation membrane material for bioreceptor immobilization, benefiting from large surface-to-volume ratio, high surface energy, catalytic activity, and adsorption ability (Bhakta et al., 2015; Zhang et al., 2017). Different modification techniques also develop with the update of membrane material, such as layer-by-layer, sol–gel, ionic liquids, and electrostatic adsorption. Compared to noncovalent immobilization, covalent immobilization is becoming more popular in recent years. Self-assembled monolayer (SAM) technique is the most widely used technique for surface immobilization of receptors, which presents increased stability, reproducible activity, and is a much easier method to control the orientation of biomolecules on the transducer interface (Levicky et al., 1998; Li et al., 2014; Pramanik et al., 2012). The interaction of thiol group and gold surface is one well-known covalent attachment procedure. Coating the transducer with a gold thin film provides an available interface for the capture of biorecognition elements containing a free thiol group, especially for thiol-functionalized probe DNA and protein with cysteine. Another common covalent binding method is based on condensation reaction, which exhibits better applicability. Carboxyl group functionalized transducer surface can be combined with amino group-containing recognition biomolecules through the reaction catalyzed by 3-(ethyliminomethyleneamino)-N,N-dimethyl-propan-1-amine (EDC) and N-hydroxysuccinimide (NHS). The immobilization of a bioreceptor is usually based on the structure and function of recognition elements. For example, antibody has a high requirement for native structure and function, which is usually immobilized on a carboxylic group functionalized surface through a condensation reaction. Compared to an antibody, there are more choices for immobilization of peptide ligand on the transducer. Additional cysteine or functional free thiol group help immobilization of peptide ligand through Au-S interaction, while carboxylic group and amino group involved in peptide ligand facilitate immobilization through covalent interaction. Moreover, chemoselectivity and supramolecular interaction are also utilized to immobilize peptide ligand based on the nature of amino acids, such as arginine and phosphate group, aromatic amino acid and cucurbituril (Gauchet et al., 2006; Yang et al., 2012).

I.1.2 Transducer

A bioreceptor for recognition and interaction with a target and a transducer for signal production make up an integrated biosensor. A transducer translates recognition information from a bioreceptor mediated biological interaction event into a readable signal for target detection. Based on different signal sources, electrical, mechanical, magnetic, and optical biosensors are most developed for biological measurements. Sufficient signal for biosensing is generally produced in a manner of either label-free or chemical labeling. For label-free detection, the analyte of interest usually has an intrinsic tag for useful signal output, otherwise chemical labeling is required for efficient distinguishing target from other interferes.

I.1.2.1 Electric Transducer

Since the first biosensor in 1962, the electrical biosensor is the most extensively studied biosensor by making use of a chemically modified electrode as a transducer. Electrochemical techniques provide various measurement modes and thus extend the applications for different biosensing purposes, including amperometry, potentiometry, electrical impedance spectroscopy, field-effect transistor, and conductometry. Specifically, amperometry traces current changes from redox reaction of an electro-active substance on the electrode surface using a conventional three-electrode system (Moreira et al., 2017). A three-electrode system consists of a working electrode, a reference electrode, and an auxiliary electrode. The common working electrodes are noble metal-based electrode (e.g., platinum electrode, gold electrode, and silver electrode) and carbon-based electrode (e.g., glassy carbon electrode, pyrolytic graphite electrode, and carbon paste electrode), which are easy and stable for modification and reproduction. A reference electrode provides a stable and well-known potential, such as saturated calomel electrode ($E = +0.241$ V) and silver chloride electrode ($E = +0.197$ V). An auxiliary electrode (e.g., platinum wire), also known as a counter electrode, constitutes a complete circuit along with a working electrode as well as balances the reaction on the surface of working electrode. The resulting current is proportional to the amount of electro-active species on the electrode surface, which is the best choice for target quantification. Potentiometry is based on potential differences between analyte and other references where there are no obvious currents for target tracing (Tarasov et al., 2016). Ion-selective electrode (ISE) is a typical transducer for such measurement by making use of an ion-selective membrane for target identification (van de Velde et al., 2016). The transducer converts ionic activity into a specific potential signal, which is proportional to the logarithm of ionic activity based on the Nernst–Donnan equation. Electrical impedance spectroscopy (EIS) is another commonly used technique in electrochemical biosensor fabrication (Bahadir and Sezginturk, 2016). EIS monitors the current flowing over samples under a given voltage, and electrical impedance is presented by voltage-to-current ratio in the range from 10 kHz to 10 MHz. In general, the binding of analyte and bioreceptor prevents electron transfer between electro-active species and electrode surface for either steric hindrance or charge changes, resulting in changes of electrical impedance. The field-effect transistor (FET) regulates the electrical behavior of the device using an electric field, which is also applied to trace ions with the assistance of an ion-selective membrane (Sarkar et al., 2014). Conductometry is able to monitor enzymatic reactions and biological recognition events through measurement of electrolytic conductivity, especially when using microelectrodes (Kucherenko et al., 2015).

I.1.2.2 Optical Transducer

An optical biosensor is another appealing biosensor today due to low signal-to-noise, excellent sensitivity, and fast response. The optical biosensor is based on the development of different spectroscopic techniques, such as UV-vis spectroscopy, fluorescent spectrometry, Raman spectroscopy, surface plasmon resonance, and nuclear magnetic resonance. The natural properties of analytes enable label-free, real-time, and simultaneous detection using optical biosensors (Feng et al., 2014; Yan et al., 2018). Fluorescence spectrometry is the most powerful technique for fabrication of an optical biosensor (Ma et al., 2016). Fluorescence spectrometry describes an optical property of a substance that emits a certain wavelength of light after absorbing energy from an external light source. Usually, the wavelength of emitted light is longer than that of excitation light, ascribed to energy loss in the luminescence process. The emitted light is called fluorescence. However, most biological analytes do not have the fluorescent properties, so labeling of fluorescent tags is necessary for biosensing. In this case, the information of a biological recognition event is converted to a fluorescent signal for quantification of the desired concentration of target molecules. Chemiluminescence is another optical method to quantify the emitted light for target measurements, but the energy resource for substrate absorption is a chemical reaction while not external light as that for fluorescence (Roda et al., 2016). Chemiluminescence is able to monitor specific biochemical reactions at a sensor surface, presenting extremely high sensitivity with the use of only a simple instrumentation. Similar to a fluorescent biosensor, chemical labeling is sometimes required to characterize the reaction between an analyte and a recognition system in chemiluminescence. Electrochemiluminescence is an electrogenerated chemiluminescence, which has attracted increasing attention in recent years (Chen et al., 2017b). Different from chemiluminescence, electrochemiluminescence absorbs the energy from an electrochemical reaction during the application of potential in the solution. It is a highly sensitive and specific technique for the combination of the advantages of chemiluminescence and electrochemical technique, which has low background signal and is easy to control. Surface plasmon resonance (SPR) originates from a particular optic phenomenon of a nonradiative electromagnetic surface wave, which is a useful, label-free, and real-time tool to identify the interaction of analyte and bioreceptor by tracing the variations in the refractive index (Masson, 2017). However, label-free detection mode limits the sensitivity of SPR biosensors. For example, SPR can only distinguish the binding with a biological molecule that is larger than 2 kDa, while smaller molecules cannot arouse sufficient signal readout after recognition event. Besides, infrared spectroscopy, nuclear magnetic resonance, and surface-enhanced Raman spectroscopy have also been active optical methods in recent years (Castro et al., 2014; Henry et al., 2016; Neubrech et al., 2017).

I.1.2.3 Magnetic Transducer

A magnetic biosensor makes use of the magnetic properties involved in interaction of the analyte and bioreceptor system (Lee et al., 2015; Zheng et al., 2016). However, only a few biological molecules possess magnetic properties except some iron-containing or metal-cluster proteins, thus chemical labeling is always needed for such a biosensor. Superparamagnetic nanoparticles are an important signal resource to detect targets of interest after labeling on a certain recognition element. Much slower spin–spin relaxation (T2) time of nanoparticles promotes the characterization for excellent contrast using magnetic resonance imaging (MRI). In the meantime, magnetoresistive material is another signal resource for the fabrication of a magnetic biosensor. The ultrathin ferromagnetic film facilitates construction of a multilayered structure by overlaying a functionalization surface. The interaction of magnetically labeled analyte and sensor surface is presented through current changes under a given magnetic field. Besides the need for chemical labeling, false-positive results from nonspecific adsorption on nanoparticles always limit the application of magnetic biosensors. Even so, a magnetic biosensor is still considered as a useful tool for biosensing as it is easily incorporated with microfluidic systems.

I.1.2.4 Mechanical Transducer

In a mechanical biosensor, mechanical force and motion are utilized to trace analyte concentration. Microcantilever and quartz crystal microbalance (QCM) are two well-known examples of mechanical biosensors (Afzal et al., 2017; Gopinath et al., 2015). Microcantilever measures the adsorption of target biomolecules by tracing either static deflection or change of oscillation frequency, which is quite similar to atomic force microscopy. However, QCM monitors frequency changes of a quartz crystal resonator upon adsorption of an analyte at a crystal surface. Although these mechanical biosensors are always limited by sensitivity, they enable specific detection in a label-free manner, and possess the advantages of easy fabrication and functionalization.

I.1.3 Signal Amplification

One big challenge for a biosensor is to detect biomarkers at an extremely low level in a given biological sample. Although good cooperation of bioreceptor and biotransducer enables direct measurement of target biological molecules through signal readout, lack of sensitivity always leads to failure in detection of targets of low abundance. Therefore, two strategies are commonly used to enhance detection sensitivity: target-based amplification and signal-based amplification (Goggins and Frost, 2016). Target-based amplification facilitates continuous generation of target molecules through a specific cycle process. For example, PCR is a typical target-based amplification, which amplifies only a few copies of target genes across several orders of magnitude through a thermal-cycle process. However, the generation process limits the extensive application of target-based amplification, which is only suitable for amplified studies of target nucleic acid to date. In contrast, signal amplification is much more universal than the target-based amplification strategy in the fabrication of a biosensor; this directly magnifies readable signals for achieving goals of ultrasensitive detection in biological samples.

I.1.3.1 Catalyst-Assisted Signal Amplification

Catalyst-assisted signal amplification is the most common signal amplification strategy in the fabrication of a biosensor (Gianneschi et al., 2005; Scrimin and Prins, 2011; Willner et al., 2008). Besides being a bioreceptor, an enzyme is also a popular element in the design of a signal amplification strategy. An enzyme is a natural catalyst involved in almost all metabolic processes in living organisms, which accelerates the conversion of thousands of substrates into a large number of products. If the product is a source of an output signal, enzymatic catalysis generates a multitude of signal molecules for enhancement of detection sensitivity. Enzyme-linked immunosorbent assays (ELISA) is a good example of enzyme-based signal amplification (Wei et al., 2016). Horseradish peroxidase (HRP)-labeled antibody is an important source of detectable signal for ELISA, which catalyzes oxidation of substrates (e.g., OPD, TMB) to generate an amplified amount of readable optical signals. As a result, ELISA achieves a quite low detection limit at pM level, which is a great progress when compared to that based on a single signal labeling. For achieving high efficiency in signal amplification, high stability, catalytic efficiency, and low cost of enzyme are usually required. However, most enzymes are not stable for modification and only work in a mild environment for the characteristics of native protein. Only a few enzymes are suitable for signal labeling at present, such as HRP and alkaline phosphatase (ALP). Besides, catalytic activity of an enzyme may decrease dramatically beyond its optimal environment (e.g., temperature and pH). Therefore, the limited choices of catalytic protein and strict requirement of reaction environment restrict the application of enzyme-based signal amplification.

Inspired by the fact that RNA (ribozymes) could work as a catalyst in a biological system, it was believed that nucleic acid might have the potential to be a catalyst. In 1994, the first single-stranded DNA with catalytic activity was reported, which was named as DNAzyme (Breaker and Joyce, 1994). Artificial DNAzyme is obtained from large-scale chemical synthesis with low cost, and exhibits high thermal stability that is not susceptible to the external environment and is able to catalyze substrates even at high temperature. The most appealing DNAzyme is reported by Sen et al., which is a hemin-binding DNA strand and could work as a mimic of a peroxidase (Travascio et al., 1998). Guanine-rich single-stranded DNA binds with hemin and thus forms a quadruple structure. Accordingly, hemin-binding DNA complex then exhibits increased peroxidase activity toward catalytical reactions. Compared to native peroxidase, DNAzymes are much more flexible and stable for structural design and surface modification, which could be divided into two or even four parts for different purposes. In this sense, artificial DNAzyme becomes a powerful substitute for native enzymes in design of catalyst-based signal amplification strategy (Kosman and Juskowiak, 2011; Liang et al., 2017b; Willner et al., 2008).

Besides, nuclease-assisted target recycling is also an appealing strategy for signal amplification in recent years (Gerasimova and Kolpashchikov, 2014; Miao et al., 2015; Yan et al., 2014). Nuclease catalyzes hydrolysis of a phosphate diester bond within nucleic acids. With the cyclic enzymatic reaction, thousands of amplified signals are produced in a few hours or even several minutes. In nuclease-assisted target recycling strategy, a target-binding event initiates nuclease-assisted cyclic digestion of reporter nucleic acid probes, releasing an increased amount of readout signal molecules for enhanced detection sensitivity. To realize the purpose, nucleases that selectively digest a specific strand of duplex DNA or single-stranded DNA are used, including Exonuclease III (Exo III), DNase I, T7 exonuclease, and nicking enzymes. Exo III is a type of exonuclease that catalyzes the removal of mononucleotides from the 3'-terminal of duplex DNA. Accordingly, a signal labeled at the 3'-terminal is released by Exo III-catalyzed digestion, while another intact strand with a redundant terminal is protected from Exo III-catalyzed digestion and then facilitates cyclic enzymatic digestion of the reporter strand (Zuo et al., 2010). Similar to Exo III, T7 exonuclease catalyzes the removal of mononucleotides from the 5'-terminal of duplex DNA, which also promotes signal amplification with recycling of intact catalyst DNA (Chen et al., 2013). DNase I is a nonspecific exonuclease that preferably digests single-stranded DNA, which usually helps to reduce background signal for biological detection (Lu et al., 2010). Unlike the abovementioned nucleases, nicking enzyme belongs to endonuclease family. Different from the traditional restriction endonuclease that cleaves duplex DNA at a recognition site; nicking enzyme only cuts one strand of duplex DNA, and leaves another intact DNA for cyclic enzymatic reaction (Cao et al., 2012). Nuclease-assisted signal amplification can be conducted at a constant temperature with quite high detection efficiency, as the detection limit of aM could be achieved within one hour. With the use of nuclease-assisted signal amplification, a target molecule is not limited to nucleic acid, as the binding of different biomolecules (protein or small molecules) to aptamer could also be dissociated for enzymatic digestion reactions.

I.1.3.2 DNA Amplification—Assisted Signal Amplification

DNA-based target amplification enhances sensitivity of bioanalysis by greatly increasing the actual amount of target DNA. Although the abovementioned PCR is the most mature technique of DNA amplification, the thermal cycle that requires an expensive instrument to control temperature for different steps always restricts its application. In this case, isothermal DNA amplification is considered as the best substitute of PCR, including loop-mediated isothermal amplification (LAMP), helicase-dependent amplification, strand displacement amplification (SDA), nicking enzyme amplification reaction (NEAR), and rolling circle amplification (RCA) (Ali et al., 2014; Feng et al., 2016; Liu et al., 2018; Zhang et al., 2013; Zhao et al., 2015). These isothermal amplification methods can greatly amplify target DNA under a mild temperature, which do not need any specific and expensive instrument. Although the replication process limits the application of target-based amplification in detection of other biomolecules, isothermal DNA amplification strategy inspires development of a DNA-based signal amplification strategy for improving detection sensitivity of a biosensor. Taking RCA as an example, RCA has emerged as an attractive strategy in the design of signal amplification. RCA is a high-efficient isothermal amplification method using a circular DNA as a template, which generates hundreds of repetitive sequences that are complementary to the template after cyclic amplification reaction (Ali et al., 2014). RCA is conducted at a mild temperature from room temperature to 37°C both in solution and on an interface, enabling generation of amplified DNA signals on the surface of a transducer. RCA is a versatile amplification technique, which satisfies different demands of target detection by designing template sequences. On one hand, RCA product is able to provide a large amount of binding sites for signal labeling through DNA hybridization. On the other hand, if the circular template is a sequence complementary of functional motif, such as DNAzyme, combination of DNA amplification and catalyst-based amplification may induce multiple signal amplification for enhancement of detection sensitivity.

Although isothermal amplification provides a better choice for signal amplification, the use of DNA polymerase and various nucleases increase the cost of experiments and also limit its reaction efficiency. Recently, nuclease-free DNA amplification strategy becomes popular in DNA-based signal amplification for a biosensor, which is based on a kinetics-controlled hybridization reaction (Dirks and Pierce, 2014; Jung and Ellington, 2014; Li et al., 2011; Liang et al., 2017a). Two typical examples of nuclease-free DNA amplification are hybridization chain reaction (HCR) and catalyzed hairpin assembly amplification strategy (CHA). HCR was firstly reported by Pierce et al. in 2004 (Dirks and Pierce, 2014). Two hairpins (H1 and H2) that are kinetically trapped are required for a cascade hybridization event. As a toehold-mediated strand displacement, HCR should be triggered by an initiator DNA. The initiator hybridizes with the single-stranded part of H1, and thus opens the stem-loop structure of H1. In this case, a single-stranded sequence that is complementary of H2 is exposed, and then opens the stem-loop structure of H2 through DNA hybridization, exposing another single-stranded complementary part to H1. With the cycle of hybridization event, a long nicked double-stranded DNA strand is formed with hundreds of repetitive sequences of H1 and H2. The long DNA product could provide increased sites for signal labeling from a hybridization event, resulting in enhancement of readable signal for a biosensor. Another example is CHA. CHA is a toehold-mediated strand displacement reaction, which is also initiated by a linear single-stranded catalyst in the presence of two kinetically trapped hairpins (H1 and H2) (Li et al., 2011). The linear catalyst is able to hybridize with a toehold at the terminal of H1 and thus destroys the secondary structure of H1. In this case, a single-stranded region within H1 that is complementary to the toehold on H2 is exposed, which further hybridizes with H2 and triggers the release and reuse of the catalyst DNA. Unlike long duplex DNA from HCR, the product of CHA is a short duplex DNA from single hybridization of H1 and H2, which may also introduce a large amount of signal labeling for amplified detection.

I.2 NANO-INSPIRED BIOSENSOR

Nowadays, nanomaterial receives widespread attention for its application in the fabrication of biosensors (Farka et al., 2017; Kurbanoglu et al., 2017; Maduraiveeran et al., 2018; Wang et al., 2017a). Nanomaterial is defined as a material with at least one dimension within nanoscale (between 1 and 100 nm). Based on the number of dimensions that exceed the nanoscale range, nanomaterial is divided into zero dimensional (0 D), one-dimensional (1 D), and two-dimensional (2 D) material. For example, carbon nanoparticles, whose three dimensions are all within 100 nm, are 0 D nanomaterials; carbon nanotubes, whose both dimensions are within 100 nm, are 1 D nanomaterials; graphene, whose one dimension is within 100 nm, is a 2 D nanomaterial. Nanomaterial is synthesized through both "top-down" and "bottom-up" procedures, and also exhibits prominent electric and optical properties due to nano-size effect. Moreover, nanomaterial has several distinct advantages over bulk material, including high surface area, electric conductivity, and catalytic activity. In fabrication of a biosensor, nanomaterial can not only be incorporated in the design of a bioreceptor, but can also facilitate signal readout and signal amplification for improving the performance of a biosensor.

I.2.1 Immobilization of Biomolecules

The immobilization of biomolecules on the surface of a transducer is a critical step in the fabrication of a biosensor, which has a direct influence on the specificity and sensitivity of biological detection. Taking the development of a gluocose biosensor as an example, immobilization of glucose oxidase (GOx) on a transducer (such as electrode) undergoes several stages for continuous improvement of the performance of a glucose biosensor. In the first-generation of a glucose biosensor, GOx was entrapped on an oxygen electrode using a semipermeable membrane similar to the first biosensor.

GOx catalyzes oxidation of β-D-glucose to produce β-D-gluconolactone, along with reduction of FAD to $FADH_2$. The conversion of $FADH_2$ to FAD then induces the reduction of O_2 to produce H_2O_2, and thus arouses an electrochemical signal on the electrode surface. The first generation of glucose biosensor has some noticeable defects. For example, the deeply buried active center of the enzyme always restricts its electron transfer with electrode surface, and electrochemical response of H_2O_2 is easily interfered with by other electro-active species. In the second generation of glucose biosensor, a synthetic mediator was used to replace O_2. The mediator facilitated regeneration of FAD, and could be regenerated on the electrode surface for production of an electrochemical response. The mediator had the advantages of high stability, low toxicity, and good electrochemical properties, which also helped to reduce the background signal. However, the use of mediator increases the cost and complexities of biological detection, especially for some toxic mediator. In third-generation biosensors, new type of membrane material works as both an immobilization medium and an electrochemical mediator for co-immobilization of membrane on the electrode surface. In this case, the third-generation biosensor has a high capacity for the immobilization procedure. Amazingly, nanomaterial offers a favorable choice for immobilization of enzyme as a coating membrane and mediator in third-generation of biosensor, especially gold nanoparticles and carbon nanotubes (Cipolatti et al., 2016; Husain, 2017; Zhang and Chen, 2017). The good compatibility of nanomaterial maintains the native bioactivity and stability of the enzyme, while the high surface area provides increased binding sites for immobilization and orientation of enzyme. What is more, excellent conductivity facilitates electron transfer between the biological receptor and a transducer and facilitates the acquisition of the electrochemical signal.

With the wide usage of nanomaterial, different immobilization techniques have been developed for the immobilization of a bioreceptor onto a transducer (Das et al., 2016; Putzbach and Ronkainen, 2013). Electrostatic adsorption is a simple and fast operation for the interaction of nanomaterial and biological recognition element; covalent binding provides highly stable interaction of nanomaterial and biological recognition element; self-assembly membrane regulates the orientation of immobilization and thus facilitates further biological recognition reactions; layer-by-layer technique increases the amount of biological recognition molecules on the surface of a transducer using repetitive use of nanomaterial for coating. Compared to the solid interface of a transducer, nanomaterial provides a friendlier interface for immobilization of biological molecules due to its high surface activity, good biocompatibility, and excellent conductivity, thereby ensuring sensitive and specific target identification after functionalization of a bioreceptor.

Magnetic nanomaterial belongs to a family of nanomaterial containing iron, nickel, and cobalt, which is able to be manipulated in magnetic fields. Fe_3O_4 nanoparticles and magnetic graphene are typical magnetic nanomaterials in biosensing. As another nanomaterial, magnetic nanomaterial is also a favorable carrier for immobilization of different biological molecules by offering a large number of binding sites on the surface. What is more important, magnetic nanomaterial provides a new method for the immobilization of bioreceptors, magnetically controlled immobilization (Zhu et al., 2014). The bioreceptor on the surface of the magnetic nanomaterial could be immobilized on a specific site of a transducer under the control of the magnetic field, which is reversible for reproduction and reuse of a transducer interface.

I.2.2 Design of Signal Probe

Surface immobilization of a bioreceptor plays an important role in the maintenance of high selectivity of a biosensor, while signal output has a direct impact on the sensitivity of a biosensor. Nanomaterial exhibits prominent chemical, physical, electrical and mechanical properties, thereby providing an alternative to design of signal probes. Multitudinous nanomaterials offer a variety of choices for almost all the demands of biosensors, which could be combined with different transducers for sensitive and specific detection of biological targets.

I.2.2.1 Electrochemical Probe

The electrochemical biosensor has been the most desired biosensor since the report of the first glucose biosensor. The conventional electrochemical signal molecules are small chemical compounds, which always have low electron transfer efficiency with the electrode surface. In order to improve the sensitivity of electrochemical detection, new types of electrochemical probes are needed for signal response and even signal amplification. Nanomaterial with redox activity is a satisfactory electrochemical probe to arouse an electrochemical signal, such as a silver nanoparticle (Cao et al., 2015). After immobilization onto an electrode surface upon a biological recognition event, a current from oxidation of silver is obtained in proportion to the amount of biological targets. In the meantime, nanomaterial with excellent catalytic activity is also an ideal electrochemical probe for production of catalytic signals, such as gold nanoparticles, platinum nanoparticles, and palladium nanoparticles (Garg et al., 2015; Tang and Tang, 2015; Wang et al., 2013). These nanoprobes catalyze conversion of a multitude of substrate molecules into a measurable electrochemical signal, which also bring in signal amplification with high efficiency. Besides, metal-containing nanoparticle is also a source of an electrochemical probe, which releases a large amount of metal ions in an acidic environment, and they are traced by stripping voltammetry (Gill et al., 2008; Valera et al., 2016; Wang et al., 2017b). What is more, since the potentials of metal atoms or catalytic reactions are different from each other, simultaneous electrochemical detection of several targets could be realized with combined utilization of different nanoprobes.

I.2.2.2 Optical Probe

Due to small size effect and quantum size effect, nanomaterial, especially semiconductor and metallic nanomaterial, presents excellent optical properties, such as particular linear absorption and strong photoluminescence emission. For example, nanomaterial with excellent fluorescent properties is able to replace fluorophore as a signal molecule for the fabrication of a fluorescent biosensor (Wegner and Hildebrandt, 2015; Wen et al., 2015; Yao et al., 2014). Quantum dots (QDs) are the best option for fluorescent labeling (Wegner and Hildebrandt, 2015). Because optoelectronic properties of QDs change with both size and shape, QDs emit different wavelengths of fluorescence by only changing the synthesis size. Large QDs (e.g., 5–6 nm) emit fluorescence with longer wavelength, while smaller ones (2–3 nm) emit fluorescence with shorter wavelength. Compared to organic dye, QD has wider absorption spectra, a narrow emission peak, and a longer lifetime as well as stronger fluorescence signal and higher signal-to-noise ratio, which allows sensitive characterization of multiple targets at the same time even inside living organisms. Meanwhile, QDs provide large surface-to-volume ratio and high surface activity, which facilitates surface functionalization for biological recognition and signal labeling. Gold nanoparticle is another popular example for colorimetric probes (Aldewachi et al., 2018). Gold nanoparticle provides a large surface area, high surface activity, and good biocompatible interface for immobilization of different biomolecules and design of various signal probes. As a plasmonic nanomaterial, plasmon resonance wavelength of gold nanoparticle changes with the size of nanoparticle as well as the distance between nanoparticles. The aggregation or growth of gold nanoparticles induce a blue shift in the UV-visible spectra, thereby promoting colorimetric detection of target analytes.

I.2.2.3 Magnetic Probe

As mentioned in the section on magnetic transducers, a magnetic biosensor is built based on the use of superparamagnetic nanoparticles as a signal source (Jamshaid et al., 2016; Lu et al., 2015). Superparamagnetism is a particular type of magnetism that is usually reflected in sufficiently small ferromagnetic or ferrimagnetic nanoparticles. The use of superparamagnetic nanoparticles is always combined with magnetic resonance imaging (MRI) or magnetic hyperthermia treatment (MHT) for medical imaging and cell tracking. Superparamagnetic nanoparticles refer to a family of iron oxide-cored inorganic nanoparticles, which could be coated by inorganic or organic material as a shell. After functionalization with biological molecules, superparamagnetic nanoparticle is an available probe for increasing magnetic resonance contrast and used for detection of disease-related biomarkers at the physiological and molecular level. On the other hand, magnetic nanomaterial offers an appealing choice for a magnetic nanoprobe instead of the traditional magnetoresistive material, which is low cost, high sensitivity, and fast response.

I.2.2.4 Mechanical Probe

The extraordinary mechanical properties of nanomaterial enable its amplification in the design of mechanical probes (Lee et al., 2016). In the mechanical biosensor, the mechanical characteristics of a sensing probe are critical to detection efficiency, such as spring constant and resonance frequency. Nanomaterial provides an opportunity to design remarkable nanoprobes as an alternative to conventional probes. Polymeric, organic, and inorganic nanomaterials are introduced for new types of mechanical nanoprobes with tunable mechanical properties and flexible surface functionalization. In the meantime, molecular self-assembly technique substitutes conventional microfabrication for enhanced performance of a nanoprobe. Using gold nanoparticles as a nucleating agent for mechanical detection using microcantilever, the growth of silver on gold nanoparticles induces a frequency shift, which is a mechanical signal for the increase of the effective mass

I.2.2.5 Signal Amplification

Nanomaterial is more frequently used in the design of a signal amplification strategy (Hasanzadeh et al., 2015; Krishnan et al., 2011; Lim and Ahmed, 2016; Pelossof et al., 2012; Tiwari et al., 2016; Wang and Dai, 2015; Wu et al., 2014). Nanomaterial provides an efficient interface for immobilization of multiple numbers of biological molecules and even a perfect scaffold to recruit functional molecules for multiple signal amplification due to high surface-to-volume ratio and surface activity. For example, Willner et al. proposed that about 96 copies of single-stranded DNA were immobilized on a single gold nanoparticle, which offered an interface for functionalization of multiple DNAzyme through DNA hybridization (Pelossof et al., 2012). Similarly, nanomaterial provided enough binding sites for signal labeling for fabrication of a hybrid electrochemical nano-labeling. Electro-active molecules were absorbed on the surface of graphene through $\pi-\pi$ interaction, and were then labeled on the recognition element to arouse amplified signal readout for increased amount of electrochemical signal tags (Wang and Dai, 2015). Moreover, aggregation of nanomaterials, such as superparamagnetic nanoparticles, changes the spin–spin relaxation time as well as SPR response and thus is used for signal amplification in both magnetic and SPR detection (Krishnan et al., 2011). Besides increasing the generation of signal molecules, magnetic nanomaterial presents another way to improve detection sensitivity. Magnetic nanomaterial is used to accumulate and separate target analytes under the control of a magnetic field. The concentration treatment increases the absolute content of a target for detection, while purification treatment facilitates a reduction of background signal, thereby dramatically promoting the increase of detection sensitivity for biosensing. For example, p19-functionalized magnetic beads achieved over 100,000-fold accumulation of target RNA and realized RNA detection at picogram levels.

I.3 BIOSENSOR APPLICATION

Since it was developed in 1980s, the most successful commercial biosensor is still the glucose meter to date, which is a medical device for quantitative determination of glucose using a small strip of blood as a sample. The portal biosensor derived from the first biosensor helps diabetic patients to monitor blood glucose level in a real-time manner, with high sensitivity, good stability, and low cost. As the concept of "point-of-care" is put forward, more types of biosensors are expected to satisfy the demand of delivering timely bioanalytical services to people in need. In this sense, extensive researches keep promoting applications in different fields relating to bioanalysis.

I.3.1 Disease Diagnosis

Diseases seriously affect human health and living quality, while early and accurate diagnosis is of great importance to disease identification and treatment. In most diseases, diagnosis is based on the determination of some specific biomarkers relating to different disease states. As a leading cause of human death, cardiovascular disease is a hot research subject for development of a biosensor. Cardiovascular disease has a close relation with the heart and blood vessels, and is caused by several factors (e.g., high blood pressure, hyperlipidemia, and smoking). Early detection and treatment is critical to prevent irreversible damage to the heart. According to guidance of the World Health Organization, the determination of cardiac biomarkers in a blood sample is a significant part in the conventional diagnosis of cardiovascular disease, providing sufficient information to guide accurate treatment and prognosis of the disease. Many biomarkers are identified for diagnosis of cardiovascular disease and becoming potent targets in fabrication of a corresponding biosensor, including C-reactive protein, myoglobin, creatine kinase-MB, creatine kinase MM, and cardiac troponins (Abdorahim et al., 2016; Rezaei et al., 2016).

Cancer diagnosis also attracts increasing attention for biosensor fabrication, since it is a leading cause of death only after cardiovascular diseases. Cancer is a large family of malignant diseases with over 200 subtypes, having the capability of infinite proliferation and distant migration. Early detection and treatment is the most effective way to improve the cure rate and survival rate of cancer patients, which brings forward a high requirement for sensitivity and accuracy of biosensors. Tumor markers, which are widely found in blood, urine, and tissue samples, are believed to predict the occurrence and development of cancers at different stages. There are several different categories of tumor markers, which may be a protein, a peptide, a small molecule or a specific DNA sequence. In this case, the main purpose of a cancer-related biosensor is to realize early diagnosis of tumors via sensitive, stable, and simultaneous determination of specific cancer biomarkers (Huang et al., 2017; Mittal et al., 2017; Peng et al., 2014; Viswambari Devi et al., 2015; Yang et al., 2018). At present, biosensors are applied in the diagnosis of several cancers, including breast cancer, lung cancer, liver cancer, and prostate cancer.

Infectious diseases that are caused by pathogenic microorganisms also have a high lethality rate worldwide. The pathogenic microorganisms are found everywhere from air to soil and include different types of tiny living things, such as bacteria, fungi, parasites, and viruses. Although great efforts have been made in the control of infectious diseases, they are still a serious threat to human health. Timely diagnosis and targeted treatment are critical to raise cure rates of diseases. Since the existing methods always suffer from drawbacks of false positivity and low sensitivity, more types of biosensors are being constantly explored to facilitate complex molecular assays in diagnosis of infectious diseases. The biosensors could sensitively target bacteria, fungi, parasites, and viruses, especially with the advance in nanotechnologies (Chen et al., 2017a; Ray et al., 2017).

I.3.2 Other Applications

There are food safety concerns with each procedure involved in handling, preparation, and storage of food, which could be directly related to human health. Food is a growth medium to reproduce and transmit pathogens for production of infectious diseases, and is also a carrier for toxic chemicals to endanger human health. Biosensors, having the capability of low cost, high sensitivity, good selectivity, and ease of use, are now being applied in monitoring food safety (Gaudin, 2017; Lan et al., 2017a; Zhang et al., 2014). For example, a biosensor is used to trace antibiotic residues that have a high risk to damage human health (Lan et al., 2017a). Although particular antibiotics are allowed to be used in food, farm, and livestock products, the amount of the drugs should be under strict supervision. Indiscriminate use of antibiotics contributes to the very serious consequence of antibiotic resistance, leading to the failing of the standard antimicrobial treatment and even the emergence of superbugs. Since antibiotic residues are usually at an extremely low level, biosensors with low detection limit and high sample throughput are becoming popular in the determination of target antibiotic residues at a trace level. Another example of application is to detect pesticides (Zhang et al., 2014). Pesticides are used to inhibit the proliferation of harmful pests in agricultural cultivation, which also do harm to other living organisms. Redundant pesticides may spread to soil, air, and water, resulting in environment pollution and biological poison. Determination of pesticide residues with high specificity is of great importance to environment protection and human health, since pesticides are highly toxic with a bioaccumulation effect, and long-term damage risk. Most pesticides are neurotoxic through irreversible inhibition of an essential enzyme system in the central nervous, acetylcholinesterase. Accordingly, biosensors, especially acetylcholinesterase-based biosensors, are widely applied for simple, fast, and sensitive determination of pesticides as a promising alternative to conventional methods.

A biosensor has the capabilities of fast response, simple operation, high sensitivity, and stability, which also meet the requirements of monitoring environmental pollutants. Environmental pollutants, such as heavy metals, small chemicals, toxins, and pathogens, are analytes for environmental monitoring. To identify different environmental analytes, various biosensors are designed based on the inner linker with bioreceptors, including nucleic acid, enzyme, cells, and even tissue (Wang et al., 2014). The detection could be realized in either a direct or indirect manner, both of which focus on avoiding false-positive or false-negative results. However, the need for continuous detection and large sample analysis bring in more challenges for biosensors in environmental monitoring. The long-term storage and high reproducibility of biosensors under different environmental conditions, especially the tough environment, are also the issues that need to be considered when developing portals and inexpensive biosensors for environmental monitoring.

I.4 OVERVIEW OF THIS BOOK

Quantitative analysis of various substances in human blood, urine, and other samples is the most important path for disease diagnosis and prognosis. The analysis of various protein markers is the most important aspect. For example, the main method for diagnosing various diseases caused by pathogenic microorganisms is to detect the antigens of pathogenic microorganisms and the corresponding antibodies produced by humans. A specific example is the diagnosis of hepatitis B. Five antigens/antibodies are adopted as the biomarkers: hepatitis B surface antigen (HBsAg), antisurface antigen antibody (HBsAb), e antigen (HBeAg), anti-e antigen antibody (HBeAb), and anticore antigen antibody (HBcAb). The early screening of various cancers also depends on the analysis of various tumor marker proteins, such as alpha-fetoprotein (AFP), carcinoembryonic antigen (CEA), prostate specific antigen (PSA), etc. Others such as cardiac troponin can be used for the diagnosis of myocardial infarction; C-reactive protein is used to indicate inflammation; and cystatin C is used for the diagnosis of renal function. In addition to this, the detection of various enzymatic activities can also be used in the diagnosis of diseases. For example, the detection of alanine aminotransferase activity in blood can be used for the analysis of liver function; glucose-6-phosphate dehydrogenase activity in blood can be used for the analysis of faba bean disease; lactate dehydrogenase activity is associated with multiple diseases such as myocardial infarction. Various new protein markers are still being discovered and used clinically. For example, in the diagnosis of osteoporosis, since the 21st century, serum osteocalcin, carboxy-terminal telopeptide of type I collagen (CTX), and amino-terminal propeptide of type I collagen (PINP) have successively replaced tartrate-resistant acid phosphatase (TRACP) and bone-specific alkaline phosphatase (BAP) to become new biomarkers. It is worth noting that all of them are proteins.

Although there are different ways to quantitatively detect proteins, such as mass spectrometry, electrophoresis etc., considering the sensitivity, specificity, cost, and other aspects, the best approach is based on molecular recognition. For nonenzymatic proteins, it is the recognition like well-known antigen-antibody couple, and aptamer recognition, supramolecular recognition, etc., which have been developed in recent years. For enzymes, it is the specific recognition between enzymes and their substrates. The detection method represented by enzyme-linked immunosorbent assay (ELISA) is widely used in hospitals, and its basis is the specific recognition between antigen and antibody. We will introduce the molecular recognition of proteins in the second part of the book.

How to convert the signals of molecular recognition into readable signals is the most critical part of protein detection. Related research is called biosensing. Various strategies and related technologies have been developed. Electrochemical analysis technology, for example, introduces electroactive molecules in the process of molecular recognition, thereby converting molecularly-recognized signals into electrochemical signals, and finally presenting detection results through electronic devices. The glucose meter is the most successful representative of electrochemical biosensing. Although it detects glucose, not protein, its principle is still in line with the universal pathway from molecular recognition (specific molecular recognition of glucose and glucose oxidase) to signal conversion (glucose oxidase catalyzes the oxidation of glucose and produces electroactive species). Immunofluorescence technology otherwise introduces fluorescent molecules in the course of immune reactions (antigen–antibody recognition), thereby converting molecularly recognized signals into fluorescent signals and presenting detection results through photosensitive elements. There are also techniques that directly output signals by measuring the physical or chemical properties of the environment during molecular recognition without the need to introduce additional signal molecules. For example, the surface plasmon resonance (SPR) technology achieves label-free target analysis by measuring the change in surface refractive index caused by interfacial molecular recognition. The development and application of various types of biosensing technology is the guarantee for the analysis of disease marker proteins. The relevant contents will be introduced in the third part of the book.

Although the above biosensing technology system from molecular recognition to signal conversion can achieve quantitative analysis of proteins, its performance (mainly sensitivity) is often not good enough. The low sensitivity on the one hand will result in the need for a large number of patient samples during the detection process. For example, assume that the detection limit of a biosensing technology for a protein is 1 μM, and that the protein content in a blood sample is exactly around 1 μM. This means that the blood sample cannot be diluted at all and may require 1 mL of blood during the test. And if the detection limit can be raised to 1 nM, the blood sample can be diluted up to 1000 times (the concentration of its protein target is reduced from 1 μM to 1 nM), thus only needing 1 μL of blood (1 μL of blood diluted 1000 times to 1 mL for testing). This not only reduces the interference of other substances in the blood, but also means that the patient's pain can be reduced during the test: 1 mL requires blood collection needle sampling, while 1 μL requires only minimally invasive blood collection needles. Or 1 mL of blood is still collected, but 1000 proteins can be analyzed theoretically: 1 mL blood is divided into 1000 parts, each 1 μL, for the detection of different proteins. Another important significance of sensitivity improvement is to help promote the discovery of new biomarkers. For a long time, an important factor constraining the discovery of new biomarkers was due to the fact that their low contents exceeded the detection limit at that time. It is also because of technological advances that the continuous improvement of detection limits has led to the discovery of various biomarkers with low abundance.

At present, one of the main ways to improve biosensing performance is the development and application of various nanomaterials and nanoprecision molecular assembly. The excellent properties of nanomaterials can bring about a qualitative improvement under reasonable design and construction. Raman spectra, for example, were discovered as early as 1928, but due to their low intensity, it was difficult to have substantial application in biochemical analysis until the application of laser light sources and the appearance of surface-enhanced Raman spectroscopy developed by Fleischmann et al. in 1974. By loading target analytes on silver nanoparticles and performing analysis, the Raman spectral intensity can be increased by 10^6 orders of magnitude, which is critical for the analysis of low-abundance materials. Gold nanoparticles, carbon nanotubes, graphene, and other nanomaterials are also excellent materials for loading biomolecules. Due to their high specific surface area, they can load a large variety of molecules. When signal molecules are loaded, signal amplification can be achieved through reasonable design. For example, in electrochemical biosensors, using various types of nanomaterials to load electroactive substances is an important way to achieve highly sensitive detection. When recognition molecules are loaded, more efficient target capture can be achieved on the one hand, and some special properties of nanomaterials can even be used to construct a completely new detection system on the other hand. For example, the initial pregnancy test was based on an ELISA strategy but was later replaced by a colloidal gold strategy, in which gold nanoparticles loaded with recognition molecules (antihuman chorionic gonadotrophin (hCG) antibody) were adopted. The red purple color developed by the deposition of gold nanoparticles at a specific location on the test paper achieves the purpose of signal output. In addition to the early pregnancy (urinary hCG), colloidal gold test strips are now widely used in the analysis of several marker proteins in HIV infection (serum HIV1/HIV2 antibody), fecal occult blood (fecal hemoglobin), renal function (urinary albumin) etc. because of their low cost and ease of use. Magnetic nanoparticle-loaded recognition molecules otherwise can achieve rapid and convenient separation and enrichment of target molecules under a magnetic field, and serve for subsequent detection and analysis. This is also a breakthrough method for the separation and enrichment of biomolecules and has been widely used. In addition to loading various functional molecules, some nanomaterials can also be used as signal reports. For example, QDs can provide higher fluorescence intensity and more stable fluorescence output than traditional organic fluorescent small molecules; upconversion nanomaterials can realize new fluorescence signal output by using an excitation light with long wavelength. By synthesizing the composite nanomaterials, it is also possible to integrate multiple properties of the nanomaterial components, thereby playing a more unique role in the biosensing system. All in all, the introduction of nanomaterials and nanoassembly has brought new horizons to biosensing, and has become a research hotspot in current biosensing. Regarding the properties of different nanomaterials and their application in biosensing, we will mainly introduce the relevant content in the first part of the book.

REFERENCES

Abdorahim, M., Rabiee, M., Alhosseini, S.N., Tahriri, M., Yazdanpanah, S., Alavi, S.H., et al., 2016. Nanomaterials-based electrochemical immunosensors for cardiac troponin recognition: an illustrated review. TrAC-Trend. Anal. Chem. 82, 337–347.

Afzal, A., Mujahid, A., Schirhagl, R., Bajwa, S., Latif, U., Feroz, S., 2017. Gravimetric viral diagnostics: QCM based biosensors for early detection of viruses. Chemosensors 5, 7.

Aldewachi, H., Chalati, T., Woodroofe, M.N., Bricklebank, N., Sharrack, B., Gardiner, P., 2018. Gold nanoparticle-based colorimetric biosensors. Nanoscale 10, 18–33.

Ali, J., Najeeb, J., Asim Ali, M., Farhan Aslam, M., Raza, A., 2017. Biosensors: their fundamentals, designs, types and most recent impactful applications: a review. J. Biosens. Bioelectron. 8, 1.

Ali, M.M., Li, F., Zhang, Z., Zhang, K., Kang, D.K., Ankrum, J.A., et al., 2014. Rolling circle amplification: a versatile tool for chemical biology, materials science and medicine. Chem. Soc. Rev. 43, 3324–3341.

Bahadir, E.B., Sezginturk, M.K., 2016. A review on impedimetric biosensors. Artif. Cell Nanomed. B. 44, 248–262.

Bhakta, S.A., Evans, E., Benavidez, T.E., Garcia, C.D., 2015. Protein adsorption onto nanomaterials for the development of biosensors and analytical devices: a review. Anal. Chim. Acta 872, 7–25.

Bock, L.C., Griffin, L.C., Latham, J.A., Vermaas, E.H., Toole, J.J., 1992. Selection of single-stranded DNA molecules that bind and inhibit human thrombin. Nature 355, 564–566.

Breaker, R.R., Joyce, G.F., 1994. A DNA enzyme that cleaves RNA. Chem. Biol. 1, 223–229.

Cao, Y., Wang, J., Xu, Y., Li, G., 2010. Combination of enzyme catalysis and electrocatalysis for biosensor fabrication: application to assay the activity of indoleamine 2,3-dioxygensae. Biosens. Bioelectron. 26, 87–91.

Cao, Y., Zhu, S., Yu, J., Zhu, X., Yin, Y., Li, G., 2012. Protein detection based on small molecule-linked DNA. Anal. Chem. 84, 4314–4320.

Cao, Y., Chen, W., Han, P., Wang, Z., Li, G., 2015. Target-driven self-assembly of stacking deoxyribonucleic acids for highly sensitive assay of proteins. Anal. Chim. Acta 890, 1–6.

Castro, C.M., Ghazani, A.A., Chung, J., Shao, H., Issadore, D., Yoon, T.J., et al., 2014. Miniaturized nuclear magnetic resonance platform for detection and profiling of circulating tumor cells. Lab Chip 14, 14–23.

Chen, C., Zhou, D., Tang, H., Liang, M., Jiang, J., 2013. A sensitive, homogeneous fluorescence assay for detection of thymine DNA glycosylase activity based on exonuclease-mediated amplification. Chem. Commun. 49, 5874–5876.

Chen, J., Andler, S.M., Goddard, J.M., Nugen, S.R., Rotello, V.M., 2017a. Integrating recognition elements with nanomaterials for bacteria sensing. Chem. Soc. Rev. 46, 1272–1283.

Chen, Y., Zhou, S., Li, L., Zhu, J.J., 2017b. Nanomaterials-based sensitive electrochemiluminescence biosensing. Nano Today 12, 98–115.

Cipolatti, E.P., Valério, A., Henriques, R.O., Moritz, D.E., Ninow, J.L., Freire, D.M.G., et al., 2016. Nanomaterials for biocatalyst immobilization – state of the art and future trends. RSC Adv. 6, 104675–104692.

Clark, L.C.J., Lyons, C., 1962. Electrode systems for continuous monitoring in cardiovascular surgery. Ann. N. Y. Acad. Sci. 102, 29–45.

Conroy, P.J., Hearty, S., Leonard, P., O'Kennedy, R.J., 2009. Antibody production, design and use for biosensor-based applications. Semin. Cell Dev. Biol. 20, 10–26.

Das, P., Das, M., Chinnadayyala, S.R., Singha, I.M., Goswami, P., 2016. Recent advances on developing 3rd generation enzyme electrode for biosensor applications. Biosens. Bioelectron. 79, 386–397.

Dhiman, A., Kalra, P., Bansal, V., Bruno, J.G., Sharma, T.K., 2017. Aptamer-based point-of-care diagnostic platforms. Sens. Actuat. B 246, 535–553.

Dirks, R.M., Pierce, N.A., 2014. Triggered amplification by hybridization chain reaction. Proc. Natl. Acad. Sci. U.S.A. 101, 15275–15278.

Ellington, A.D., Szostak, J.W., 1990. In vitro selection of RNA molecules that bind specific ligands. Nature 346, 818–822.

Farka, Z., Jurik, T., Kovar, D., Trnkova, L., Skladal, P., 2017. Nanoparticle-based immunochemical biosensors and assays: recent advances and challenges. Chem. Rev. 117, 9973–10042.

Feng, C., Dai, S., Wang, L., 2014. Optical aptasensors for quantitative detection of small biomolecules: a review. Biosens. Bioelectron. 59, 64–74.

Feng, C., Mao, X., Yang, Y., Zhu, X., Yin, Y., Li, G., 2016. Rolling circle amplification in electrochemical biosensor with biomedical applications. J. Electroanal. Chem. 781, 223–232.

Fu, J., Liu, M., Liu, Y., Woodbury, N.W., Yan, H., 2012. Interenzyme substrate diffusion for an enzyme cascade organized on spatially addressable DNA nanostructures. J. Am. Chem. Soc. 134, 5516–5519.

Gao, T., Liu, F., Yang, D., Yu, Y., Wang, Z., Li, G., 2015. Assembly of selective biomimetic surface on an electrode surface: a design of nano-bio interface for biosensing. Anal. Chem. 87, 5683–5689.

Garg, B., Bisht, T., Ling, Y.C., 2015. Graphene-based nanomaterials as efficient peroxidase mimetic catalysts for biosensing applications: an overview. Molecules 20, 14155–14190.

Gauchet, C., Labadie, G.R., Poulter, C.D., 2006. Regio- and chemoselective covalent immobilization of proteins through unnatural amino acids. J. Am. Chem. Soc. 128, 9274–9275.

Gaudin, V., 2017. Advances in biosensor development for the screening of antibiotic residues in food products of animal origin - a comprehensive review. Biosens. Bioelectron. 90, 363–377.

Gerasimova, Y.V., Kolpashchikov, D.M., 2014. Enzyme-assisted target recycling (EATR) for nucleic acid detection. Chem. Soc. Rev. 43, 6405–6438.

Gianneschi, N.G., Nguyen, S.N., Mirkin, C.M., 2005. Signal amplification and detection via a supramolecular allosteric catalyst. J. Am. Chem. Soc. 127, 1644–1645.

Gill, R., Zayats, M., Willner, I., 2008. Semiconductor quantum dots for bioanalysis. Angew. Chem. Int. Ed. 47, 7602–7625.

Goggins, S., Frost, C.G., 2016. Approaches towards molecular amplification for sensing. Analyst 141, 3157–3218.

Gopinath, P.G., Anitha, V.R., Mastani, S.A., 2015. Microcantilever based biosensor for disease detection applications. J. Med. Bioeng. 4, 307–311.

Gui, Q., Lawson, T., Shan, S., Yan, L., Liu, Y., 2017. The application of whole cell-based biosensors for use in environmental analysis and in medical diagnostics. Sensors 17, 1623.

Hasanzadeh, M., Shadjou, N., de la Guardia, M., 2015. Iron and iron-oxide magnetic nanoparticles as signal-amplification elements in electrochemical biosensing. TrAC-Trend. Anal. Chem. 72, 1–9.

Henry, A.I., Sharma, B., Cardinal, M.F., Kurouski, D., Van Duyne, R.P., 2016. Surface-enhanced raman spectroscopy biosensing: in vivo diagnostics and multimodal imaging. Anal. Chem. 88, 6638–6647.

Huang, X., Liu, Y., Yung, B., Xiong, Y., Chen, X., 2017. Nanotechnology-enhanced no-wash biosensors for in vitro diagnostics of cancer. ACS Nano 11, 5238–5292.

Husain, Q., 2017. Nanomaterials as novel supports for the immobilization of amylolytic enzymes and their applications: a review. Biocatalysis 3, 37–53.

Jamshaid, T., Neto, E.T.T., Eissa, M.M., Zine, N., Kunita, M.H., El-Salhi, A.E., et al., 2016. Magnetic particles: from preparation to lab-on-a-chip, biosensors, microsystems and microfluidics applications. TrAC-Trend. Anal. Chem. 79, 344–362.

Jung, C., Ellington, A.D., 2014. Diagnostic applications of nucleic acid circuits. Acc. Chem. Res. 47, 1825–1835.

Kokkinos, C., Economou, A., Prodromidis, M.I., 2016. Electrochemical immunosensors: critical survey of different architectures and transduction strategies. TrAC-Trend. Anal. Chem. 79, 88–105.

Kosman, J., Juskowiak, B., 2011. Peroxidase-mimicking DNAzymes for biosensing applications: a review. Anal. Chim. Acta 707, 7–17.

Krishnan, S., Mani, V., Wasalathanthri, D., Kumar, C.V., Rusling, J.F., 2011. Attomolar detection of a cancer biomarker protein in serum by surface plasmon resonance using superparamagnetic particle labels. Angew. Chem. Int. Ed. 50, 1175–1178.

Kubota, R., Hamachi, I., 2015. Protein recognition using synthetic small-molecular binders toward optical protein sensing in vitro and in live cells. Chem. Soc. Rev. 44, 4454–4471.

Kucherenko, I., Soldatkin, O., Kasap, B.O., Kirdeciler, S.K., Kurc, B.A., Jaffrezic-Renault, N., et al., 2015. Nanosized zeolites as a perspective material for conductometric biosensors creation. Nanoscale Res. Lett. 10, 209.

Kurbanoglu, S., Ozkan, S.A., Merkoci, A., 2017. Nanomaterials-based enzyme electrochemical biosensors operating through inhibition for biosensing applications. Biosens. Bioelectron. 89, 886–898.

Lan, L., Yao, Y., Ping, J., Ying, Y., 2017a. Recent advances in nanomaterial-based biosensors for antibiotics detection. Biosens. Bioelectron. 91, 504–514.

Lan, L., Yao, Y., Ping, J., Ying, Y., 2017b. Recent progress in nanomaterial-based optical aptamer assay for the detection of food chemical contaminants. ACS Appl. Mater. Interfaces 9, 23287–23301.

Lee, H., Shin, T.H., Cheon, J., Weissleder, R., 2015. Recent developments in magnetic diagnostic systems. Chem. Rev. 115, 10690–10724.

Lee, J.S., Song, J., Kim, S.O., Kim, S., Lee, W., Jackman, J.A., et al., 2016. Multifunctional hydrogel nano-probes for atomic force microscopy. Nat. Commun. 7, 11566.

Levicky, R.L., Herne, T.H., Tarlov, M.T., Satija, S.S., 1998. Using self-assembly to control the structure of DNA monolayers on gold: a neutron reflectivity study. J. Am. Chem. Soc. 120, 9787–9792.

Li, B., Ellington, A.D., Chen, X., 2011. Rational, modular adaptation of enzyme-free DNA circuits to multiple detection methods. Nucleic Acids Res. 39, e110.

Li, J., Sun, C.L., Shen, R., Cao, X.Y., Zhou, B., Bai, D.C., et al., 2014. An electrochemically switched smart surface for peptide immobilization and conformation control. J. Am. Chem. Soc. 136, 11050–11056.

Liang, C.L., Ma, P.M., Liu, H.L., Guo, X.G., Yin, B.Y., Ye, B.Y., 2017a. Rational engineering of a dynamic, entropy-driven DNA nanomachine for intracellular microRNA imaging. Angew. Chem. Int. Ed. 56, 9077–9081.

Liang, G., Man, Y., Li, A., Jin, X., Liu, X., Pan, L., 2017b. DNAzyme-based biosensor for detection of lead ion: a review. Microchem. J. 131, 145–153.

Lim, S.A., Ahmed, M.U., 2016. Electrochemical immunosensors and their recent nanomaterial-based signal amplification strategies: a review. RSC Adv. 6, 24995–25014.

Liu, M., Yin, Q., McConnell, E.M., Chang, Y., Brennan, J.D., Li, Y., 2018. DNAzyme feedback amplification: relaying molecular recognition to exponential DNA amplification. Chem. Eur. J. 24, 4473–4479.

Liu, Q., Wang, J., Boyd, B.J., 2015. Peptide-based biosensors. Talanta 136, 114–127.

Low, P.L., Henne, W.H., Doorneweerd, D.D., 2008. Discovery and development of folic-acid-based receptor targeting for imaging and therapy of cancer and inflammatory diseases. Acc. Chem. Res. 41, 120–129.

Lu, C.H., Li, J., Lin, M.H., Wang, Y.W., Yang, H.H., Chen, X., et al., 2010. Amplified aptamer-based assay through catalytic recycling of the analyte. Angew. Chem. Int. Ed. 49, 8454–8457.

Lu, L., Wang, X., Xiong, C., Yao, L., 2015. Recent advances in biological detection with magnetic nanoparticles as a useful tool. Sci. China: Chem. 58, 793–809.

Ma, F., Li, Y., Tang, B., Zhang, C.Y., 2016. Fluorescent biosensors based on single-molecule counting. Acc. Chem. Res. 49, 1722–1730.

Maduraiveeran, G., Sasidharan, M., Ganesan, V., 2018. Electrochemical sensor and biosensor platforms based on advanced nanomaterials for biological and biomedical applications. Biosens. Bioelectron. 103, 113–129.

Masson, J.F., 2017. Surface plasmon resonance clinical biosensors for medical diagnostics. ACS Sens. 2, 16–30.

Mauriz, E., García-Fernández, M.C., Lechuga, L.M., 2016. Towards the design of universal immunosurfaces for SPR-based assays: a review. TrAC-Trend. Anal. Chem. 79, 191–198.

Mi, J., Liu, Y., Rabbani, Z.N., Yang, Z., Urban, J.H., Sullenger, B.A., et al., 2010. In vivo selection of tumor-targeting RNA motifs. Nat. Chem. Biol. 6, 22–24.

Miao, P., Tang, Y., Wang, B., Yin, J., Ning, L., 2015. Signal amplification by enzymatic tools for nucleic acids. TrAC-Trend. Anal. Chem. 67, 1–15.

Mittal, S., Kaur, H., Gautam, N., Mantha, A.K., 2017. Biosensors for breast cancer diagnosis: a review of bioreceptors, biotransducers and signal amplification strategies. Biosens. Bioelectron. 88, 217–231.

Mohanty, S.P., Kougianos, E., 2006. Biosensors: a tutorial review. IEEE Potentials 25, 35–40.

Moreira, F.T.C., Sale, M.G.F., Di Lorenzo, M., 2017. Towards timely alzheimer diagnosis: a self-powered amperometric biosensor for the neurotransmitter acetylcholine. Biosens. Bioelectron. 87, 607–614.

Neubrech, F., Huck, C., Weber, K., Pucci, A., Giessen, H., 2017. Surface-enhanced infrared spectroscopy using resonant nanoantennas. Chem. Rev. 117, 5110–5145.

Pavan, S., Berti, F., 2012. Short peptides as biosensor transducers. Anal. Bioanal. Chem. 402, 3055–3070.

Pelossof, G., Tel-Vered, R., Willner, I., 2012. Amplified surface plasmon resonance and electrochemical detection of Pb2+ ions using the Pb2+-dependent DNAzyme and hemin/G-quadruplex as a label. Anal. Chem. 84, 3703–3709.

Peng, F., Su, Y., Zhong, Y., Fan, C., Lee, S., He, Y., 2014. Silicon nanomaterials platform for bioimaging, biosensing, and cancer therapy. Acc. Chem. Res. 47, 512–623.

Perumal, V., Hashim, U., 2014. Advances in biosensors: principle, architecture and applications. J. Appl. Biomed. 12, 1–15.

Pode, Z., Peri-Naor, R., Georgeson, J.M., Ilani, T., Kiss, V., Unger, T., et al., 2017. Protein recognition by a pattern-generating fluorescent molecular probe. Nat. Nanotechnol. 12, 1161–1168.

Pramanik, D., Sengupta, K., Mukherjee, S., Dey, S.G., Dey, A., 2012. Self-assembled monolayers of Abeta peptides on Au electrodes: an artificial platform for probing the reactivity of redox active metals and cofactors relevant to Alzheimer's disease. J. Am. Chem. Soc. 134, 12180–12189.

Puiu, M., Bala, C., 2018. Peptide-based biosensors: from self-assembled interfaces to molecular probes in electrochemical assays. Bioelectrochemistry 120, 66–75.

Putzbach, W., Ronkainen, N.J., 2013. Immobilization techniques in the fabrication of nanomaterial-based electrochemical biosensors: a review. Sensors 13, 4811–4840.

Ray, M., Ray, A., Dash, S., Mishra, A., Achary, K.G., Nayak, S., et al., 2017. Fungal disease detection in plants: traditional assays, novel diagnostic techniques and biosensors. Biosens. Bioelectron. 87, 708–723.

Reverdatto, S.R., Burz, D.B., Shekhtman, A.S., 2015. Peptide aptamers: development and applications. Curr. Top. Med. Chem. 15, 1082–1101.

Rezaei, B., Ghani, M., Shoushtari, A.M., Rabiee, M., 2016. Electrochemical biosensors based on nanofibres for cardiac biomarker detection: a comprehensive review. Biosens. Bioelectron. 78, 513–523.

Roda, A., Mirasoli, M., Michelini, E., Di Fusco, M., Zangheri, M., Cevenini, L., et al., 2016. Progress in chemical luminescence-based biosensors: a critical review. Biosens. Bioelectron. 76, 164–179.

Saidur, M.R., Abdul Aziz, A.R., Basirun, W.J., 2017. Recent advances in DNA-based electrochemical biosensors for heavy metal ion detection: a review. Biosens. Bioelectron. 90, 125–139.

Sarkar, D., Liu, D., Xie, J., Anselmo, A.C., Mitragotri, S., Banerjee, K., 2014. MoS2 field-effect transistor for next generation label-free biosensors. ACS Nano 8, 3992–4003.

Scott, D.E., Bayly, A.R., Abell, C., Skidmore, J., 2016. Small molecules, big targets: drug discovery faces the protein-protein interaction challenge. Nat. Rev. Drug Discov. 15, 533–550.

Scrimin, P., Prins, L.J., 2011. Sensing through signal amplification. Chem. Soc. Rev. 40, 4488–4505.

Sefah, K., Shangguan, D., Xiong, X., O'Donoghue, M.B., Tan, W.H., 2010. Development of DNA aptamers using Cell-SELEX. Nat. Protoc. 5, 1169–1185.

Su, Y., Hickey, S.F., Keyser, S.G., Hammond, M.C., 2016. In vitro and in vivo enzyme activity screening via RNA-based fluorescent biosensors for S-Adenosyl-l-homocysteine (SAH). J. Am. Chem. Soc. 138, 7040–7047.

Tang, J., Tang, D., 2015. Non-enzymatic electrochemical immunoassay using noble metal nanoparticles: a review. Microchim. Acta 182, 2077–2089.

Tarasov, A., Gray, D.W., Tsai, M.Y., Shields, N., Montrose, A., Creedon, N., et al., 2016. A potentiometric biosensor for rapid on-site disease diagnostics. Biosens. Bioelectron. 79, 669–678.

Tiwari, J.N., Vij, V., Kemp, K.C., Kim, K.S., 2016. Engineered carbon-nanomaterial-based electrochemical sensors for biomolecules. ACS Nano 10, 46–80.

Travascio, P.T., Li, Y.L., Sen, D.S., 1998. DNA-enhanced peroxidase activity of a DNA aptamer-hemin complex. Chem. Biol. 5, 505–517.

Trilling, A.K., Beekwilder, J., Zuilhof, H., 2013. Antibody orientation on biosensor surfaces: a mini review. Analyst 138, 1619–1627.

Turek, C., Gold, L., 1990. Systematic evolution of ligands by exponential enrichment: RNA ligands to bacteriophage T4 DNA polymerase. Science 249, 505–510.

Valera, E., Hernández-Albors, A., Marco, M.P., 2016. Electrochemical coding strategies using metallic nanoprobes for biosensing applications. TrAC-Trend. Anal. Chem. 79, 9–22.

van de Velde, L., d'Angremont, E., Olthuis, W., 2016. Solid contact potassium selective electrodes for biomedical applications - a review. Talanta 160, 56–65.

Viswambari Devi, R., Doble, M., Verma, R.S., 2015. Nanomaterials for early detection of cancer biomarker with special emphasis on gold nanoparticles in immunoassays/sensors. Biosens. Bioelectron. 68, 688–698.

Wang, L., Lei, J., Ma, R., Ju, H., 2013. Host-guest interaction of adamantine with a beta-cyclodextrin-functionalized AuPd bimetallic nanoprobe for ultrasensitive electrochemical immunoassay of small molecules. Anal. Chem. 85, 6505–6510.

Wang, L., Xiong, Q., Xiao, F., Duan, H., 2017a. 2D nanomaterials based electrochemical biosensors for cancer diagnosis. Biosens. Bioelectron. 89, 136–151.

Wang, X., Lu, X., Chen, J., 2014. Development of biosensor technologies for analysis of environmental contaminants. TrAC-Trend. Anal. Chem. 2, 25–32.

Wang, Z., Dai, Z., 2015. Carbon nanomaterial-based electrochemical biosensors: an overview. Nanoscale 7, 6420–6431.

Wang, Z., Han, P., Mao, X., Yin, Y., Cao, Y., 2017b. Sensitive detection of glutathione by using DNA-templated copper nanoparticles as electrochemical reporters. Sens. Actuat. B 238, 325–330.

Wang, Z.G., Wilner, O.I., Willner, I., 2009. Self-assembly of aptamer-circular DNA nanostructures for controlled biocatalysis. Nano Lett. 9, 4098–4102.

Wegner, K.D., Hildebrandt, N., 2015. Quantum dots: bright and versatile in vitro and in vivo fluorescence imaging biosensors. Chem. Soc. Rev. 44, 4792–4834.

Wei, T., Du, D., Zhu, M.J., Lin, Y., Dai, Z., 2016. An improved ultrasensitive enzyme-linked immunosorbent assay using hydrangea-like antibody-enzyme-inorganic three-in-one nanocomposites. ACS Appl. Mater. Interfaces 8, 6329–6335.

Wen, J., Xu, Y., Li, H., Lu, A., Sun, S., 2015. Recent applications of carbon nanomaterials in fluorescence biosensing and bioimaging. Chem. Commun. 51, 11346–11358.

Willner, I., Shlyahovsky, B., Zayats, M., Willner, B., 2008. DNAzymes for sensing, nanobiotechnology and logic gate applications. Chem. Soc. Rev. 37, 1153–1165.

Wilson, G.W., Hu, Y.H., 2000. Enzyme-based biosensors for in vivo measurements. Chem. Rev. 100, 2693–2704.

Wu, L., Xiong, E., Zhang, X., Zhang, X., Chen, J., 2014. Nanomaterials as signal amplification elements in DNA-based electrochemical sensing. Nano Today 9, 197–211.

Yan, L., Zhou, J., Zheng, Y., Gamson, A.S., Roembke, B.T., Nakayama, S., et al., 2014. Isothermal amplified detection of DNA and RNA. Mol. BioSyst. 10, 970–1003.

Yan, X., Li, H., Su, X., 2018. Review of optical sensors for pesticides. TrAC-Trends Anal. Chem. Available from: https://doi.org/10.1016/j.trac.2018.03.004.

Yang, L., Gomez-Casado, A., Young, J.F., Nguyen, H.D., Cabanas-Danes, J., Huskens, J., et al., 2012. Reversible and oriented immobilization of ferrocene-modified proteins. J. Am. Chem. Soc. 134, 19199–19206.

Yang, Y., Yang, X., Yang, Y., Yuan, Q., 2018. Aptamer-functionalized carbon nanomaterials electrochemical sensors for detecting cancer relevant biomolecules. Carbon 129, 380–395.

Yao, J., Yang, M., Duan, Y., 2014. Chemistry, biology, and medicine of fluorescent nanomaterials and related systems: new insights into biosensing, bioimaging, genomics, diagnostics, and therapy. Chem. Rev. 114, 6130–6178.

Ye, D.K., Zuo, X.L., Fan, C., 2018. DNA nanotechnology enabled interfacial engineering for biosensor development. Annu. Rev. Anal. Chem. 11, 3.1–3.25.

Yin, Y., Cao, Y., Xu, Y., Li, G., 2010. Colorimetric immunoassay for detection of tumor markers. Int. J. Mol. Sci. 11, 5077–5094.

Zhang, H., Li, F., Dever, B., Li, X.F., Le, X.C., 2013. DNA-mediated homogeneous binding assays for nucleic acids and proteins. Chem. Rev. 113, 2812–2841.

Zhang, R., Chen, W., 2017. Recent advances in graphene-based nanomaterials for fabricating electrochemical hydrogen peroxide sensors. Biosens. Bioelectron. 89, 249–268.

Zhang, S., Geryak, R., Geldmeier, J., Kim, S., Tsukruk, V.V., 2017. Synthesis, assembly, and applications of hybrid nanostructures for biosensing. Chem. Rev. 117, 12942–13038.

Zhang, W., Asiri, A.M., Liu, D., Du, D., Lin, Y., 2014. Nanomaterial-based biosensors for environmental and biological monitoring of organophosphorus pesticides and nerve agents. TrAC-Trend. Anal. Chem. 54, 1–10.

Zhao, Y., Chen, F., Li, Q., Wang, L., Fan, C., 2015. Isothermal amplification of nucleic acids. Chem. Rev. 115, 12491–12545.

Zheng, Z., Sun, H., Hu, C., Li, G., Liu, X., Chen, P., et al., 2016. Using "On/Off" ^{19}F NMR/magnetic resonance imaging signals to sense tyrosine kinase/phosphatase activity in vitro and in cell lysates. Anal. Chem. 88, 3363–3368.

Zhou, W., Huang, P.J., Ding, J., Liu, J., 2014. Aptamer-based biosensors for biomedical diagnostics. Analyst 139, 2627–2640.

Zhu, X., Feng, C., Ye, Z., Chen, Y., Li, G., 2014. Fabrication of magneto-controlled moveable architecture to develop reusable electrochemical biosensors. Sci. Rep. 4, 4169.

Zuo, X.Z., Xia, F.X., Xiao, Y.X., Plaxco, K.P., 2010. Sensitive and selective amplified fluorescence DNA detection based on exonuclease III-aided target recycling. J. Am. Chem. Soc. 132, 1816–1818.

Part I

Nanomaterials for Protein Assay

Following the sensing principle described above and the clinical application of nano-inspired biosensors in protein assays, in part II, we mainly summarize the design and fabrication of biosensors based on the employment of different kinds of nanomaterials, such as carbon nanomaterials, metal nanomaterials, quantum dots, and nanoclusters, etc. Thanks to the special functions and properties of the employed nanomaterials, many kinds of novel sensing systems have been created by using the recognition mechanism, such as materials–proteins, protein–proteins, and aptamer–proteins, etc. Furthermore, various signal output strategies and signal amplification approaches have been possible to be designed and utilized. In this part, some typical examples showing the applications of these biosensing systems for protein assays are presented, by using different techniques for signal output, such as colorimetry, fluorescence, surface plasmon resonance, surface-enhanced Raman scattering, and electrochemistry, etc.

Chapter 1

Carbon Nanomaterials

Jingjing Xu[1] and Lei Wang[2]

[1]Center for Molecular Recognition and Biosensing, School of Life Sciences, Shanghai University, Shanghai, P. R. China, [2]State Key Laboratory of Pharmaceutical Biotechnology, Department of Biochemistry, Nanjing University, Nanjing, P. R. China

Chapter Outline

1.1 INTRODUCTION

Carbon is a very important element in the natural world. Due to its various electron orbital features (sp, sp2, sp3), there are many unique structures and properties for carbon (Bhattacharya and Samanta, 2016; Tan et al., 2017). With the rapid development of nanotechnology, the use of carbon nanomaterials has been increasingly widespread. After experiencing an upsurge of interest over the past several decades, researches on carbon nanomaterials

Nano-inspired Biosensors for Protein Assay with Clinical Applications.
DOI: https://doi.org/10.1016/B978-0-12-815053-5.00001-5

represented by fullerenes, carbon nanotubes, and graphene have now come into a steady development stage. With the deepening of research, it has been shown that carbon nanomaterials have more and more irreplaceable effects on human production and life (Yang et al., 2015).

Carbon nanomaterials have excellent physical properties like mechanical, optical, and electronic properties, as well as good chemical versatility and biocompatibility. They have become a promising platform especially in the sensing field (Yang et al., 2010). Owing to the advantages such as high surface to volume ratio, high conductivity, chemical stability, and strong mechanical strength, carbon nanomaterials-based biosensors are good choices for protein assay.

Some chemical, optical, and immunological methods have been used to detect protein macromolecules. Biosensors have received much attention due to their abilities of quick detection of specific proteins (Liu et al., 2015b). A wide variety of carbon nanomaterials can be combined with these characteristics to prepare a corresponding carbon-based sensor, which allows for more sensitive detection of proteins. As graphene and graphene oxide are excellent fluorescence acceptors with outstanding performance, a large number of immunosensors have been designed by means of fluorescence resonance energy transfer where graphene and graphene oxide were chosen as quenchers (Chang et al., 2010; Lu et al., 2011; Shi et al., 2015; Wang et al., 2011a; Wang et al., 2012). Other carbon nanomaterials, like carbon quantum dots, have unique optical properties and are prepared as elements of carbon quantum dot-based optical sensors (Sidhu et al., 2017). In addition, owing to the excellent electrical conductivity, good mechanical strength, and high surface area-to-weight ratio, graphene, graphene oxide (GO), and carbon nanotubes have been universally used to fabricate electrochemistry biosensors for protein. These biosensors based on carbon nanomaterials are generally constructed in two forms: one is directly modified on an electrode or other interfaces (Majd and Salimi, 2018; Munge et al., 2010; Sardesai et al., 2011), and the other can act as carriers for signaling labels (Feng et al., 2014; Guo et al., 2011). Besides, carbon nanomaterials can also be used as biomolecule recognition probes to bind protein by covalent or noncovalent methods to achieve the aim of protein detection (Georgakilas et al., 2012; Zheng et al., 2003a). Compared with ordinary sensors, there is still much room for improvement of carbon-based sensors, such as reducing the detection cost, improving the detection performance of biological samples, and realizing high-throughput composite detection.

In this review, the recent progress of carbon nanomaterials-based biosensors, which are used to detect protein, are overviewed. We introduce the characteristics of carbon nanomaterials, such as fullerenes, graphene and its derivatives, and carbon nanotubes, and their applications on protein detection by preparing corresponding sensors with their unique electrochemical and optical properties.

1.2 FULLERENE

1.2.1 The Fundamental Properties of Fullerene

In 1985 Kroto found a new form of molecular carbon, C_{60}. It has a definite composition, 60 carbon atoms constituting a football-like 32 panel body, including 20 hexagons and 12 pentagons (Kroto, 1997). This structure was inspired by the architect Buckminster Fuller. Fuller has designed a hexagonal and pentagonal spherical shell structure, which was named as Fullerene (Krätschmer et al., 1990). In the 1990s, theories and applications of fullerene were developed rapidly. At present, researches and applications of fullerene have been involved in many disciplines like chemistry and life science. There exists great potential for its applications. The three-dimensional structure and numerous double bonds of fullerenes provide great room for fullerene development and indicate many new properties and functions to be discovered. The structure of fullerene is shown in Fig. 1.1.

After the discovery of fullerene (C_{60}), a drastic change in the research of biosensors took place. With the unique topological attributions and electrochemical properties, such as broad light absorption in the UV−vis region, photothermal effect, structural angle strain, the ability to accommodate multiple electrons and endohedral metal atoms, long-living triplet state, singlet oxygen production, as well as the ability to act as an electron acceptor with a dual nature of electrophilic and nucleophilic characteristics, fullerene has

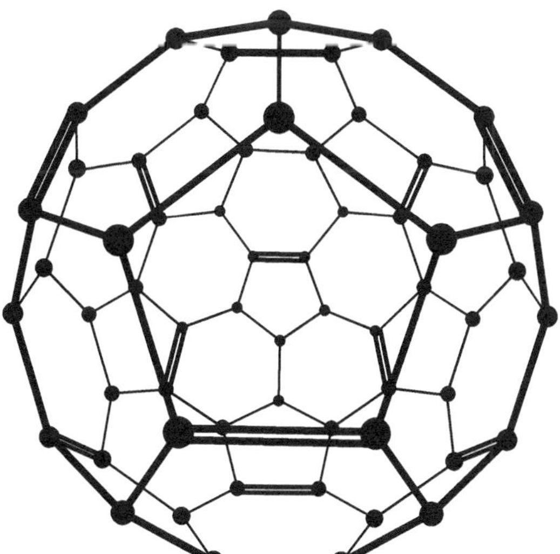

FIGURE 1.1 Schematic representation of the structure of fullerene. *Reprinted from Haufler, R.E., Conceicao, J., Chibante, L.P.F., Chai, Y., Byrne, N.E., Flanagan, S., et al., 1990. Efficient production of C_{60} (buckminsterfullerene), $C_{60}H_{36}$, and the solvated buckide ion. J. Phys. Chem. (USA), 94, 8634–8636.*

promoted continued interest for researchers to investigate the possibilities of its applications on biosensor devices (Biju, 2014). The outcomes are inspiring, since fullerene has been successfully used in developing biosensors for protein detection (Afreen et al., 2015).

1.2.2 The Fullerene-Modified Electrodes for Protein Assay

The elemental fullerene alone is not an effective material for constructing a highly sensitive biosensor. It acts as a mediator in biosensing systems. The effective mediator of a biosensor should be hydrophilic and possess some reactive functional groups, so that it could establish a link between the recognition site of the target molecule and the electrode surface, which could therefore promote subsequent biochemical reactions. However, elemental fullerenes are hydrophobic and insoluble in polar solvents, making it difficult to couple with many biologically molecules such as proteins, DNA, and the like. Once some suitable functionalized functional groups, such as carboxyl groups, hydroxyl groups, and amine groups, are introduced, fullerenes become hydrophilic in nature and they are characterized by reactive functional groups or free radicals. Therefore, it is a great challenge for researchers to obtain potential derivatives of fullerenes by modifications on them for certain proposes. The function of fullerene as a mediator can be shown in the Fig. 1.2.

It has been reported that numerous studies of C60-based sensors could auxiliarily diagnose different types of clinical diseases by detecting proteins that are abnormal in the human body. There have been many strategies for developing immunosensors based on fullerene derivatives or immobilized fullerenes because they can strongly adsorb proteins. Proteins such as hemoglobin and myoglobin contain abundant NH_2 groups that can chemically bond with fullerene molecules to form immobilized C60-Hb and C60-Mb. Chang et al. developed a surface SAW system for detection of

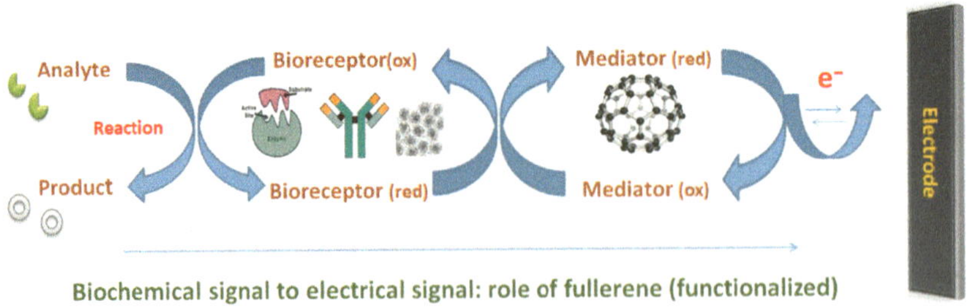

FIGURE 1.2 Function of fullerene at the interface of recognition site and electrode. *Reprinted from Afreen, S., Muthoosamy, K., Manickam, S., Hashim, U., 2015. Functionalized fullerene (C60) as a potential nanomediator in the fabrication of highly sensitive biosensors. Biosens. Bioelectron. 63, 354–364.*

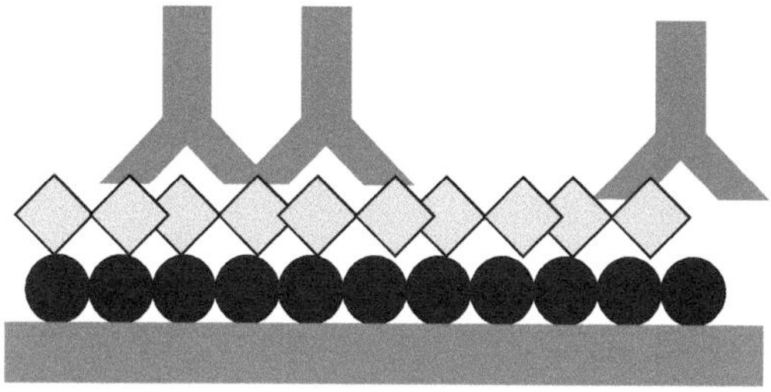

FIGURE 1.3 Diagram of the C_{60}-Hb-coated $LiTaO_3$ SAW crystal for anti-Hb antibody. *Reprinted from Chang, H.-W., Shih, J.-S., 2007. Surface acoustic wave immunosensors based on immobilized C60-proteins. Sens. Actuators B-Chem. 121, 522–529.*

antihemoglobin (anti-Hb) using C60-Hb and C60-Mb coatings (see Fig. 1.3) (Chang and Shih 2007).

Sheng et al. introduced fullerene–nitrogen-doped carbon nanotubes and chitosan (C60-NCNTs/CHIT) composite matrix. They used cyclic voltammetry and electrochemical impedance spectroscopy to characterize modified electrodes, demonstrating that using NCNT-modified fullerenes could achieve dramatic improvements on the properties of fullerenes involving electronic conductivity, stability, and biocompatibility. The direct electrochemistry of C60-NCNTs/CHIT composite matrix immobilized hemoglobin (Hb) was demonstrated. Simultaneously, C60-NCNT/CHIT modified electrode can catalyze the reduction of H_2O_2 for sensitive determination of H_2O_2 in biological and pharmaceutical samples (Sheng et al., 2013).

Yuan et al. constructed a novel sandwich-type biosensor that could detect α2,3-sialic acid-Gs using 4-mercaptophenylboronic acid (4-MPBA). They synthesized amino-functionalized fullerenes and palladium–platinum bimetallic alloy nanocrystals (n-C_{60}-PdPt) to modify the surface of the glassy carbon electrode, using 4-MPBA having a thiol group that can bind to the PdPt alloy by strong adsorption. The 4-MPBA is immobilized on n-C_{60}-PdPt, and a molecular recognition system is formed by the coordination of the boron atom of 4-MPBA with the amide group of Neu5Ac in α2,3-sial-Gs structure. Mackia amurensis lectin (MAL) could act as a signal amplification component by being covalently immobilized on Au-poly (Au-PMB) and be used to identify α2,3-sial-Gs. Because n-C_{60} nanomaterial provides a large surface area and excellent electron transfer ability for in situ reduction of bimetallic alloy nanoparticles, the detection limit of the sensor is much lower (see Fig. 1.4; Yuan et al., 2018).

As is shown in Fig. 1.5, Zhu et al. developed a novel self-assembly method to construct a linear bicontinuous donor–acceptor (H2BCPP) n-C_{60} array on an indium tin oxide electrode. This (H2BCPP) n-C_{60} array can enhance the photocurrent generation capability and therefore provide a highly effective and sensitive photoelectrochemical immunosensing platform

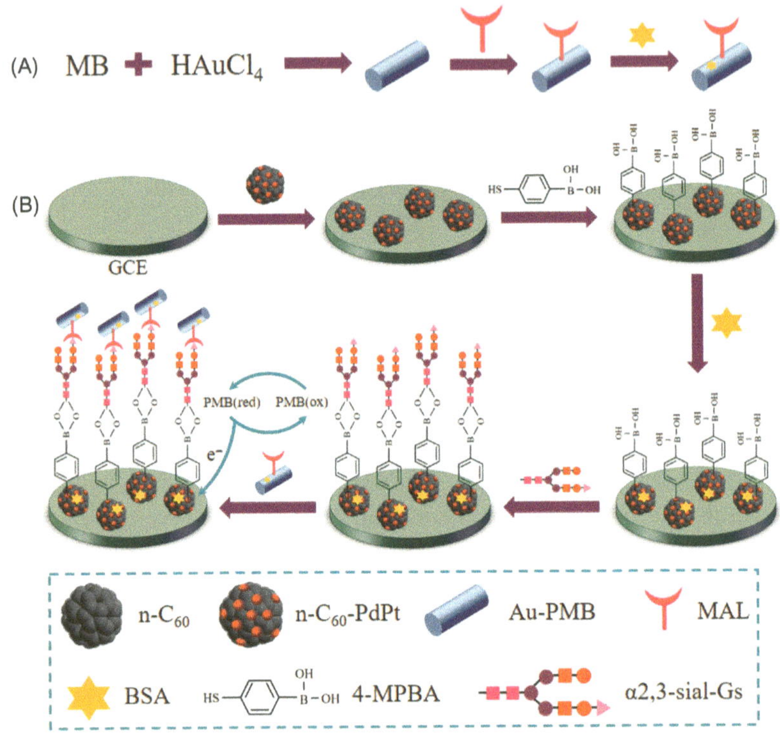

FIGURE 1.4 Sandwich-type biosensor for α2,3-sialylated glycan detection based on fullerene–palladium–platinum alloy. *Reprinted from Yuan, Q., He, J., Niu, Y., Chen, J., Zhao, Y., Zhang, Y., et al., 2018. Sandwich-type biosensor for the detection of α2, 3-sialylated glycans based on fullerene-palladium-platinum alloy and 4-mercaptophenylboronic acid nanoparticle hybrids coupled with Au-methylene blue-MAL signal amplification. Biosens. Bioelectron. 102, 321–327.*

for carcinoembryonic antigen (CEA) detection. Quantitative measurements of CEA were based on a decrease in photocurrent intensity of the (H_2BCPP) n-C_{60} array which resulted from competitions between CEA and CEA-CdTe. The photocurrent drop was linear with the CEA concentration in the range of 0.01–60 ng mL^{-1} and the detection limitation was 3.4 pg mL^{-1} (Zhu et al., 2015).

1.2.3 Fullerene-Based Piezoelectric Quartz Crystal for Protein Assay

Piezoelectric quartz crystal sensor is a new product of sensing technology. Quartz crystal oscillation frequency is highly sensitive to the changes of crystal surface mass loading and reaction system physical properties such as density, viscosity, conductivity, and so on, with sub-ng levels of mass

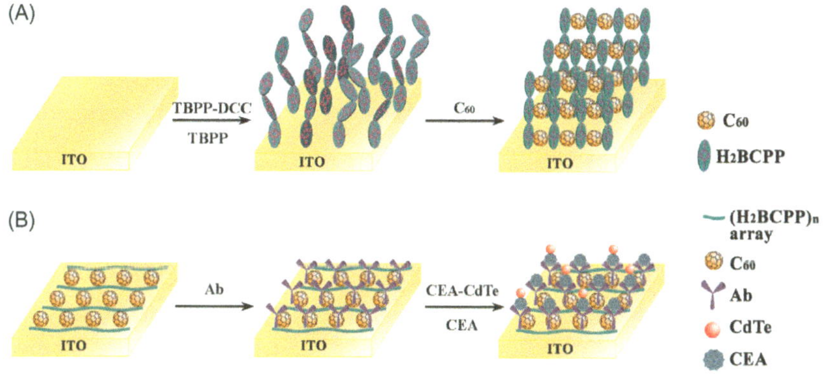

FIGURE 1.5 Schematic molecular structure of 5,15-bis(4-carboxyphenyl)10,20-bis(mesityl) porphyrin (H_2BCPP). *Reprinted from Zhu, P., Wang, P., Kan, L., Sun, G., Zhang, Y., Yu, J., 2015. An enhanced photoelectrochemical immunosensing platform: supramolecular donor–acceptor arrays by assembly of porphyrin and C60. Biosens. Bioelectron. 68, 604–610.*

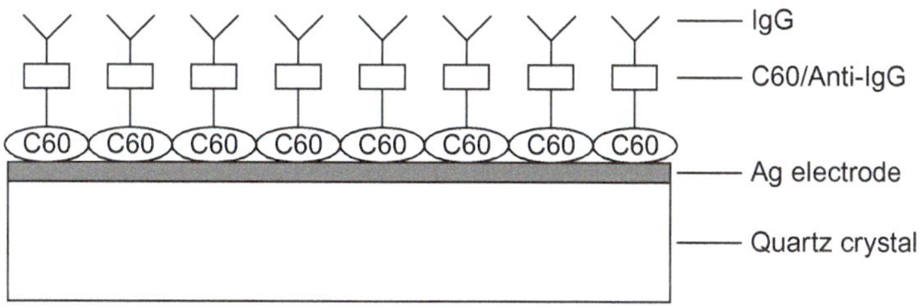

FIGURE 1.6 Scheme of the C60-antihuman IgG coated piezoelectric (PZ) crystal detection system and quartz crystal electrode for IgG. *Reprinted from Pan, N.-Y., Shih, J.-S., 2004. Piezoelectric crystal immunosensors based on immobilized fullerene C60-antibodies. Sens. Actuators B-Chem. 98, 180–187.*

detection. As is shown in Fig. 1.6, Pan and Shih constructed the C_{60}-antihuman IgG and C_{60}-anti-Hb coated piezoelectric (PZ) quartz crystal immunosensors and immobilized fullerene to detect immunoglobulin G (IgG) and hemoglobin (Hb) in aqueous solutions using antibodies. This process can be used to indirectly diagnose disorders of the immune system. The C_{60}-antibody-coated quartz crystals were coated with C_{60} on quartz crystals which were bound with silver (Ag)-metal electrodes followed by antibody adsorption (Pan and Shih, 2004).

A similar observation on C_{60}-myoglobin (C_{60}-Mb), C_{60}-Hb, and C_{60}-gliadin-coated piezoelectric (PZ) quartz crystal immunosensors was reported by Liao and Shih. The optimum frequency response for these antiprotein PZ-immunosensors was observed at pH 7.0 with a temperature around 30°C. Sensitivity of C_{60}-Mb, C_{60}-Hb, and C_{60}-gliadin PZ-immunosensors were 1.43×10^3, 2.59×10^3, and 8.05×10^3 Hz/(mg/mL) and their detection limits

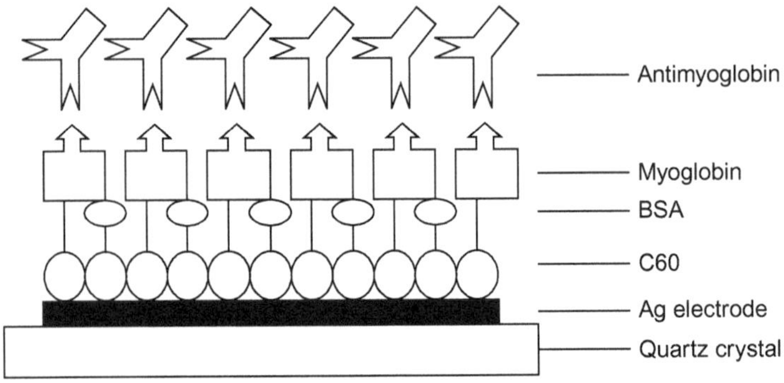

FIGURE 1.7 The scheme of immobilized C_{60}-myoglobin on quartz crystal for adsorption of antimyoglobin in the sample solution (BSA = Bovine serum Albumn). *Reprinted from Liao, Y. H., Shih, J.S., 2013. Piezoelectric quartz crystal anti-protein immunosensors based on immobilized fullerene C60-proteins. J. Chin. Chem. Soc. 60, 1387–1395.*

were 4.36×10^{-3}, 3.23×10^{-3}, and 1.98×10^{-3} mg mL^{-1} respectively. The interference induced by other biological species found in the human blood like cysteine, tyrosine, urea, glucose, ascorbic acid, and metal ions was negligible (see Fig. 1.7) (Liao and Shih, 2013).

1.3 GRAPHENE AND GRAPHENE OXIDE

1.3.1 The Fundamental Properties of Graphene and Graphene Oxide

Graphene is a monatomic-thick two-dimensional planar sheet of sp2-bonded honeycomb crystal (see Fig. 1.8A; Geim and Novoselov, 2007). Graphene oxide (GO) possesses a similar planar sheet lattice but contains oxygen atoms such as carboxyl on its edges and esters, hydroxyl groups, and epoxides on planes (see Fig. 1.8B and C; Du et al., 2011; Lerf et al., 1998; Robinson et al., 2008). Due to those oxygen-containing functional groups, GO has a high affinity to water molecules and can be dissolved in water. Besides, GO has good mechanical strength that can extend tens of nanometers to tens of micrometers. With these functional groups, GO can be easily decorated with biomolecules (Loh et al., 2010; Pumera 2011; Wang et al., 2011b). Recently, these two-dimensional nanomaterials have attracted more attention and research (Compton and Nguyen, 2010; Loh et al., 2010; Park and Ruoff, 2009).

FRET is a phenomenon where photoexcited energy from a donor fluorescence group transfers to a receptor molecule. Graphene can be used as a high efficient fluorescent quenching agent because of its electronic properties, as well as similarities and dissimilarities with metal surfaces. In fact, a lot of experimental results have shown that GO can act as the best FRET acceptor compared with other carbon structures, such as graphite, carbon

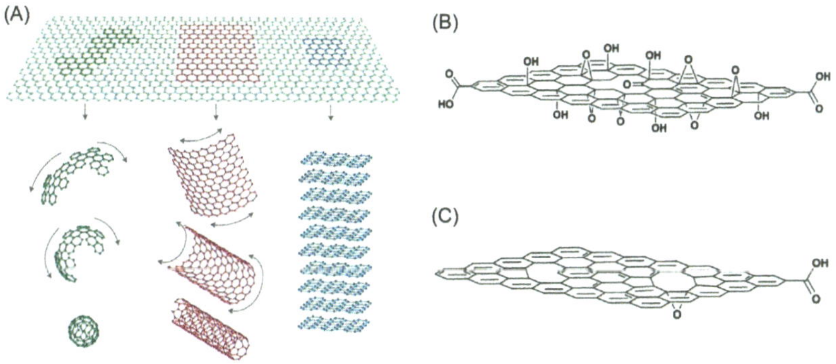

FIGURE 1.8 The structure representations of graphenes and GO. (A) The different carbon allotropes based on graphene from left to right: fullerene, nanotube, and graphite. Copyright 2007, Nature Publishing Group. (B) Sketch of GO and (C) reduced GO. *Figure 1.8 (A): Reprinted from Geim, A.K., Novoselov, K.S., 2007. The rise of graphene. Nat. Mater. 6, 183–191. Figure 1.8 (B) and (C): Reprinted from Compton, O.C., & Nguyen, S.T., 2010. Graphene oxide, highly reduced graphene oxide, and graphene: versatile building blocks for carbon-based materials. Small, 6, 711–723.*

nanotubes, and carbon nanofibers (Morales-Narváez et al., 2012). Therefore, GO has been widely exploited as a highly efficient long-rang quencher (Morales-Narvaez and Merkoci, 2012).

1.3.2 The Highly Efficient Quencher of Graphene and Graphene Oxide for Protein Assay

The oxygenated lattice of GO can specifically recognize biomolecules through electrostatic interaction, $\pi-\pi$ stacking bond, and hydrogen bonding (Chen et al., 2011). For example, GO can strongly interact with single-stranded DNA by $\pi-\pi$ stacking interaction (Zheng et al., 2003b). Lu and his colleagues were the first to establish a biosensor platform with GO performing as a high quencher in FRET (Lu et al., 2009).

In terms of protein detection, Liu et al. have exploited GO sheets modified with capture antibodies by hydrothermal treatment and quantum dots modified by detection antibodies. After adding analytes (a-fetoprotein, AFP), the emission of quantum dots was quenched through forming sandwich immune complexes (Liu et al., 2010; see Fig. 1.9).

Graphene and graphene oxide have been successfully developed to sensitively detect various proteins based on their high efficiency of energy-transfer in FRET. Han et al. reported a new method to detect Glypican-3 (GPC3), a prognostic indicator in early-stage cancer, using graphene oxide as a high fluorescence quenching agent (Han et al., 2017). As is shown in Fig. 1.10, αGCC specifically recognized the C-terminal domain of the GPC3 core protein, and αGCN specifically inhibited the remaining part. First, GO was individually precoated with αGCC and αGCN. In the presence of

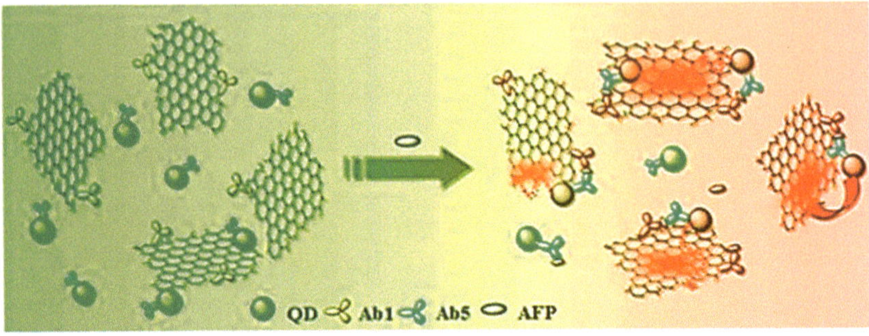

FIGURE 1.9 Schematic representation of the change of quantum dots fluorescence based on self-assembled QD and nanoscale-graphene in immunoassay. *Reprinted from Liu, M., Zhao, H., Quan, X., Chen, S., Fan, X., 2010. Distance-independent quenching of quantum dots by nanoscale-graphene in self-assembled sandwich immunoassay. Chem. Commun. (Camb) 46, 7909–7911.*

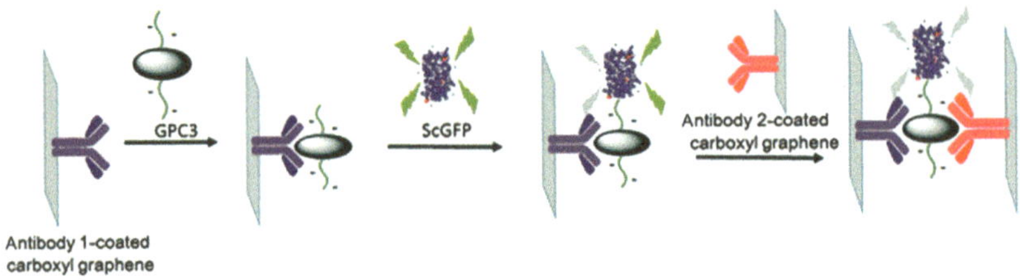

FIGURE 1.10 The construction process of homogeneous sandwich method for GPC3 detection. *Reprinted from Han, N., Wang, W., Lu, D., Wang, H., Ma, X., Fan, X., et al., 2017. A novel, rapid, and sensitive homogeneous sandwich detection method of Glypican-3 as a serum marker for hepatocellular carcinoma. Chem. Commun. (Camb) 53, 12209–12212.*

GPC3, the αGCC–coated graphene complexes and the αGCN–coated graphene complexes could specifically capture GPC3 molecules. After adding ScGFP, GPC3 would combine the HS side chain of ScGFP through electrostatic interaction. Finally, the fluorescence of ScGFP was completely quenched by graphene oxide. The concentration of GPC3 was positively correlated with the fluorescence quenching, and the detection limit was 15.6 pg mL^{-1}.

In the majority of studies on graphene-based FRET sensors, graphene and its derivative materials are quenchers. Graphene oxide, an important derivative of graphene, is a precursor of graphene prepared by chemical reduction. The presence of oxygen-containing groups makes GO an ideal material with better water solubility and can be applied to biological systems. Due to their large area of conjugated structure, graphene and GO can be used as energy receptors to quench the fluorescence of various organic dyes and quantum dots. It is a versatile fluorescence quencher. Compared with the traditional quencher, graphene material has a higher quenching efficiency, which could

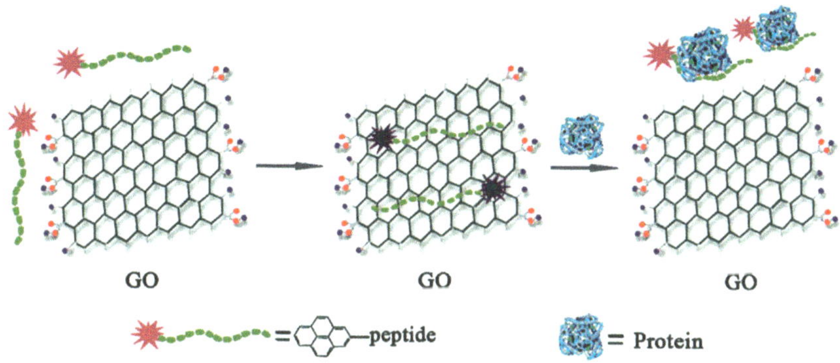

FIGURE 1.11 General approach for monitoring peptide–protein interactions based on graphene–peptide complex. *Reprinted from Lu, C.-H., Li, J., Zhang, X.-L., Zheng, A.-X., Yang, H.-H., Chen, X., et al., 2011. General approach for monitoring peptide–protein interactions based on graphene–peptide complex. Anal. Chem. 83, 7276–7282.*

endow FRET sensors distinctive features such as low background, high signal to noise ratio, and multiple detections (Su et al., 2011). As is shown in Fig. 1.11, Lu et al. designed an ultrasensitive fluorescence sensor to monitor peptide–protein interactions. The peptide was labeled by helium and then strongly adsorbed on GO surface by hydrophobic interactions and π-π interactions. When GO is close to helium, fluorescence of helium can be effectively quenched. In the presence of the target protein, the binding force of target protein to peptide is greater than that of peptide to GO, resulting in the separation of peptide from GO and restoration of fluorescence signal (Lu et al., 2011).

Based on the fluorescence quenching effect of GO, Wang et al. also constructed a fluorescence sensor to detect caspase-3 (see Fig. 1.12). The peptide chain is labeled with a fluorophore and attached to the GO surface by the succinimide coupling (EDC–NHS) reaction. Caspase-3 is able to specifically hydrolyze the peptide, which makes it possible to keep the fluorophore away from GO, eventually leading to a recovery of the fluorescence signal (Wang et al., 2011a).

Chang et al. developed a fluorescence resonance energy transfer (FRET) aptamer sensor for thrombin detection. They labeled the aptamer with FAM and assembled noncovalently with graphene, which quenched FAM fluorescence through FRET. When thrombin was added, it competed with graphene to bind to an aptamer, which allowed fluorescence to recover. This graphene aptamer sensor can detect thrombin in buffer and serum with exceptional sensitivity and specificity (Chang et al., 2010; see Fig. 1.13).

GO is not only an outstanding fluorescence quencher, but also a good chemiluminescence and color quenching agent. Chemiluminescence resonance energy transfer (CRET) is a nonradiative dipole–dipole whose energy

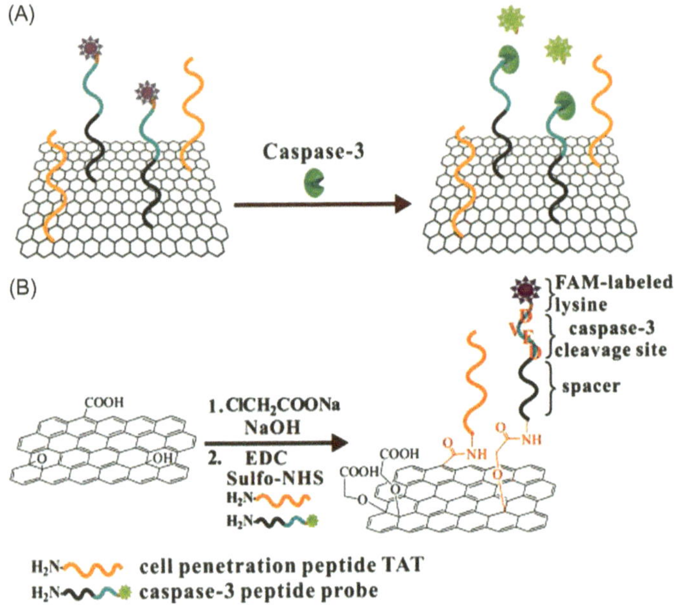

FIGURE 1.12 (A) Caspase-3 detection using GO–peptide conjugate and caspase-3-mediated cleavage of the peptide. (B) The construction of GO–peptide conjugate. *Reprinted from Wang, H., Zhang, Q., Chu, X., Chen, T., Ge, J., Yu, R., 2011a. Graphene oxide–peptide conjugate as an intracellular protease sensor for caspase-3 activation imaging in live cells. Angew. Chem., Int. Ed. 50, 7065–7069.*

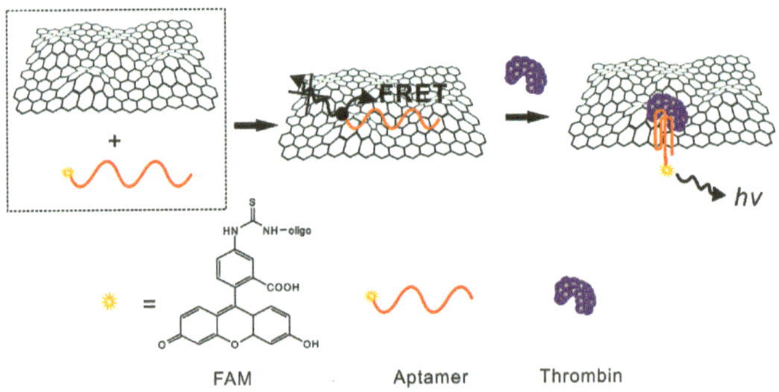

FIGURE 1.13 Schematic demonstration of graphene FRET aptasensor and the detection mechanism for Thrombin. *Reprinted from Chang, H., Tang, L., Wang, Y., Jiang, J., Li, J., 2010. Graphene fluorescence resonance energy transfer aptasensor for the thrombin detection. Anal. Chem. 82, 2341–2346.*

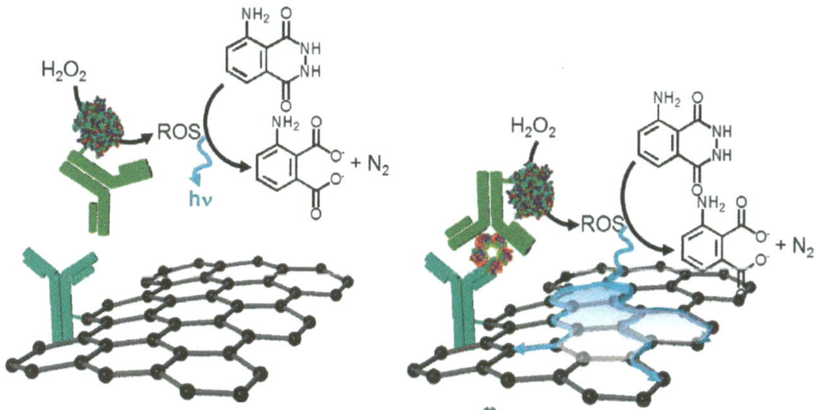

FIGURE 1.14 Schematic illustration of the biosensor for C-reactive protein (CRP) through a graphene-based chemiluminescence resonance energy transfer (CRET). Anti-CRP antibody-conjugated graphene was chosen as an acceptor and luminol excited by horseradish peroxidase (HRP)-catalyzed oxidation are used as a donor. *Reprinted from Lee, J.S., Min-Gon Kim, H.-A.J., Park, C.B., 2012. Graphene-based chemiluminescence resonance energy transfer for homogeneous immunoassay. ACS Nano 6, 2978–2983.*

is transferred from a chemiluminescent donor to an acceptor. CTET is produced by specific oxidation of luminous substrate without an external excitation source; thus, it can avoid the drawbacks of FRET like needing external photoexcitation. Lee et al. developed a CRET immunosensor platform based on graphene acceptor and chemiluminescence donors (Lee et al., 2012). As is shown in Fig. 1.14, they used horseradish peroxidase (HRP) to catalyze luminol/hydrogen peroxide CL reaction. Simply, HRP can catalyze hydrogen peroxide to generate reactive oxygen species which could react with luminol to form high energy species. Then the high energy species decomposed with loss of nitrogen and emitted a strong blue fluorescence at 430 nm. They used C-reactive protein (CRP) as a model analyte for biomolecules recognition. It plays a key role in human inflammation and cardiovascular diseases. The detection limit of CRP in human serum is 0.93 ng mL^{-1}.

Li et al. reported a new signal amplification scheme consisting of small chromotropic dyes and antibodies modified carboxyl graphene oxide (cGO) (see Fig. 1.15; Li et al., 2016). Carboxygraphene oxide was used as the carrier and quenching agent of small molecule chromotropic dyes. This simple, sensitive, and multicolor detection method was called allochroic-cGO linked immunosorbent assay (ALISA). When just adding acidic or alkaline water to the test solution, the chromatic dye dissociated quickly from cGO. When the target protein was added, the antibody, target, and graphene complex would form a sandwich structure. After the PH of the detection system decreased, the additional malachite green carbinol base (MGCB) dissociated with the

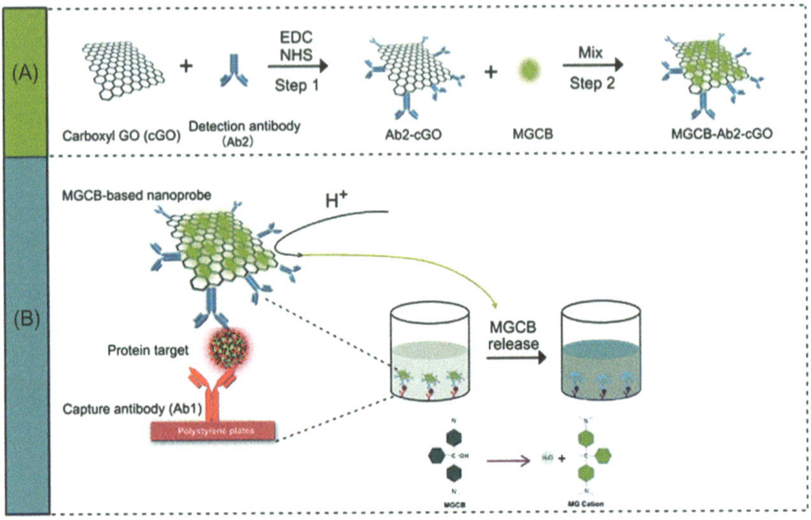

FIGURE 1.15 Schematic illustration of the allochroic-cGO linked immunosorbent assay (ALISA) for protein detection manipulated in 96-well polystyrene (PS) plates. *Reprinted from Li, C., Yang, Y., Wu, D., Li, T., Yin, Y., Li, G., 2016. Improvement of enzyme-linked immunosorbent assay for the multicolor detection of biomarkers. Chem. Sci. 7, 3011–3016.*

cGO nanosheet and the color of the MGCB was recovered. Due to the high adsorption amount and wide selection range of dye molecules, the detection limit for protein biomarkers was successfully achieved at picogram level for the first time.

Yuan et al designed a graphene oxide cationic conjugated polymer probe to detect calmodulin (CaM) conformation changes induced by Ca^{2+} (Yuan et al., 2015). The flexible CaM was transformed into rigid α-helical structure when binding with Ca^{2+}, which exposed the surface of CaM to be more hydrophobic and with less negative charges (see Fig. 1.16). Therefore, Ca^{2+} (apoCaM) would lead to more CaM adsorption to the surface of GO. Hence, the distance between poly[(9,9-bis(6′-*N*,*N*,*N*-trimethylammonium)hexyl)-fluorenylene phenylene dibromide] (PFP) and green fluorescent protein labeled at the *N*-terminus of CaM was farther and leads to a much weaker FRET. The conformation changes of CaM and the assembly of CaM with GO and PFP could be observed intuitively under ultraviolet irradiation.

In recent years, fluorescent carbon nanoparticles (or carbon dots) have been synthesized and applied in protein assay due to their low-cost, unique optical properties, and good biocompatibility (Baker and Baker, 2010; Esteves da Silva and Gonçalves, 2011; Guo et al., 2013). A fluorescence "off-to-on" approach for detecting different biological species such as protein is exploited. The biosensor is based on the functionalized AgNPs and highly fluorescent n-butylamine functionalized-grapheme oxides (GO-NHBu) nanosheets (Mei and Zhang, 2012; see Fig. 1.17). The platform of carbon dots could be used to detect peptides, proteins, and DNAs. The fluorescence

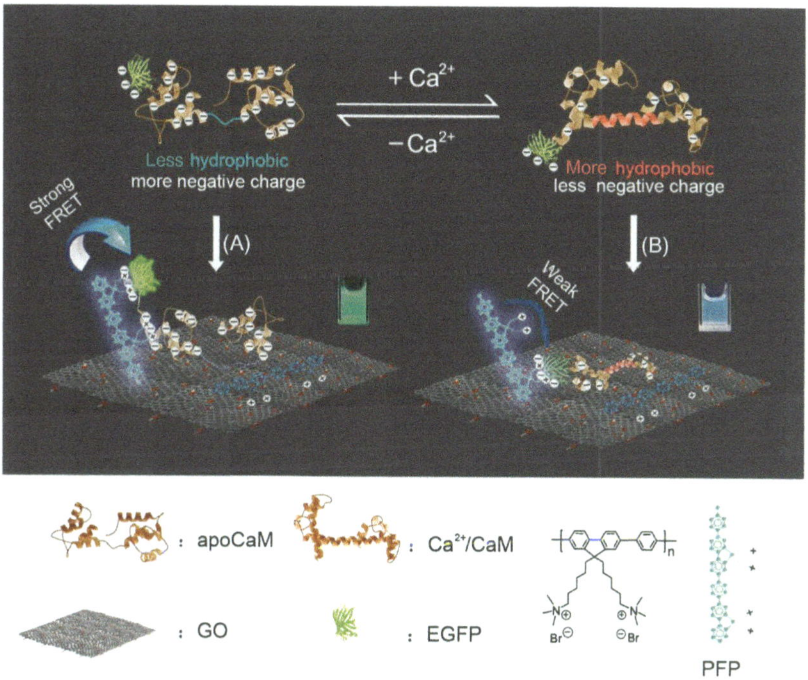

FIGURE 1.16 Schematic illustration of the detection for conformation changes of calmodulin based on GO-PFP hybrid probe. *Reprinted from Yuan H., Qi J., Xing C., An H., Niu R., Zhan Y., et al., 2015. Graphene-oxide-conjugated polymer hybrid materials for calmodulin sensing by using FRET strategy, Adv. Funct. Mater. 25, 4412–4418*

of GO-NHBu nanosheets could be strongly quenched by AgNPs through FRET or charge-transfer process. In the presence of the corresponding analytes, AgNPs could specifically recongnize analytes via donor–acceptor interactions, leading to the dissociation between AgNPs and the GO-NHBu nanosheets. Then the fluorescence of GO-NHBu nanosheets would recover immediately. This method provides a good platform to assay other proteins by using corresponding specific aptamers (Xu et al., 2012).

A fluoroimmunosensing strategy based on regulation of the interaction between graphene (Gr) and graphene quantum dots (GQDs) was designed for the detection of human immunoglobulin G (IgG, antigen) (see Fig. 1.18; Zhao et al., 2013a,b). In the sensor, graphene acted as an acceptor and mouse antihuman immunoglobulin G (mIgG, antibody)-conjugated GQDs (mIgGGQDs) as donors. Upon the addition of Gr to the mIgG-GQDs solution, the fluorescence of GQDs would be quenched due to FRET between Gr and GQDs and the nonspecific binding interaction between mIgG and the Gr surface. IgG could be successfully detected through the change of the fluorescence and the detection limit was 10 ng mL^{-1}. The method could also be used to detect other proteins only by replacing the mIgG with other antibodies functionalized GQDs.

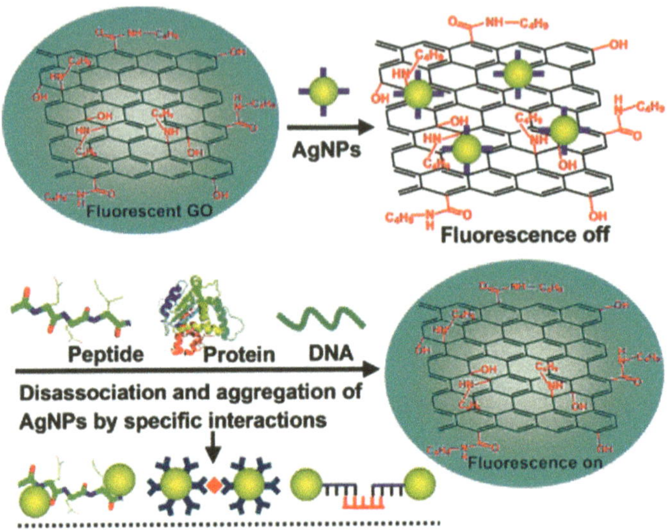

FIGURE 1.17 Schematic illustration of the "off-to-on" mechanism of GO-NHBu nanosheets for biological assays. The functionalized AgNPs are adsorbed on photoluminescent GO nanosheets, making the GO fluorescence extinguishing. The addition of target species results in the disassociation and aggregation of AgNPs, and the fluorescence of GO is recovered again. *Reprinted from Mei, Q., Zhang, Z., 2012. Photoluminescent graphene oxide ink to print sensors onto microporous membranes for versatile visualization bioassays. Angew. Chem. Int. Ed. 51, 5602–5606.*

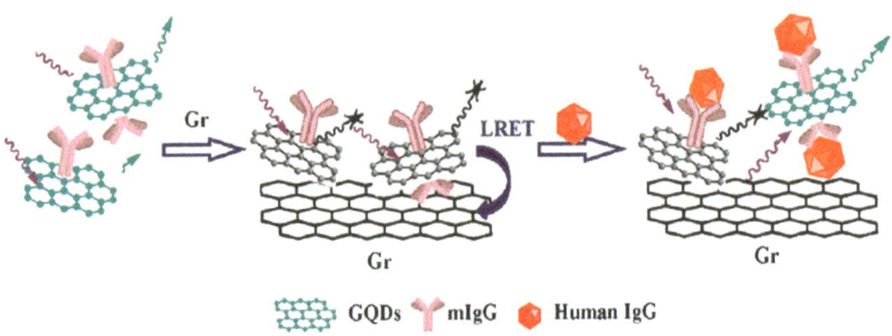

FIGURE 1.18 Schematic representation of a universal immunosensing platform based on regulation of the interaction between Gr and GQDs. *Reprinted from Zhao, H., Chang, Y., Liu, M., Gao, S., Yu, H., & Quan, X., 2013. A universal immunosensing strategy based on regulation of the interaction between graphene and graphene quantum dots. Chem. Commun., 49, 234–236.*

1.3.3 Graphene and Graphene Oxide-Modified Electrodes for Protein Assay

Because graphene and graphene oxide are very sensitive to the changes of electrons, the detection of proteins by electrochemistry based on carbon nanomaterials often achieves high sensitivity (Zhang et al., 2013).

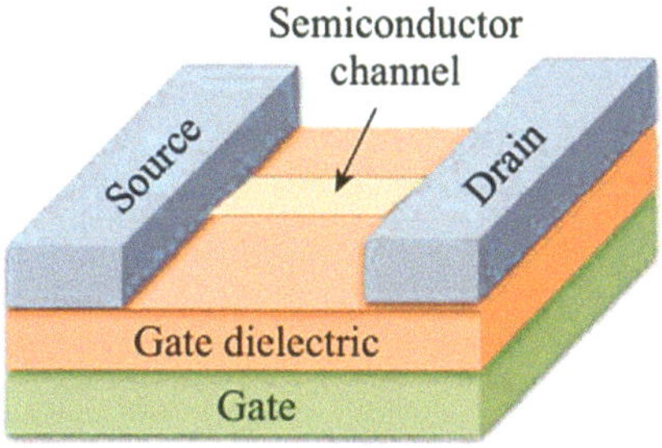

FIGURE 1.19 Device considerations for development of conductance-based biosensors. *Reprinted from Zhu, Z., 2017. An overview of carbon nanotubes and graphene for biosensing applications. Nano-Micro Lett. 9, 25.*

A field effect transistor (FET) is a semiconductor device that controls the output circuit current by controlling the electric field effect of the input circuit, and has a high input resistance, low noise, low power consumption, large dynamic range, easy integration, no secondary breakdown phenomenon, wide safe working area, etc. The working principle is as shown in Fig. 1.19, the drain current flows through the channel between the drain and source, and the drain current is controlled by the reversed-biased formed by the PN junction between the gate and the channel (Zhu, 2017).

Graphene has been developed to fabricate biosensors for protein detection based on FET (He et al., 2011; Kwon et al., 2012; Mao et al., 2010; Mao et al., 2011; Park et al., 2012; Stine et al., 2013). Thermally reduced graphene oxide (TRGO) sheets were chosen to construct a highly sensitive and selective FET biosensor for IgG (Mao et al., 2010; Fig. 1.20). TRGO sheets were deposited on the electrodes and AuNPs-antibody were absorbed to TRGO sheets via noncovalent interactions. After the adding of the target proteins (i.e., IgG) the significant changes in the electrical characteristics of the biosensor would occur because of the antigen–antibody interaction. Through the FET of graphene and direct current measurements, the changes of the electrical characteristics could be observed. The limit of detection of the novel method was 2 ng mL^{-1} (<13 pM) for IgG.

FET-based nitrogen-doped graphene was also used to fabricate a sensitive biosensor for the detection of vascular endothelial growth factor (VEGF) (Kwon et al., 2012; Fig. 1.21). Polypyrrole was deposited on the Cu substrate through vapor deposition and then was grown on Cu substrate by deposition polymerization. The VEGF aptamer was immobilized on the side plane of graphene by glutaraldehyde-conjugated 1,5-diaminonaphthalene, which could attach to the graphene via $\pi-\pi$ stacking interaction. In the present of VEGF the current of the biosensor would change, which was used to detect VEGF as low as 100 fM.

Many studies show that screen-printed graphene electrodes are useful tools for constructing highly sensitive protein sensors. The sensors for hCG

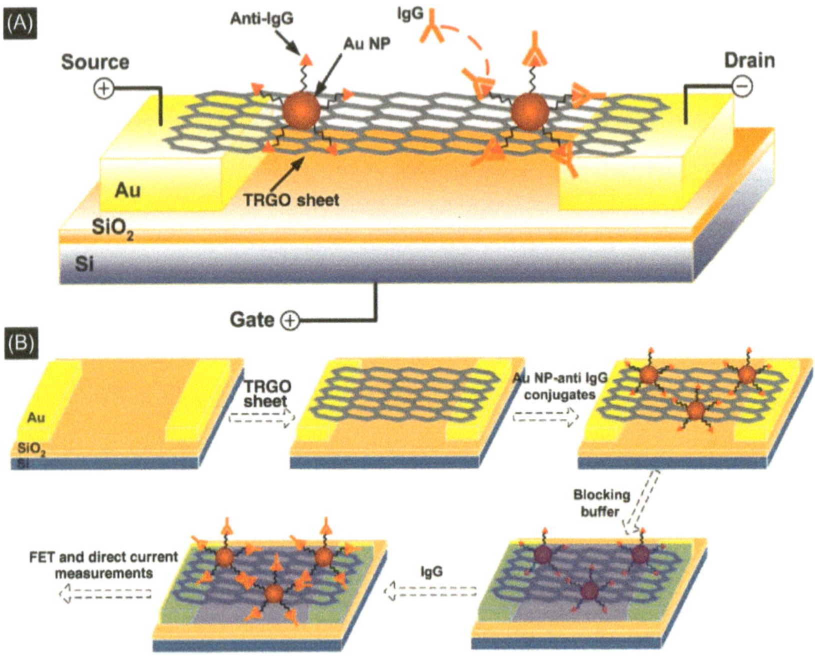

FIGURE 1.20 (A) Schematic representation of a TRGO FET. Anti-IgG is fixed on the TRGO sheet surface based on AuNPs and a specific function group for the IgG binding. The electrical detection of protein binding (IgG to anti-IgG) is achieved through FET and direct current measurements. (B) Schematic representation of the fabrication procedure for TRGO FET biosensor. TRGO sheets were firstly dispersed on the electrodes and then modified with Au. *Reprinted from Mao, S., Lu, G., Yu, K., Bo, Z., Chen, J., 2010. Specific protein detection using thermally reduced graphene oxide sheet decorated with gold nanoparticle-antibody conjugates. Adv. Mater. 22, 3521–3526.*

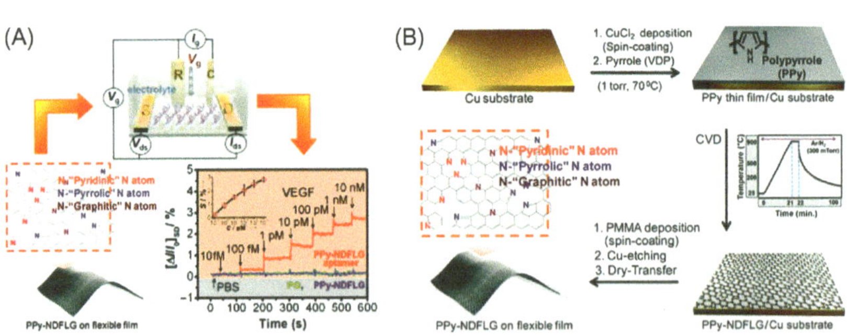

FIGURE 1.21 (A) Schematic representation of the nitrogen-doped graphene-based FET biosensor for VEGF. (B) Synthetic protocol of PPy-NDFLG on flexible substrate. *Reprinted from Kwon, O.S., Park, S.J., Hong, J.-Y., Han, A.R., Lee, J.S., Lee, J.S., et al., 2012. Flexible FET-type vegf aptasensor based on nitrogen-doped graphene converted from conducting polymer. ACS Nano 6, 1486–1493.*

detection based on screen-printed graphene electrodes was developed. hCG is a diagnostic marker for pregnancy, ovarian, and testicular cancers. Through immobilizing anti-hCG on commercially available screen-printed graphene electrodes via physical adsorption, Ahmed and coworkers constructed an hCG biosensor and the detection limit was 5 pg mL^{-1} (Lim et al., 2014). On the other hand, Sales and coworkers designed a label-free hCG biosensor based on a screen-printed graphene electrode coated with poly(aniline) film (Teixeira et al., 2014b). This sensor could detect hCG in a linear range from 0.001 to 50 ng mL^{-1} in a urine sample and the limit of detection was 0.286 pg mL^{-1} through impedimetric assay.

Reduced graphene oxide (rGO) is also combined with paper-based microfluidics to construct immunosensors for protein biomarkers (Kumar et al., 2015b; Wu et al., 2014; Wu et al., 2013). Wu et al. fabricated microfluidic electrochemical devices on a cellulose paper by photolithography, in which eight working electrodes, reference electrode, and counter electrode were screen printed with carbon ink (Wu et al., 2013). GO was modified on the working electrodes via drop-cast dispersion followed by electrochemical reduction to rGO in order for further modification with antibodies. According to the principle immunosensor, four kinds of antibodies for AFP, CEA, CA125, and CA153 were immobilized on the working electrodes. This sensor showed good voltammetry responses for AFP, CEA, CA125, and CA153 in the concentration ranges of 0.001–100, 0.005–100, 0.001–100, and 0.005–100 ng mL^{-1}, respectively. Other biosensors based on electrodes modified with several graphene derivatives including N-doped rGO (Li et al., 2013), epitaxially grown multilayered GO (Teixeira et al., 2014a), and three-dimensional macroporous GO foams (Liu et al., 2015a) were also developed to improve the performance of protein assay.

1.3.4 Graphene and Graphene Oxide as Signaling Labels in Electrochemistry for Protein Assay

Graphene, due to its large surface area and easy surface modification, is often used as a scaffold for the construction of electrochemical immunosensor signal labels (Gao et al., 2014; Li et al., 2015). Ma and colleagues developed a biosensor for simultaneous detection of CEA and AFP based on carbonylated GO sheets modified with toluidine blue (TB), anti-CEA or PB, and anti-AFP (Chen et al., 2013). The redox signals of modified GO sheets in DPV were related to the concentration of analytes. The detection limit of these sensors for CEA and AFP was down to 0.1 and 0.05 ng mL^{-1}, respectively. They also developed different signaling labels such as GO/PB/Au/ionic liquid nanocomposites and GO/CuS/Cu_2S redox couple for the detection of AFP.

Graphene has been also utilized to fabricate electrogenerated chemiluminescence (ECL) immunosensor. Cao et al. designed an electrochemiluminescence (ECL) aptamer sensor based on GO@AuNRs linking streptavidin (SA) and glucose oxidase (GOD) for the detection of a specific antigen (PSA). In their principle, first, GOD and SA-biotin-DNA were loaded on GO@AuNR, and this conjugate was used as a signal probe. Second, gold (DpAu) was deposited on the glassy carbon electrode to stimulate ECL, and s1-PSA aptamer-dsDNA attached to a modified electrode. After that, the signal probe is connected to the electrode through the DNA complementary pairing. In this way, the GOD loaded on the probe can catalyze the in situ generation of H_2O_2 in the glucose, and then the AuNR catalyzes the reaction of the luminol reagent with H_2O_2 to generate reactive oxygen species (ROS). This cascade of chemical reactions amplifies the ECL signal in the ECL system and improves the sensitivity of biosensing. In the presence of PSA, signal probes can be separated from the electrodes due to the competitive binding of protein and DNA to aptamers, resulting in the inability to generate the abovementioned cascade reaction and no ECL signal (see Fig. 1.22; Cao et al., 2018).

Zhao et al. designed an ultrasensitive electrochemiluminescence biosensor based on GO/AgNPs/luminol complex signal amplification to measure DNA adenine methylation (MTase). First, they prepared the GO/AgNPs/luminol complex which is used as a signal probe. Second, the capture probe DNA on the surface of the gold electrode is hybridized with the

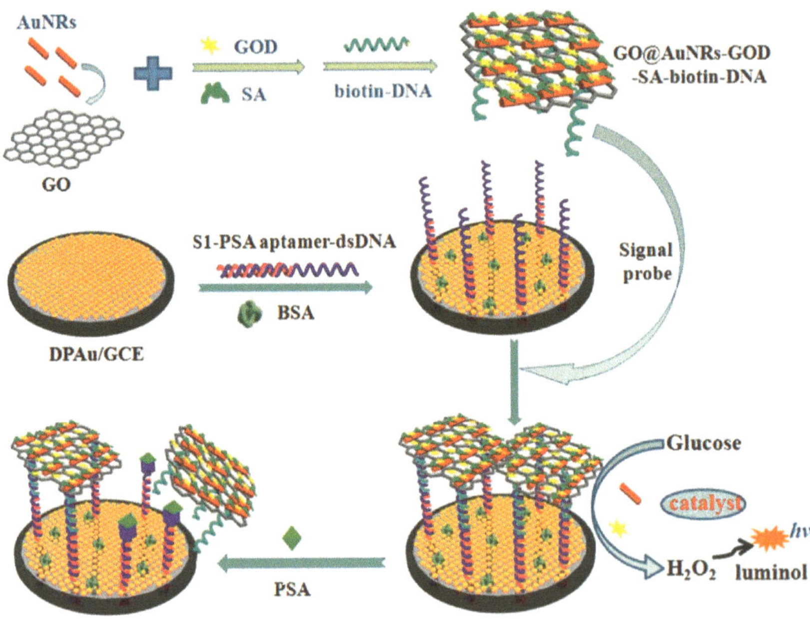

FIGURE 1.22 Schematic illustration of the ECL biosensor for detection of PSA. *Reprinted from Cao, J.-T., Yang, J.-J., Zhao, L.-Z., Wang, Y.-L., Wang, H., Liu, Y.-M., et al., 2018. Graphene oxide@gold nanorods-based multiple-assisted electrochemiluminescence signal amplificationstrategy for sensitive detection of prostate specific antigen. Biosens. Bioelectron. 99, 92–98.*

azide-attached DNA. Then, the signaling probes are fixed to the azide-terminated double-stranded DNA by click chemistry, resulting in a high electrochemiluminescence (ECL) signal. In the presence of Dam MTase, DNA hybrids are methylated and further cleaved by Dpn I endonucleases, signaling probes are released from the electrode surface into solution, resulting in a dramatic reduction in ECL signal. The quantification of DNA adenine methylation (Dam) methyltransferase (MTase) can be accomplished by measuring change in ECL signal (see Fig. 1.23; Zhao et al., 2015).

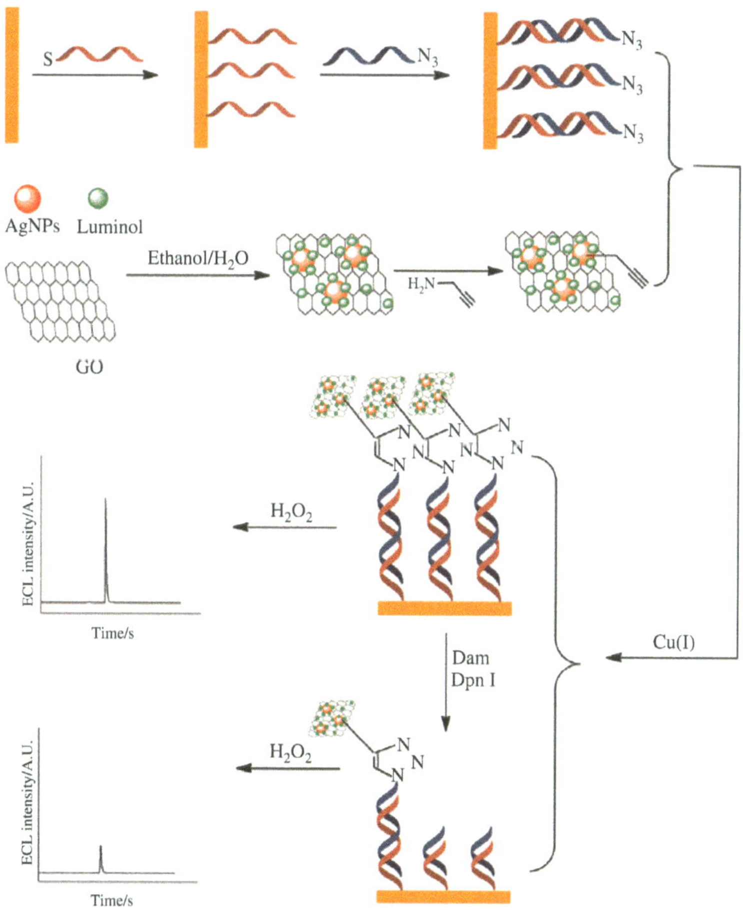

FIGURE 1.23 Schematic illustration of the preparation of GO/AgNPs/luminol composites and the procedures of the MTase activity assay. *Reprinted from Zhao, H.-F., Liang, R.-P., Wang, J.-W., Qiu, J.-D., 2015. One-pot synthesis of GO/AgNPs/luminol composites with electrochemiluminescence activity for sensitive detection of DNA methyltransferase activity. Biosens. Bioelectron. 63, 458–464.*

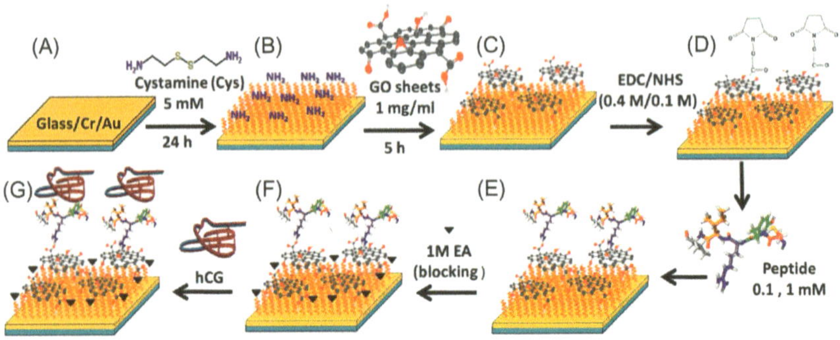

FIGURE 1.24 The GO–peptide-based SPR biochip experimental conditions. (A) The BK7 substrate deposited with a 2-nm chromium (Cr) adhesion layer and a 47-nm gold (Au) film. (B) The Au film was modified with a thiol self-assembled monolayer (SAM) of cystamine (Cys). (C) The fabrication of the GO sheets on the Au film. (D) The activation of the carboxyl end groups on the surfaces of the GO sheets. (E) A peptide aptamer probe was immobilized on the GO sheet surface using covalent bonds. (F) The unreacted carboxyl groups on the GO sheet surfaces were blocked by ethanolamine. (G) The target hCG proteins was captured by the peptide probe. *Reprinted from Chiu, N.-F., Kuo, C.-T., Lin, T.-L., Chang, C.-C., Chen, C.-Y., 2017b. Ultra-high sensitivity of the non-immunological affinity of graphene oxide-peptide-based surface plasmon resonance biosensors to detect human chorionic gonadotropin. Biosens. Bioelectron. 94, 351–357.*

1.3.5 Graphene-Based Surface Plasmon Resonance (SPR) for Protein Assay

Graphene oxide composites can form biocompatible surfaces on sensing films based on surface plasmon resonance (SPR). Chiu et al. prepared a SPR biosensor based on a GO sheet bound to a specific peptide aptamer for detection of human hCG protein (see Fig. 1.24; Chiu et al., 2017b).

Chiu et al. constructed an immunosensor based on carboxyl-functionalized graphene oxide (GO-COOH) in SPR. Graphene which underwent carboxylation could adjust its visible spectrum, thereby improving and controlling plasma coupling mechanism (see Fig. 1.25; Chiu et al., 2017a).

1.4 CARBON NANOTUBES

1.4.1 The Fundamental Properties of CNT

Carbon nanotubes (CNTs) are carbon molecules with a cylindrical hollow structure whose walls are formed by one-atom-thick sheets of sp^2-hybridized carbon. The diameter of CNTs is in the range of 0.5–50 nm and the length is usually in micrometer range. In fact, CNTs with 18.5 cm in length have also been reported (Wang et al., 2009). According to the number of layers of sheets in the wall, CNTs can be divided into single-walled carbon nanotubes (SWCNTs) and multiwalled carbon nanotubes (MWCNTs). CNTs are widely

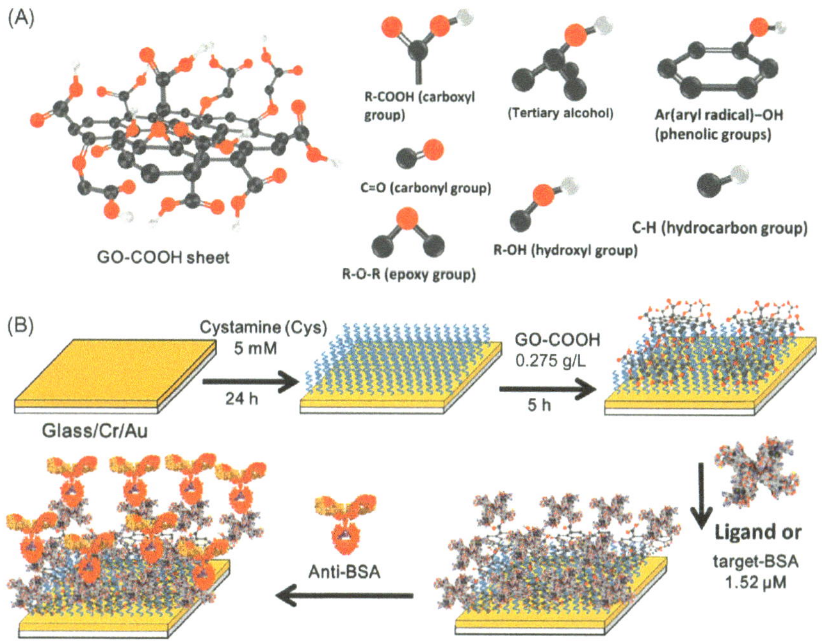

FIGURE 1.25 (A) The molecular structure of carboxyl-functionalized GO sheet and its oxygen-containing groups. (B) The process of fabrication of SPR chip with bio-molecular immobilization on the surface of carboxyl-functionalized GO film. *Reprinted from Chiu, N.-F., Fan, S.-Y., Yang, C.-D., Huang, T.-Y., 2017a. Arboxyl-functionalized graphene oxide composites as SPR biosensors with enhanced sensitivity for immunoaffinity detection. Biosens. Bioelectron. 89, 370–376.*

exploited to construct nanodevices including biosensors because of their high mechanical strength and electrical and thermal conductivity (Braik et al., 2016; Huang et al., 2007; Katz and Willner 2004).

SWNTs can be deemed to be one rolled-up graphene sheet, while MWNTs are concentric tubes separated by about 0.34 nm of two or more rolled-up graphene sheets. SWNTs have very distinctive electrical properties, because of the chirality, the wrap CNT can be modified with different chemical groups such as amine and carboxyl via both covalent and noncovalent which makes CNTs biocompatible and hence makes them a suitable candidate for fabricating a biosensor (Kumar et al., 2015a). Functionalization of CNTs can also promote the direct electron transfer between active sites of the biological element and the electrode.

1.4.2 CNT Based on Optical Spectrum for Protein Assay

1.4.2.1 CNT-Based Fluorescence

In the past few years, the unique optical properties of carbon nanotubes have attracted widespread attention, and a large number of them have been used as optical biosensors to detect proteins. Previous studies have shown that carbon nanotubes can be used as energy receptors in fluorescence resonance energy transfer, which is a good fluorescence quencher. Unlike common organic quenchers, carbon nanotubes have a broad absorption spectrum, making them highly efficient in quenching fluorescence. Using this feature of carbon nanotubes, scientists have constructed many fluorescent biosensors for detecting proteins.

Ouyang et al. built a fluorescent biosensor based on single-walled carbon nanotubes for protein detection. In the absence of the probe DNA (lysozyme aptamer), single-walled carbon nanotubes (SWNTs) existed as a solution. After the aptamer was added, the polymerized SWNT bound to the probe DNA and dispersed it, and the Eu^{3+}/BHHT complex can be more easily adsorbed onto the aptamer-dispersed SWNT surface, at which time the fluorescence was quenched. Later, when the target lysozyme was added, the specific binding capacity of lysozyme to its aptamer was greater than that of aptamer and SWNTs, and the aptamer and SWNTs were separated, resulting in a decrease in the dispersibility of the SWNTs and agglomeration at the bottom, followed by ultracentrifugation to remove SWNTs at the bottom. At this time, the Eu^{3+}/BHHT complex was added to the solution, and the fluorescence of the complex could not be quenched by the SWNTs (Li et al., 2017).

SWNTs also can be utilized as NIR fluorescent tags for selective cell probing and imaging. In the work reported by Welsher et al. (see Fig. 1.26), antibodies bound to polyethylene glycol (PEG)-modified SWNTs, such as Rituxan antibodies that specifically recognized CD20 cell surface receptors on B cells and had small nonspecificity for negative cells and Herceptin. Heterologous binding identified HER2/neu positive breast cancer cells. Cell imaging was performed by detecting the intrinsic NIR photoluminescence of the nanotubes (Welsher et al., 2008).

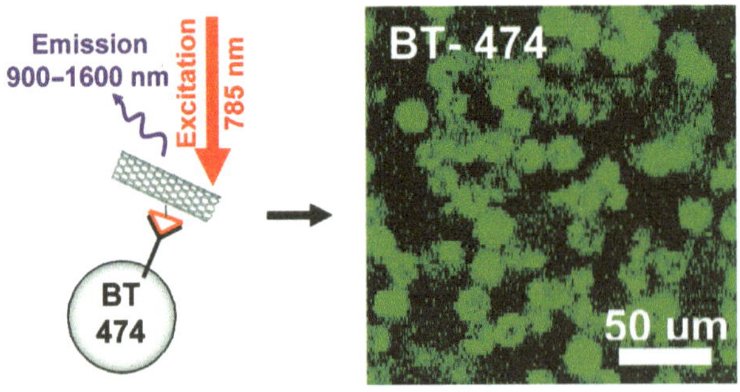

FIGURE 1.26 Selective probing and imaging of cells with single walled carbon nanotubes as near-infrared fluorescent molecules. *Reprinted from Welsher, K., Liu, Z., Daranciang, D., Dai, H., 2008. Selective probing and imaging of cells with single walled carbon nanotubes as near-infrared fluorescent molecules. Nano Lett. 8, 586–590.*

1.4.2.2 CNT-Based Raman Scattering

Another important optical property for SWNTs is that they exhibit strong Raman scattering. Chen et al. designed a multicolor Raman array for highly sensitive detection of multiple proteins (see Fig. 1.27). The array employed antibody-modified SWNTs as polychromatic Raman probes to immobilize human IgG and mouse IgG in two groups, and each set had three 400 nm diameter spots on a gold-coated slide. ^{12}C and ^{13}C isotope SWNTs were synthesized and linked to antimouse immunoglobulin G (GaM-IgG) and antihuman immunoglobulin G (GaH-IgG), respectively. Incubating these two biological mixtures of SWNTs on a sensing platform allows them to efficiently bind specifically to mouse or human IgG. From the G-mode Raman scattering spectra, a red shift was observed at the G-peak position for the

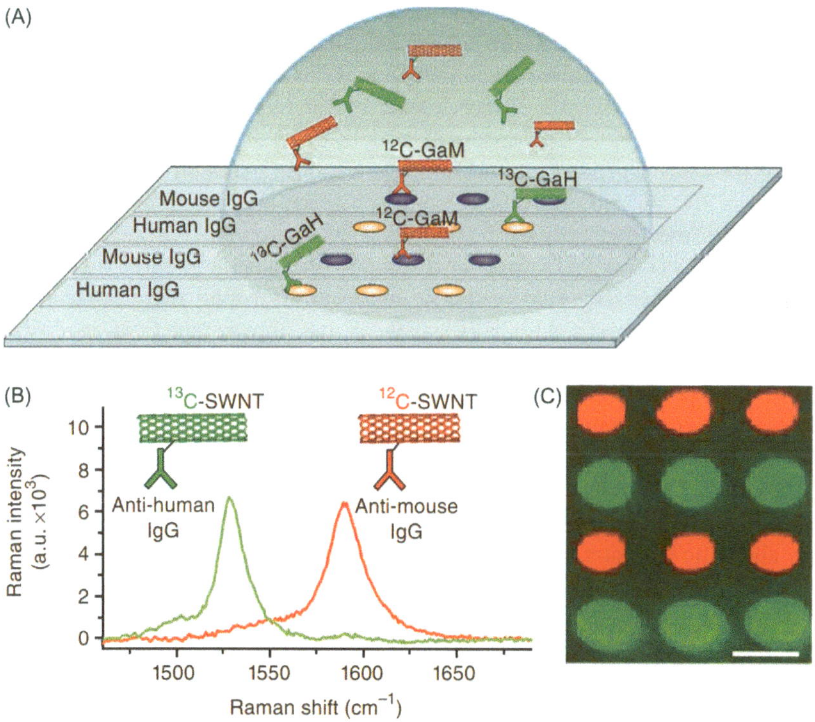

FIGURE 1.27 The detection for multiplexed protein using multi-color SWNT Raman labels. (A) Two-layer, direct, microarray-format protein detection with excellent Raman labels based on pure 12C and 13C SWNT tags. (B) The G-mode Raman scattering spectra of 12C (red) and 13C (green) SWNT Raman tags excited simultaneously with a 785 nm laser. (C) Raman scattering map of integrated 12C (red) and 13C (green) SWNT G-mode scattering. Scale bar, 500 μm. *Reprinted from Chen, Z., Tabakman, S.M., Goodwin, A.P., Kattah, M.G., Daranciang, D., Wang, X., et al., 2008. Protein microarrays with carbon nanotubes as multicolor Raman labels. Nat. Biotechnol. 26, 1285–1292.*

^{13}C SWNTs biological mixture due to the isotope effect, which allows the simultaneous detection of both types of IgG (Chen et al., 2008).

1.4.3 CNT-Modified Electrodes for Protein Assay

Due to the improved surface area conductivity and easily chemically modified surfaces, CNT-modified electrodes have been widely used for designing high-performance biosensors for the detection of cancer biomarkers. Screen-printed SWCNT electrodes were chosen for constructing label-free immunosensors for hCG by modifying the electrode surface with an anti-hCG antibody (Teixeira et al., 2014c). The concentration of hCG correlated with the response of the sensors in the range from 0.01 to 100 ng mL^{-1}. Screen-printed MWCNTs electrodes were used to prepare aptamer sensors for the breast cancer marker mucine (MUC1) (Nawaz et al., 2016). The response of the aptamer sensors depended on the MUC1 in a range of 0.1–2 U mL^{-1}. In addition, a carbonylated CNTs-coated conductive paper was also prepared for detecting CEA (Kumar et al., 2015c). CEA antibody (anti-CEA) was covalently modified on the surface of the conductive paper by carbodiimide coupling to fabricate CEA sensors. The detection range of this paper-based sensor for CEA depended on impedimetric response in the physiological situation was 2–15 ng mL^{-1}. Another approach based on voltammetry detection for matrix metalloproteinase-3 (MMP-3) was proposed using vertically aligned SWCNT (or CNT forests) to modify the electrode surface (Munge et al., 2010). This sensor showed a detection limit of 4 pg mL^{-1} based on a sandwich immunoassay with HRP labels. It has shown that CNT-modified field effect transistors (FETs) are promising biosensor platforms (Allen et al., 2007). A CNT network sensor for the prostate cancer marker osteopontin (OPN) was fabricated through chemical vapor deposition on the surface of a silicon wafer to construct CNT-modified FETs (Lerner et al., 2012). The CNT-modified FETs exhibited a wide concentration range of OPN from 0.001 to 1000 ng mL^{-1}. One of the advantages of FET sensors is the ability to measure the electrochemical responses of samples under dry conditions. A biosensor for C-reactive protein (CRP) was established by Justino and coworkers using CNT-modified FETs (Justino et al., 2013). In this scheme the drain current was measured under dry conditions after the sensors were exposed to CRP solutions to eliminate the potential effects of ionic species in sample solutions. Sharma designed a label-free SWCNT-based immunosensor for detecting the biomarker cTnI (see Fig. 1.28). This sensor contained two pairs of electrodes: one for the dielectrophoretic concentration of cTnI and the other for the electrical detection of the concentrated protein via resistance measurements (Sharma et al., 2016). A monoclonal antibody (mAb) against cTnI was immobilized on the surface of SWCNT through 1-ethyl-3-(3-dimethylaminopropyl) carbodiimide (EDC)

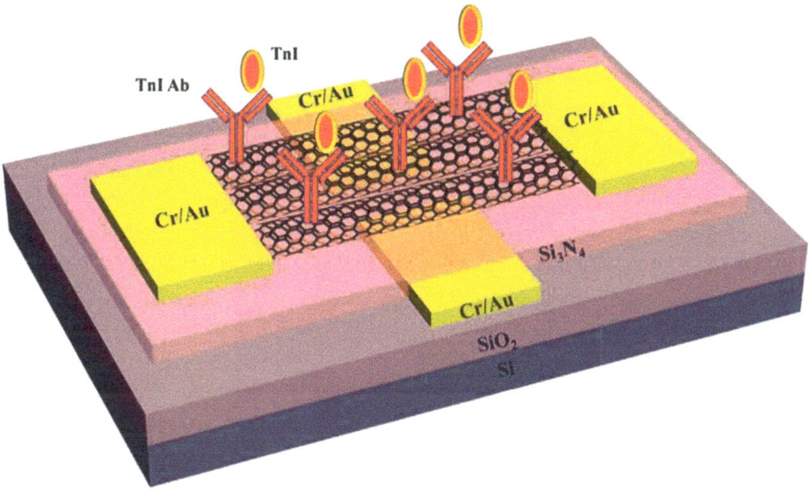

FIGURE 1.28 A schematic illustration of the SWCNT immunosensor with the concentration and detection electrodes (Cr/Au). SWCNTs were aligned between the detection electrodes by dielectrophoretic (DEP) and modified with cTnI antibodies. The cTnI was concentrated onto the SWCNTs by alternating current (AC)-DEP while the concentration electrodes imbedded between two dielectric layers to generate high electric fields. *Reprinted from Sharma, A., Han, C.H., Jang, J., 2016. Rapid electrical immunoassay of the cardiac biomarker troponin I through dielectrophoretic concentration using imbedded electrodes. Biosens. Bioelectron. 82, 78–84.*

and N-hydroxysuccinimide (NHS) reaction. The relative resistance change (RRC) depending on cTnI concentration was recorded in human serum and the detection range varied from 1 pg mL^{-1} to 100 ng mL^{-1}.

Ji et al. developed a sensitive and simple biosensor for early detection of prostate cancer, which was based on PSA antibody which wwa modified on MWCNTs immobilized on micropore filter paper. And its detection limit of 1.18 ng mL^{-1} was still superior to the available ELISA kit of 51.0 pg mL^{-1} (Ji et al., 2018).

1.4.4 CNTs as Signaling Labels in Electrochemistry for Protein Assay

CNTs can also be used as a carrier of signaling labels due to their high surface area-to-volume ratio. MWCNT labels modified with ferritin and a secondary antibody were used to detect carbohydrate antigen 15-3 (CA15-3), a breast cancer marker. And this biosensor could selectively detect CA15-3 as low as 0.01 ± 0.07 U mL^{-1} in human serum samples (Akter et al., 2016).

In a similar method, an immunosensor was constructed based on anti-PSA and HRP covalently cross-linking with carbonylated MWCNTs as signal tag.(Akter et al., 2012). The linear response range for PSA concentration was 1 pg mL^{-1} to 10 ng mL^{-1} and the limit of detection was 0.4 pg mL^{-1} (Yang et al., 2014). Carbon nanohorns (CNHs) were long cone-shaped tip carbon nanomaterials with a tubular structure similar to SWCNTs. A biosensor for AFP including CNH as labels was reported. CNH labels were designed by modifying anti-AFP, HRP, and Gox on carbonylated CNHs by carbodiimide coupling (Zhao et al., 2013a,b).

An aptamer is a single-stranded DNA/RNA oligonucleotide that can specifically bind to target molecules such as proteins. Guo et al. developed an electrochemical biosensor for detecting thrombin using aptamer-CNT as electrochemical labels (Fig. 1.29) (Guo et al., 2011). Alkanethiol monolayer 16-mercaptohexadecanoic acid (MHA) was modified on gold electrode to prevent the electron transfer between the electrode surface and redox species. Aptamer was wrapped on the sidewall of CNTs through $\pi-\pi$ interaction. In the presence of thrombin, aptamer was separated from the CNTs through the antibody–antigen interaction. Then the CNTs can be assembled on the MHA-modified electrode to promote electron transfer between the electrode and electroactive species. Additionally, the current was positively correlated with the concentration of target protein, and the detection limit was 50 pM.

Sánchez-Tirado and coworkers reported a novel method for the preparation of viologen SWCNT hybrids (V-Phe-SWCNTs) carrier tags for signal amplification to detect transforming growth factor β1 (TGF-β1) (see Fig. 1.30; Sánchez-Tirado et al., 2017). TGF-β1 is considered as a reliable

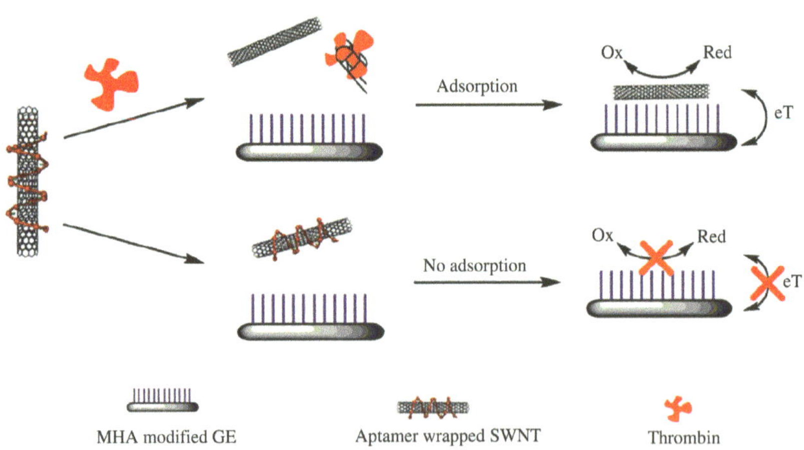

FIGURE 1.29 A schematic illustration of electrochemical biosensor for thrombin using aptamer-wrapped SWNTs as electrochemical labels. *Reprinted from Guo, K., Wang, Y., Chen, H., Ji, J., Zhang, S., Kong, J., et al., 2011. An aptamer–SWNT biosensor for sensitive detection of protein via mediated signal transduction. Electrochem. Commun. 13, 707–710.*

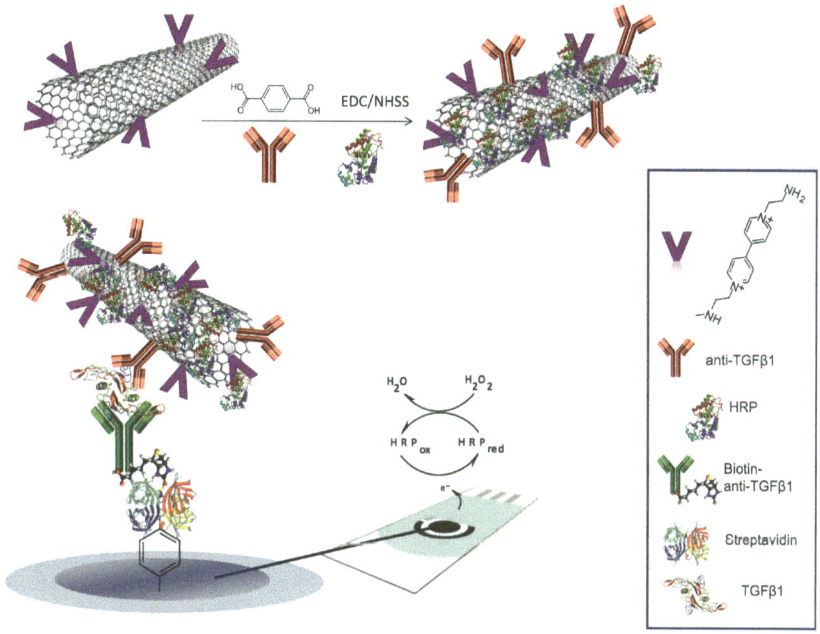

FIGURE 1.30 Schematic display of the multiple steps involved in the fabrication of an amperometry immunosensor for TGF-β1 using V-Phe-SWCNT hybrids. *Reprinted from Sánchez-Tirado, E., Arellano, L.M., González Cortés, A., Yáñez-Sedeño, P., Langa, F. and Pingarrón, J.M., 2017. Viologen-functionalized single-walled carbon nanotubes as carrier nanotags for electrochemical immunosensing. application to TGF-β1 cytokine. Biosens. Bioelectron. 98, 240–247.*

biomarker in several human diseases. The limit of detection for TGF-β1 is 0.95 pg mL^{-1}.

Wan et al. constructed an electrochemical immunosensor array to detect both PSA and interleukin-8 (IL-8) at the same time (Wan et al., 2011). As shown in Fig. 1.31, the 16-channel disposable SPCE array was achieved by modifying mouse monoclonal anti-PSA antibody (PSA mAb) or mouse monoclonal anti-IL-8 antibody (IL-8 mAb). Then PSA or IL-8 was then fixed on the sensor platform based on antibody–antigen interaction, followed by the immobilization of rabbit polyclonal signal anti-PSA antibodies (PSA pAb) or rabbit polyclonal anti-IL-8 antibodies (IL-8 pAb). Finally, HRP and goat antirabbit IgG (Ab2)-modified MWNTs were modified on the electrode to provide an amperometry readout. The detection limits were as low as 5 pg mL^{-1} of PSA and 8 pg mL^{-1} of IL-8 with this electrochemical immunosensor.

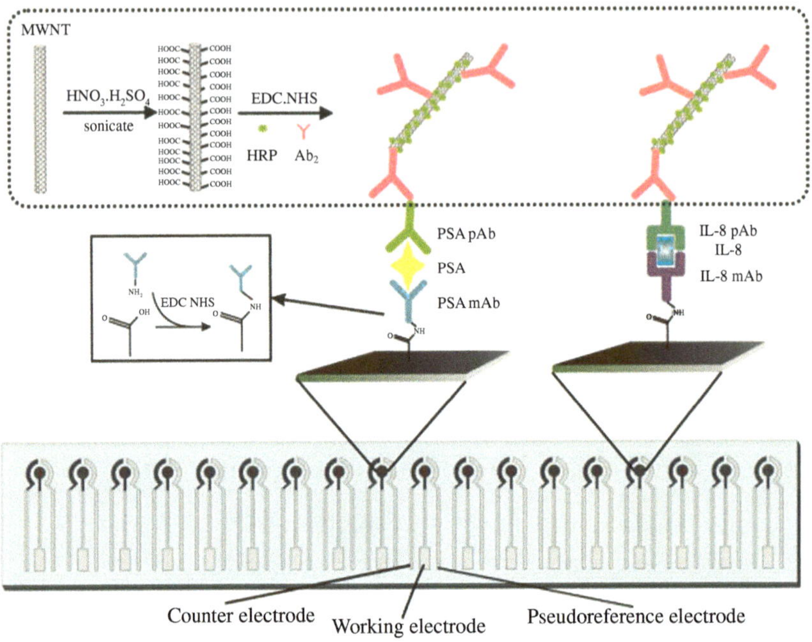

FIGURE 1.31 Schematic display of the "sandwich" type strategy electrochemical immunosensor. *Reprinted from Wan, Y., Deng, W., Su, Y., Zhu, X., Peng, C., Hu, H., et al., 2011. Carbon nanotube-based ultrasensitive multiplexing electrochemical immunosensor for cancer biomarkers. Biosens. Bioelectron. 30, 93–99.*

1.5 CONCLUSION

Carbon nanomaterials have been widely used in the preparation of electrochemical or spectral biosensors for the detection of protein biomarkers by means of the distinctive optical/electrochemical properties of carbon materials. Carbon nanomaterials as FRET receptors are good fluorescence quenching agents. When carbon nanomaterials are modified on the electrode surface, it can accelerate the electron transfer between solution and electrode surface, thus improving the output signal of the biosensor. In addition, using carbon nanomaterials as signal labels can also enhance the electrochemical signals. Furthermore, the high surface area-to-weight ratio and facile surface modification of the carbon nanomaterials are also advantages to be exploited. Therefore, it is attractive to use carbon nanomaterials with antibodies to develop a variety of biosensors for protein detection. It is demonstrated that the detection limit of biosensors for protein can achieve extremely low concentration ranges, including picograms per milliliter. One drawback of most biosensors is that lots of complex steps are needed to fabricate a sensor platform. Therefore, how to develop a protein biosensor based on just a simpler protocol is a challenge. Finally, if the carbon nanomaterials-based biosensors were to be used in clinical laboratories and hospitals, their performance characteristics such as reusability, stability, and compatibility with biological fluids should be further improved.

REFERENCES

Afreen, S., Muthoosamy, K., Manickam, S., Hashim, U., 2015. Functionalized fullerene (C60) as a potential nanomediator in the fabrication of highly sensitive biosensors. Biosens. Bioelectron. 63, 354–364.

Akter, R., Rahman, M.A., Rhee, C.K., 2012. Amplified electrochemical detection of a cancer biomarker by enhanced precipitation using horseradish peroxidase attached on carbon nanotubes. Anal. Chem. 84, 6407–6415.

Akter, R., Jeong, B., Choi, J.-S., Rahman, M.A., 2016. Ultrasensitive Nanoimmunosensor by coupling non-covalent functionalized graphene oxide platform and numerous ferritin labels on carbon nanotubes. Biosens. Bioelectron. 80, 123–130.

Allen, B.L., Kichambare, P.D., Star, A., 2007. Carbon nanotube field-effect-transistor-based biosensors. Adv. Mater. 19, 1439–1451.

Baker, S.N., Baker, G.A., 2010. Luminescent carbon nanodots: emergent nanolights. Angew. Chem. Int. Edit. 49, 6726–6744.

Bhattacharya, S., Samanta, S.K., 2016. Soft-nanocomposites of nanoparticles and nanocarbons with supramolecular and polymer gels and their applications. Chem. Rev. 116, 11967–12028.

Biju, V., 2014. Chemical modifications and bioconjugate reactions of nanomaterials for sensing, imaging, drug delivery and therapy. Chem. Soc. Rev. 43, 744–764.

Braik, M., Barsan, M.M., Dridi, C., Ben Ali, M., Brett, C.M.A., 2016. Highly sensitive amperometric enzyme biosensor for detection of superoxide based on conducting polymer/CNT modified electrodes and superoxide dismutase. Sens. Actuators B-Chem. 236, 574–582.

Cao, J.-T., Yang, J.-J., Zhao, L.-Z., Wang, Y.-L., Wang, H., Liu, Y.-M., et al., 2018. Graphene oxide@ gold nanorods-based multiple-assisted electrochemiluminescence signal amplification strategy for sensitive detection of prostate specific antigen. Biosens. Bioelectron. 99, 92–98.

Chang, H., Tang, L., Wang, Y., Jiang, J., Li, J., 2010. Graphene fluorescence resonance energy transfer aptasensor for the thrombin detection. Anal. Chem. 82, 2341–2346.

Chang, H.-W., Shih, J.-S., 2007. Surface acoustic wave immunosensors based on immobilized C60-proteins. Sens. Actuators B-Chem. 121, 522–529.

Chen, J.-L., Yan, X.-P., Meng, K., Wang, S.-F., 2011. Graphene oxide based photoinduced charge transfer label-free near-infrared fluorescent biosensor for dopamine. Anal. Chem. 83, 8787–8793.

Chen, X., Jia, X., Han, J., Ma, J., Ma, Z., 2013. Electrochemical immunosensor for simultaneous detection of multiplex cancer biomarkers based on graphene nanocomposites. Biosens. Bioelectron. 50, 356–361.

Chen, Z., Tabakman, S.M., Goodwin, A.P., Kattah, M.G., Daranciang, D., Wang, X., et al., 2008. Protein microarrays with carbon nanotubes as multicolor Raman labels. Nat. Biotechnol. 26, 1285–1292.

Chiu, N.-F., Fan, S.-Y., Yang, C.-D., Huang, T.-Y., 2017a. Arboxyl-functionalized graphene oxide composites as SPR biosensors with enhanced sensitivity for immunoaffinity detection. Biosens. Bioelectron. 89, 370–376.

Chiu, N.-F., Kuo, C.-T., Lin, T.-L., Chang, C.-C., Chen, C.-Y., 2017b. Ultra-high sensitivity of the non-immunological affinity of graphene oxide-peptide-based surface plasmon resonance biosensors to detect human chorionic gonadotropin. Biosens. Bioelectron. 94, 351–357.

Compton, O.C., Nguyen, S.T., 2010. Graphene oxide, highly reduced graphene oxide, and graphene: versatile building blocks for carbon-based materials. Small 6, 711–723.

Du, D., Wang, L., Shao, Y., Wang, J., Engelhard, M.H., Lin, Y., 2011. Functionalized graphene oxide as a nanocarrier in a multienzyme labeling amplification strategy for ultrasensitive electrochemical immunoassay of phosphorylated p53 (S392). Anal. Chem. 83, 746–752.

Esteves da Silva, J.C.G., Gonçalves, H.M.R., 2011. Analytical and bioanalytical applications of carbon dots. TrAC Trends in Anal. Chem. 30, 1327–1336.

Feng, D., Li, L., Han, X., Fang, X., Li, X., Zhang, Y., 2014. Simultaneous electrochemical detection of multiple tumor markers using functionalized graphene nanocomposites as non-enzymatic labels. Sens. Actuators B-Chem. 201, 360–368.

Gao, Q., Liu, N., Ma, Z., 2014. Prussian blue–gold nanoparticles-ionic liquid functionalized reduced graphene oxide nanocomposite as label for ultrasensitive electrochemical immunoassay of alpha-fetoprotein. Anal. Chim. Acta 829, 15–21.

Geim, A.K., Novoselov, K.S., 2007. The rise of graphene. Nat. Mater. 6, 183–191.

Georgakilas, V., Otyepka, M., Bourlinos, A.B., Chandra, V., Kim, N., Kemp, K.C., et al., 2012. Functionalization of graphene: covalent and non-covalent approaches, derivatives and applications. Chem. Rev. 112, 6156–6214.

Guo, K., Wang, Y., Chen, H., Ji, J., Zhang, S., Kong, J., et al., 2011. An aptamer–SWNT biosensor for sensitive detection of protein via mediated signal transduction. Electrochem. Commun. 13, 707–710.

Guo, Y., Wang, Z., Shao, H., Jiang, X., 2013. Hydrothermal synthesis of highly fluorescent carbon nanoparticles from sodium citrate and their use for the detection of mercury ions. Carbon 52, 583–589.

Han, N., Wang, W., Lu, D., Wang, H., Ma, X., Fan, X., et al., 2017. A novel, rapid, and sensitive homogeneous sandwich detection method of Glypican-3 as a serum marker for hepatocellular carcinoma. Chem. Commun. (Camb) 53, 12209–12212.

Haufler, R.E., Conceicao, J., Chibante, L.P.F., Chai, Y., Byrne, N.E., Flanagan, S., et al., 1990. Efficient production of C_{60} (buckminsterfullerene), $C_{60}H_{36}$, and the solvated buckide ion. J. Phys. Chem. (USA) 94, 8634–8636.

He, Q., Wu, S., Gao, S., Cao, X., Yin, Z., Li, H., et al., 2011. Transparent, flexible, all-reduced graphene oxide thin film transistors. ACS Nano 5, 5038–5044.

Huang, J., Song, Z., Li, J., Yang, Y., Shi, H., Wu, B., et al., 2007. A highly-sensitive l-lactate biosensor based on sol-gel film combined with multi-walled carbon nanotubes (MWCNTs) modified electrode. Mater. Sci. Eng., C 27, 29–34.

Ji, S., Lee, M., Kim, D., 2018. Detection of early stage prostate cancer by using a simple carbon nanotube@paper biosensor. Biosens. Bioelectron. 102, 345–350.

Justino, C.I.L., Freitas, A.C., Amaral, J.P., Rocha-Santos, T.A.P., Cardoso, S., Duarte, A.C., 2013. Disposable immunosensors for C-reactive protein based on carbon nanotubes field effect transistors. Talanta 108, 165–170.

Katz, E., Willner, I., 2004. Biomolecule-functionalized carbon nanotubes: applications in nanobioelectronics. Chemphyschem 5, 1084–1104.

Krätschmer, W., Lamb, L.D., Fostiropoulos, K., Huffman, D.R., 1990. Solid C60: a new form of carbon. Nature 347, 354.

Kroto, H., 1997. Symmetry, space, stars, and C60 (nobel lecture). Angew. Chem., Int. Ed. 36, 1578–1593.

Kumar, S., Ahlawat, W., Kumar, R., Dilbaghi, N., 2015a. Graphene, carbon nanotubes, zinc oxide and gold as elite nanomaterials for fabrication of biosensors for healthcare. Biosens Bioelectron 70, 498–503.

Kumar, S., Kumar, S., Srivastava, S., Yadav, B.K., Lee, S.H., Sharma, J.G., et al., 2015b. Reduced graphene oxide modified smart conducting paper for cancer biosensor. Biosens. Bioelectron. 73, 114–122.

Kumar, S., Willander, M., Sharma, J.G., Malhotra, B.D., 2015c. A solution processed carbon nanotube modified conducting paper sensor for cancer detection. J. Mater. Chem. B 3, 9305–9314.

Kwon, O.S., Park, S.J., Hong, J.-Y., Han, A.R., Lee, J.S., Lee, J.S., et al., 2012. Flexible FET-type vegf aptasensor based on nitrogen-doped graphene converted from conducting polymer. ACS Nano 6, 1486–1493.

Lee, J.S., Min-Gon Kim, H.-A.J., Park, C.B., 2012. Graphene-based chemiluminescence resonance energy transfer for homogeneous immunoassay. ACS Nano 6, 2978–2983.

Lerf, A., He, H., Forster, M., Klinowski, J., 1998. Structure of graphite oxide revisited. J. Mater. Chem. B 102, 4477–4482.

Lerner, M.B., D'Souza, J., Pazina, T., Dailey, J., Goldsmith, B.R., Robinson, M.K., et al., 2012. Hybrids of a genetically engineered antibody and a carbon nanotube transistor for detection of prostate cancer biomarkers. ACS Nano 6, 5143–5149.

Li, C., Yang, Y., Wu, D., Li, T., Yin, Y., Li, G., 2016. Improvement of enzyme-linked immunosorbent assay for the multicolor detection of biomarkers. Chem. Sci. 7, 3011–3016.

Li, D., Wang, C., Sun, G., Senapati, S., Chang, H.-C., 2017. A shear-enhanced CNT-assembly nanosensor platform for ultra-sensitive and selective protein detection. Biosens. Bioelectron. 97, 143–149.

Li, H., He, J., Li, S., Turner, A.P.F., 2013. Electrochemical immunosensor with N-doped graphene-modified electrode for label-free detection of the breast cancer biomarker CA 15-3. Biosens. Bioelectron. 43, 25–29.

Li, L., Zhang, L., Yu, J., Ge, S., Song, X., 2015. All-graphene composite materials for signal amplification toward ultrasensitive electrochemical immunosensing of tumor marker. Biosens. Bioelectron. 71, 108–114.

Liao, Y.H., Shih, J.S., 2013. Piezoelectric quartz crystal anti-protein immunosensors based on immobilized fullerene C60-proteins. J. Chin. Chem. Soc. 60, 1387–1395.

Lim, S.A., Yoshikawa, H., Tamiya, E., Yasin, H.M., Ahmed, M.U., 2014. A highly sensitive gold nanoparticle bioprobe based electrochemical immunosensor using screen printed graphene biochip. RSC Adv. 4, 58460–58466.

Liu, J., Wang, J., Wang, T., Li, D., Xi, F., Wang, J., et al., 2015a. Three-dimensional electrochemical immunosensor for sensitive detection of carcinoembryonic antigen based on monolithic and macroporous graphene foam. Biosens. Bioelectron. 65, 281–286.

Liu, R., Wan, L., Liu, S., Pan, L., Wu, D., Zhao, D., 2015b. An interface-induced co-assembly approach towards ordered mesoporous carbon/graphene aerogel for high-performance supercapacitors. Adv. Funct. Mater. 25, 526–533.

Liu, M., Zhao, H., Quan, X., Chen, S., Fan, X., 2010. Distance-independent quenching of quantum dots by nanoscale-graphene in self-assembled sandwich immunoassay. Chem. Commun. (Camb) 46, 7909–7911.

Loh, K.P., Bao, Q., Eda, G., Chhowalla, M., 2010. Graphene oxide as a chemically tunable platform for optical applications. Nat. Chem. 2, 1015–1024.

Lu, C.-H., Yang, H.-H., Zhu, C.-L., Chen, X., Chen, G.-N., 2009. A graphene platform for sensing biomolecules. Angew. Chem. Int. Ed. 48, 4785–4787.

Lu, C.-H., Li, J., Zhang, X.-L., Zheng, A.-X., Yang, H.-H., Chen, X., et al., 2011. General approach for monitoring peptide–protein interactions based on graphene–peptide complex. Anal. Chem. 83, 7276–7282.

Majd, S.M., Salimi, A., 2018. Ultrasensitive flexible FET-type aptasensor for CA 125 cancer marker detection based on carboxylated multiwalled carbon nanotubes immobilized onto reduced graphene oxide film. Anal. Chim. Acta 1000, 273–282.

Mao, S., Lu, G., Yu, K., Bo, Z., Chen, J., 2010. Specific protein detection using thermally reduced graphene oxide sheet decorated with gold nanoparticle-antibody conjugates. Adv. Mater. 22, 3521–3526.

Mao, S., Yu, K., Lu, G., Chen, J., 2011. Highly sensitive protein sensor based on thermally-reduced graphene oxide field-effect transistor. Nano Res. 4, 921.

Mei, Q., Zhang, Z., 2012. Photoluminescent graphene oxide ink to print sensors onto microporous membranes for versatile visualization bioassays. Angew. Chem. Int. Ed. 51, 5602–5606.

Morales-Narvaez, E., Merkoci, A., 2012. Graphene oxide as an optical biosensing platform. Adv. Mater. 24, 3298–3308.

Morales-Narváez, E., Pérez-López, B., Pires, L.B., Merkoçi, A., 2012. Simple Förster resonance energy transfer evidence for the ultrahigh quantum dot quenching efficiency by graphene oxide compared to other carbon structures. Carbon 50, 2987–2993.

Munge, B.S., Fisher, J., Millord, L.N., Krause, C.E., Dowd, R.S., Rusling, J.F., 2010. Sensitive electrochemical immunosensor for matrix metalloproteinase-3 based on single-wall carbon nanotubes. Analyst 135, 1345–1350.

Nawaz, M.A., Rauf, S., Catanante, G., Nawaz, M.H., Nunes, G., Marty, J.L., et al., 2016. One step assembly of thin films of carbon nanotubes on screen printed interface for electrochemical aptasensing of breast cancer biomarker. Sensors (Basel) 16, 1651.

Pan, N.-Y., Shih, J.-S., 2004. Piezoelectric crystal immunosensors based on immobilized fullerene C60-antibodies. Sens. Actuators B-Chem. 98, 180–187.

Park, S., Ruoff, R.S., 2009. Chemical methods for the production of graphenes. Nat. Nanotech. 4, 217–224.

Park, S.J., Kwon, O.S., Lee, S.H., Song, H.S., Park, T.H., Jang, J., 2012. Ultrasensitive flexible graphene based field-effect transistor (FET)-type bioelectronic nose. Nano Lett. 12, 5082–5090.

Pumera, M., 2011. Graphene in biosensing. Mater. Today 14, 308–315.

Robinson, J.T., Perkins, F.K., Snow, E.S., Wei, Z., Sheehan, P.E., 2008. Reduced graphene oxide molecular sensors. Nano Lett. 8, 3137–3140.

Sánchez-Tirado, E., Arellano, L.M., González-Cortés, A., Yáñez-Sedeño, P., Langa, F., Pingarrón, J.M., 2017. Viologen-functionalized single-walled carbon nanotubes as carrier nanotags for electrochemical immunosensing. application to TGF-β1 cytokine. Biosens. Bioelectron. 98, 240–247.

Sardesai, N.P., Barron, J.C., Rusling, J.F., 2011. Carbon nanotube microwell array for sensitive electrochemiluminescent detection of cancer biomarker proteins. Anal. Chem. 83, 6698–6703.

Sharma, A., Han, C.H., Jang, J., 2016. Rapid electrical immunoassay of the cardiac biomarker troponin I through dielectrophoretic concentration using imbedded electrodes. Biosens. Bioelectron. 82, 78–84.

Sheng, Q., Liu, R., Zheng, J., 2013. Fullerene–nitrogen doped carbon nanotubes for the direct electrochemistry of hemoglobin and its application in biosensing. Bioelectrochemistry 94, 39–46.

Shi, J., Guo, J., Bai, G., Chan, C., Liu, X., Ye, W., et al., 2015. A graphene oxide based fluorescence resonance energy transfer (FRET) biosensor for ultrasensitive detection of botulinum neurotoxin A (BoNT/A) enzymatic activity. Biosens. Bioelectron. 65, 238–244.

Sidhu, J.S., Singh, A., Garg, N., Singh, N., 2017. Carbon dot based, naphthalimide coupled FRET pair for highly selective ratiometric detection of thioredoxin reductase and cancer screening. ACS Appl. Mater. Interfaces 9, 25847–25856.

Stine, R., Mulvaney, S.P., Robinson, J.T., Tamanaha, C.R., Sheehan, P.E., 2013. Fabrication, optimization, and use of graphene field effect sensors. Anal. Chem. 85, 509–521.

Su, C.-Y., Lu, A.-Y., Xu, Y., Chen, F.-R., Khlobystov, A.N., Li, L.-J., 2011. High-quality thin graphene films from fast electrochemical exfoliation. ACS Nano 5, 2332–2339.

Tan, C., Cao, X., Wu, X.-J., He, Q., Yang, J., Zhang, X., et al., 2017. Recent advances in ultrathin two-dimensional nanomaterials. Chem. Rev. 117, 6225–6331.

Teixeira, S., Burwell, G., Castaing, A., Gonzalez, D., Conlan, R.S., Guy, O.J., 2014a. Epitaxial graphene immunosensor for human chorionic gonadotropin. Sens. Actuators B-Chem. 190, 723–729.

Teixeira, S., Conlan, R.S., Guy, O.J., Sales, M.G.F., 2014b. Label-free human chorionic gonadotropin detection at picogram levels using oriented antibodies bound to graphene screen-printed electrodes. J. Mater. Chem. B 2, 1852–1865.

Teixeira, S., Conlan, R.S., Guy, O.J., Sales, M.G.F., 2014c. Novel single-wall carbon nanotube screen-printed electrode as an immunosensor for human chorionic gonadotropin. Electrochim. Acta 136, 323–329.

Wan, Y., Deng, W., Su, Y., Zhu, X., Peng, C., Hu, H., et al., 2011. Carbon nanotube-based ultrasensitive multiplexing electrochemical immunosensor for cancer biomarkers. Biosens. Bioelectron. 30, 93–99.

Wang, X., Li, Q., Xie, J., Jin, Z., Wang, J., Li, Y., et al., 2009. Fabrication of ultralong and electrically uniform single-walled carbon nanotubes on clean substrates. Nano Lett. 9, 3137–3141.

Wang, H., Zhang, Q., Chu, X., Chen, T., Ge, J., Yu, R., 2011a. Graphene oxide–peptide conjugate as an intracellular protease sensor for caspase-3 activation imaging in live cells. Angew. Chem., Int. Ed. 50, 7065–7069.

Wang, Y., Li, Z., Wang, J., Li, J., Lin, Y., 2011b. Graphene and graphene oxide: biofunctionalization and applications in biotechnology. Trends Biotechnol. 29, 205–212.

Wang, Q., Cui, X., Chen, J., Zheng, X., Liu, C., Xue, T., et al., 2012. Well-dispersed palladium nanoparticles on graphene oxide as a non-enzymatic glucose sensor. RSC Adv. 2, 6245–6249.

Welsher, K., Liu, Z., Daranciang, D., Dai, H., 2008. Selective probing and imaging of cells with single walled carbon nanotubes as near-infrared fluorescent molecules. Nano Lett. 8, 586–590.

Wu, Y., Xue, P., Kang, Y., Hui, K.M., 2013. Paper-based microfluidic electrochemical immunodevice integrated with nanobioprobes onto graphene film for ultrasensitive multiplexed detection of cancer biomarkers. Anal. Chem. 85, 8661–8668.

Wu, Y., Xue, P., Hui, K.M., Kang, Y., 2014. A paper-based microfluidic electrochemical immunodevice integrated with amplification-by-polymerization for the ultrasensitive multiplexed detection of cancer biomarkers. Biosens. Bioelectron. 52, 180–187.

Xu, B., Zhao, C., Wei, W., Ren, J., Miyoshi, D., Sugimoto, N., et al., 2012. Aptamer carbon nanodot sandwich used for fluorescent detection of protein. Analyst 137, 5483–5486.

Yang, F., Han, J., Zhuo, Y., Yang, Z., Chai, Y., Yuan, R., 2014. Highly sensitive impedimetric immunosensor based on single-walled carbon nanohorns as labels and bienzyme biocatalyzed precipitation as enhancer for cancer biomarker detection. Biosens. Bioelectron. 55, 360–365.

Yang, W., Ratinac, K.R., Ringer, S.P., Thordarson, P., Gooding, J.J., Braet, F., 2010. Carbon nanomaterials in biosensors: should you use nanotubes or graphene? Angew. Chem., Int. Ed. 49, 2114–2138.

Yang, Y., Li, P., Wu, S., Li, X., Shi, E., Shen, Q., et al., 2015. Hierarchically designed three-dimensional macro/mesoporous carbon frameworks for advanced electrochemical capacitance storage. Chem. Eur. J 21, 6157–6164.

Yuan, H., Qi, J., Xing, C., An, H., Niu, R., Zhan, Y., et al., 2015. Graphene-oxide-conjugated polymer hybrid materials for calmodulin sensing by using FRET strategy. Adv. Funct. Mater. 25, 4412–4418.

Yuan, Q., He, J., Niu, Y., Chen, J., Zhao, Y., Zhang, Y., et al., 2018. Sandwich-type biosensor for the detection of α2, 3-sialylated glycans based on fullerene-palladium-platinum alloy and 4-mercaptophenylboronic acid nanoparticle hybrids coupled with Au-methylene blue-MAL signal amplification. Biosens. Bioelectron. 102, 321–327.

Zhang, Y., Guo, Y., Xianyu, Y., Chen, W., Zhao, Y., Jiang, X., 2013. Nanomaterials for ultrasensitive protein detection. Adv. Mater. 25, 3802–3819.

Zhao, C., Lin, D., Wu, J., Ding, L., Ju, H., Yan, F., 2013a. Nanogold-enriched carbon nanohorn label for sensitive electrochemical detection of biomarker on a disposable immunosensor. Electroanalysis 25, 1044–1049.

Zhao, H., Chang, Y., Liu, M., Gao, S., Yu, H., Quan, X., 2013b. A universal immunosensing strategy based on regulation of the interaction between graphene and graphene quantum dots. Chem. Commun. 49, 234–236.

Zhao, H.-F., Liang, R.-P., Wang, J.-W., Qiu, J.-D., 2015. One-pot synthesis of GO/AgNPs/luminol composites with electrochemiluminescence activity for sensitive detection of DNA methyltransferase activity. Biosens. Bioelectron. 63, 458–464.

Zheng, M., Jagota, A., Semke, E.D., Diner, B.A., McLean, R.S., Lustig, S.R., et al., 2003a. DNA-assisted dispersion and separation of carbon nanotubes. Nat. Mater. 2, 338.

Zheng, M., Jagota, A., Strano, M.S., Santos, A.P., Barone, P., Chou, S.G., et al., 2003b. Structure-based carbon nanotube sorting by sequence-dependent DNA assembly. Science 302, 1545–1548.

Zhu, P., Wang, P., Kan, L., Sun, G., Zhang, Y., Yu, J., 2015. An enhanced photoelectrochemical immunosensing platform: supramolecular donor–acceptor arrays by assembly of porphyrin and C60. Biosens. Bioelectron. 68, 604–610.

Zhu, Z., 2017. An overview of carbon nanotubes and graphene for biosensing applications. Nano-Micro Lett. 9, 25.

Chapter 2

Metal Nanomaterials

Nandi Zhou[1] and Hai Shi[2]
[1]*School of Biotechnology and the Key Laboratory of Industrial Biotechnology, Ministry of Education, Jiangnan University, Wuxi, P. R. China,* [2]*State Key Laboratory of Pharmaceutical Biotechnology, Department of Biochemistry, Nanjing University, Nanjing, P. R. China*

Chapter Outline

2.1 INTRODUCTION

In the last few decades, metal nanomaterials (MNMs) have attracted significant attention due to their excellent physicochemical properties that result from surface effect (Liz-Marzán, 2006), small size effect (Zhang, 1997), quantum size effect (Daniel and Astruc, 2004), and macroscopic quantum tunneling effect (Ralph et al., 1997). These advantages endow MNMs with incomparable features in electrical, thermal, optical, chemical, or even catalytic activities, which emerge in innovative technologies that have improved the way we live (Li et al., 2015).

Two main strategies named "top-down" and "bottom-up" are widely employed for MNMs synthesis (Heath, 1999). Usually, the "top-down" strategy is implemented by dividing large pieces of MNMs into nanometer-scaled small pieces of different shapes, by means of lithography, cutting, grinding, etching, etc.; whereas the "bottom-up" strategy is performed by

Nano-inspired Biosensors for Protein Assay with Clinical Applications.
DOI: https://doi.org/10.1016/B978-0-12-815053-5.00002-7

using metal ions as precursors to generate metal atom nuclei which can grow into monodispersed MNM colloids subsequently (Wang and Xia, 2004; Choi et al., 2008; Nuraje et al., 2009). Compared to the "top-down" strategy, the "bottom-up" strategy costs less and endows MNMs with higher flexibility, thus drawing more attention in recent years (Heath, 1999; Xia et al., 2013).

According to their composition, MNMs can be classified into single-component MNMs and multiple-components MNMs (Raveendran et al., 2006; Chen et al., 2005; Li and Yamauchi, 2013). So far, various approaches for controlled MNMs synthesis, such as size, shape, composition, and architecture-controlled synthesis, have already been well explored (Xia et al., 2009). For instance, MNMs have been successfully molded into various shapes, such as nanospheres, nanorods, nanowires, nanoribbons, nanoplates, nanosheets, nanopolyhedra, nanotubes, nanomultipods, and nanoframes (Fan and Zhang, 2016; Howes et al., 2014a,b; Busbee et al., 2003; Diao et al., 2003; Huang et al., 2011). Generally, different shapes or compositions endow MNMs with unique properties that expand their application range, e.g., gold nanoparticles (AuNPs) for peroxidase-mimicking catalysis (Zheng et al., 2011), gold nanorods (AuNRs) for surface enhanced Raman scattering, and silver nanoparticles (AgNPs) for signal amplification (Orendorff et al., 2006; Lee et al., 2011).

The real application of MNMs in bioassays, especially in protein assays, is an exciting and challenging task worldwide. It is well known that the concentration of protein biomarkers exhibits certain physiological or pathological statuses of living system. Therefore, rapid and accurate monitoring of these proteins is extremely useful in early diagnosis of diseases, drug screening, and clinical therapy (Kingsmore, 2006; Nolan, 2007; Tessler et al., 2009; He et al., 2010). The main challenge lies at the practicability and interference-resistance of assays in harsh conditions, because target proteins are usually distributed in serum, saliva, urine, or cell extract, whose composition can be very complicated and a large number of impurities can influence the detection results. In addition, a very low limit of detection (LOD) is often required since biomarkers usually appear at an extremely low concentration. For example, LOD at pg mL^{-1} level is often required when the concentration of biomarker protein in patients is at ng mL^{-1} level (Rusling et al., 2014). To achieve such low LOD, the characteristics of MNMs used in detection systems have been studied and improved, including stability, dispersibility, and biocompatibility, by addition of surfactants, cyclodextrin, bovine serum albumin (BSA), or thiol compounds into the reaction system during the synthesis of MNMs. These reagent-protected MNMs showed excellent performance for protein detection in real samples (Mei et al., 2009; Balasubramanian et al., 2010; Gao et al., 2012; García et al., 2015). Moreover, MNMs can be easily modified with proteins, peptides, or oligonucleotides due to the property of the metal surface and their high surface-to-volume ratio. Then various signal amplification strategies can be adopted to greatly increase the sensitivity of protein assays (Lei and Ju, 2012; Wang et al., 2015a; Wang et al., 2015b; Ding et al., 2015; Fenzl et al., 2016; Zhou et al., 2018).

In this review, we highlight some typical design principles and advantages of MNMs-based protein assays and pay particular attention to their application for detecting proteins in real samples.

2.2 NUCLEIC ACID-FUNCTIONALIZED METAL NANOMATERIALS FOR PROTEIN ASSAYS

Metal nanomaterials are naturally suitable to integrate with biomolecules, and thus to perform exquisite strategies. For example, nucleic acids or oligonucleotides are easy to be immobilized on MNMs through diverse linkages, and then efficient amplification methods for nucleic acids can be employed for protein assays that enable extremely high sensitivity. Generally, an initiator oligonucleotide strand is immobilized onto the surface of MNMs, so as to trigger signal amplification in the presence of targets.

One representative example of signal amplification techniques is on-nanoparticle rolling-circle amplification (nanoRCA) strategy, which was proposed by Yan et al. (2010) and shown in Fig. 2.1, which represented the first

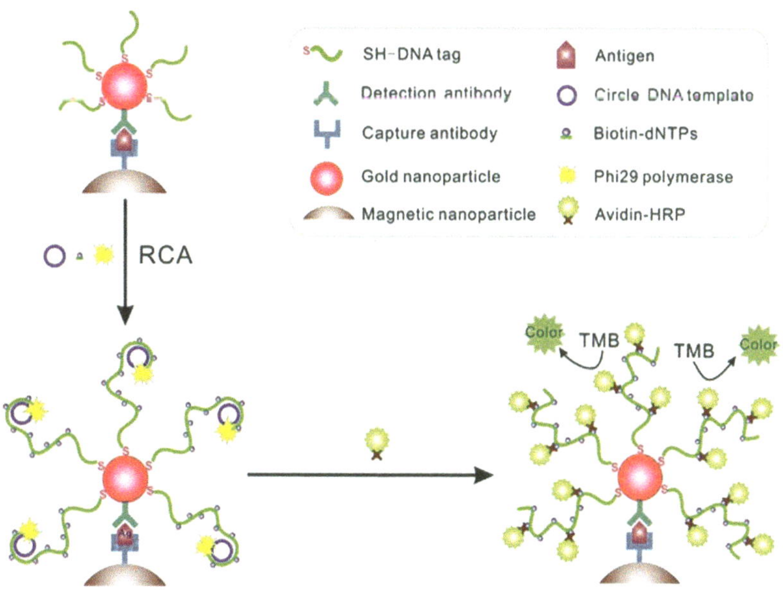

FIGURE 2.1 Schematic illustration of the nanoRCA strategy. *Reprinted from Yan, J., Song, S. P., Li, B., Zhang, Q., Huang, Q.A., Zhang, H., et al., 2010. An on-nanoparticle rolling-circle amplification platform for ultrasensitive protein detection in biological fluids. Small 6, 2520–2525.*

nanoRCA-immunoassay for protein detection. Briefly, this MNMs-based biosensing system consisted of capture antibody-conjugated magnetic nanoparticles (MNPs) and DNA primer-modified AuNPs. Performing like a typical sandwich immunoassay, the functionalized MNPs and AuNPs simultaneously recognized the different epitopes of carcinoembryonic antigen, which led to the formation of a sandwich complex. Meanwhile, the DNA primer-initiated RCA, resulting in intensive signal amplification, which allowed the visual detection of 30 molecules of carcinoembryonic antigen. The assay was successfully performed in serum and saliva samples.

In another research, Zhang et al. (2012) reported an AuNPs-based electrochemical immunoassay for the detection of IgG via hybridization chain reaction (HCR)-signal amplification. As shown in Fig. 2.2, the biosensing system consisted of AuNPs-labeled signal probe, magnetic immunosensing probe, and two different hairpin DNAs. AuNPs were conjugated with two different types of biomolecules, which are signal antibodies and initiator DNA strands of HCR, respectively. During the detection process, capture antibody-conjugated magnetic beads recognized the target IgG and captured it onto the surface magnetic beads. Afterwards, the signal antibody-conjugated AuNPs could recognize the target on the magnetic beads, and further bound to the beads to form sandwich-type complex. Then the initiator DNA strands on AuNPs opened the hairpin DNA structures and triggered HCR to form a

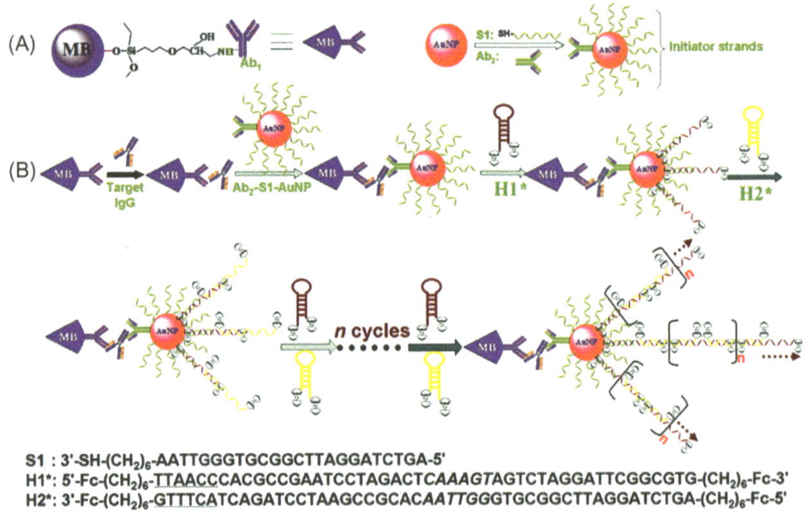

FIGURE 2.2 (A) The principle of anti-human IgG-functionalized magnetic beads (Ab1-MBs) and antihuman IgG/initiator strands-modified AuNPs (Ab2-S1-AuNPs). (B) Schematic illustration of the sandwiched immunoassays with the hybridization chain reaction. *Reprinted from Zhang, B., Liu, B.Q., Tang, D.P., Niessner, R., Chen, G.N., Knopp, D., 2012. DNA-based hybridization chain reaction for amplified bioelectronic signal and ultrasensitive detection of proteins. Anal. Chem. 84, 5392–5399.*

nicked double-helix. Using ferrocene as signal tag, the LOD of this MNMs-based IgG detection was found to be as low as 0.1 fg mL^{-1}. This example indicates that antibodies and oligonucleotides modified on MNMs could correctly recognize target proteins at very low concentration.

Recently, an interesting MNMs-based DNA machine for protein detection has been reported. The nanomachine utilized specific target binding to trigger assembly of separate DNA components which could not be spontaneously assembled. Typically, three-dimensional DNA tentacles were densely constructed on AuNPs through functionalization with hundreds of single-stranded oligonucleotides (Zhang et al., 2015a,b; Chen et al., 2017; Li et al., 2017; Jiang et al., 2017; Zheng et al., 2018). For example, Zhang et al. (Zhang et al., 2015a,b) reported an innovative DNA nanomachine that was activated upon protein binding (Fig. 2.3). The performance of this new DNA nanomachine relied on molecular recognition-induced DNA assembly. Upon the binding of the target molecule, those separate DNA components could be assembled spontaneously. Taking advantage of the high loading capacity of AuNPs, this high-density 3D DNA nanomachine could specifically and sensitively detect different targets, including streptavidin, platelet-derived growth factor, and the smallpox gene.

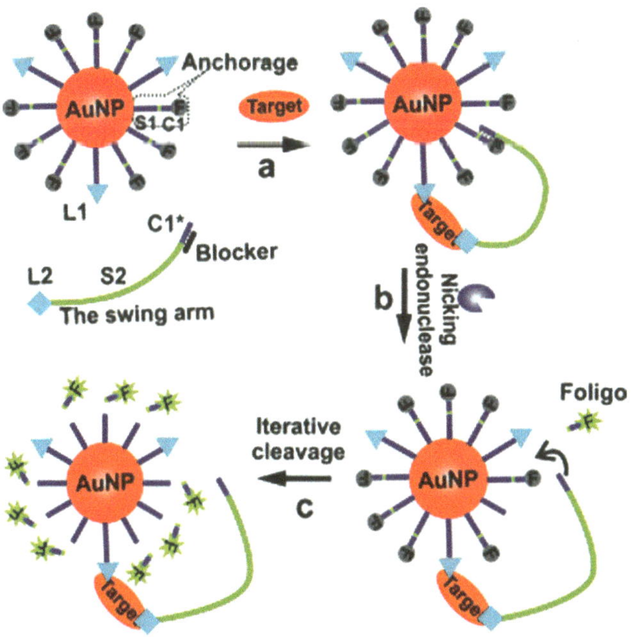

FIGURE 2.3 Schematic illustration of molecular recognition-induced DNA nanomachine. *Reprinted from Zhang, H.Q., Lai, M.D., Zuehlke, A., Peng, H.Y., Li, X.F., Le, X.C., 2015b. Binding-induced DNA nanomachines triggered by proteins and nucleic acids. Angew. Chem. Int. Ed. 54, 14326–14330.*

2.3 SURFACE PLASMON RESONANCE FEATURED METAL NANOMATERIALS FOR PROTEIN ASSAYS

Surface plasmon resonance (SPR) has been employed in research for many years. It is also one of the most widely used technologies for the study of biomolecular interactions. The unique SPR property of metal nanomaterials, especially AuNPs, give them great potential for ultrasensitive detection of trace proteins in clinical diagnosis (Liang et al., 2015; Wang et al., 2007; Rosi and Mirkin, 2005; He et al., 2005). Generally speaking, SPR generates as the result of the collective oscillation of electrons stimulated by an incident light. Specifically, the surface plasmon on metal nanoparticles is activated by an optical excitation, leading to nanoscale spatial confinement of electromagnetic fields that trigger localized surface plasmon resonance (LSPR). Thereafter, if the wavelength of the incident light matches the LSPR frequency, the collective oscillation of the local field will increase to the maximum. As a result, this LSPR generates a distinct absorbance peak in the visible frequency range and a strong electromagnetic field at the particle surface, which polarizes the local volume around the nanoparticles. Some conditions may interfere with the final result of resonance, including absorption, extinction, color change, or scattering of the nanoparticles in solution, leading to a shift of LSPR (Eustis and El-Sayed, 2006; Willets and Van Duyne, 2007; Howes et al., 2014a,b).

Typically, when the SPR of spherical and well-dispersed AuNPs generates, the solution color becomes brilliant red. However, when the morphology of AuNPs-based nanostructures is modified, the SPR wavelength changes, which leads to the change of solution color as well (Hu et al., 2006; Jung et al., 2015). In the case of gold nanorods (AuNRs), the free electrons can oscillate along the long and short axes, generating two plasmon wavelengths corresponding to their transverse and longitudinal directions. By changing the aspect ratio of AuNRs, it is possible to adjust the longitudinal plasmon wavelength into the near infrared region (NIR), which makes them applicable for biosensors in the biological window (700–900 nm). For example, the absorbance of hemoglobin and water is negligible, which allows optimal penetration of light through blood and tissue (Tsai et al., 2013; Pissuwan et al., 2007; Mayer et al., 2008; Kumar et al., 2008). Thus, AuNRs exhibit stronger SPR signal in NIR range, which makes them effective for protein detection in real samples.

The optical properties of plasmonic nanoparticles strongly depend on intermolecular distances. When the intermolecular distance is short, plasmon coupling and associated perturbation occurs in the LSPR band, which leads to a red-shift of the absorbance peak. In other words, when disperse particles in solution become aggregated; there is a conspicuous color shift from red to blue. Such system provides the possibility of "one-step" assay with a simple colorimetric output (Aldewachi et al., 2018). Therefore, various protein assays have been established by using the mechanism of aggregated plasmonic nanoparticles that may induce colorimetric shifts in the presence/absence of target proteins.

2.3.1 SPR for Colorimetric Detection of Proteins

Recently, a colorimetric approach for protein detection based on peptide-modified AuNPs has been reported (Fig. 2.4). It is well known that Cucurbit [8]uril (CB[8]), a pumpkin-shaped macrocyclic molecule, can selectively combine two peptides containing N-terminal aromatic residues, regardless of the sequence. On this basis, CB[8] was used to induce the aggregation of this kind of peptide-modified AuNPs, so as to conduct a colorimetric assay. In more detail, in the absence of the target, vascular endothelial growth factor receptor 1 (Flt-1), the aggregation of peptide-modified AuNPs was caused by CB[8]-induced dimerization of N-terminal aromatic residues. However, in the presence of the protein target, the peptides bound to the target, the aromatic residue sites would be occupied, resulting in the failure of CB[8] function and the aggregation of AuNPs. The colorimetric assay reached the LOD of 0.2 nM and was performed in serum samples.

Chang et al. (2016) developed a label-free approach for fibronectin detection using the aggregation of unmodified AuNPs-induced color change (Fig. 2.5). Fibronectin is important for cell attachment and differentiation, and is also an important biomarker for renal interstitial fibrosis and preterm

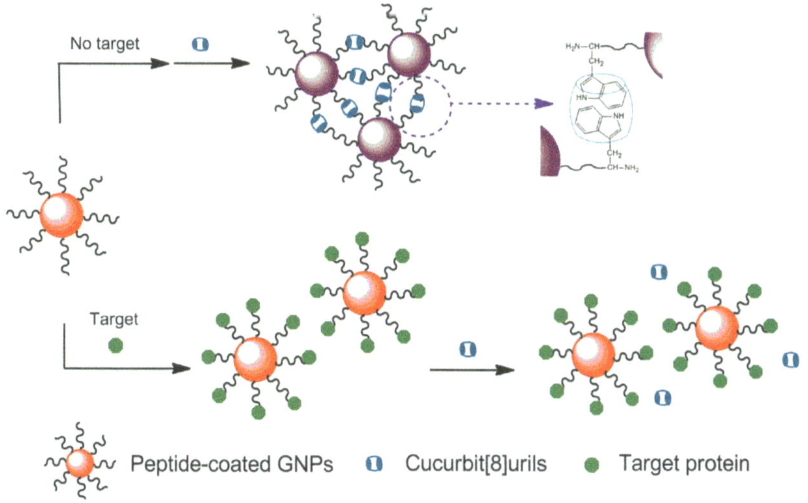

FIGURE 2.4 Schematic illustration of the peptide-coated AuNPs-based protein assay. *Reprinted from Wei, L.M., Wang, X.Y., Li, C., Li, X.X., Yin, Y.M., Li, G.X., 2015. Colorimetric assay for protein detection based on "nano-pumpkin" induced aggregation of peptide-decorated gold nanoparticles. Biosens. Bioelectron. 71, 348–352.*

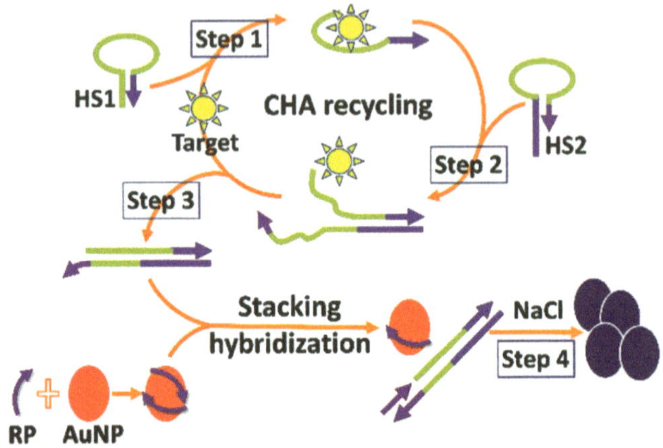

FIGURE 2.5 Schematic illustration of the colorimetric detection of fibronectin. *Reprinted from Chang, C.C., Chen, C.P., Chen, C.Y., Lin, C.W., 2016. DNA base-stacking assay utilizing catalytic hairpin assembly-induced gold nanoparticle aggregation for colorimetric protein sensing. Chem. Commun. 52, 4167–4170.*

birth. This approach was composed of a catalytic unit and a reporting probe (RP). The catalytic unit contained two metastable DNA hairpin subunits (HS1 and HS2). HS1 was designed as the catalytic hairpin assembly (CHA) initiator sequence, which formed a hairpin DNA by hybridization with intramolecular aptamer sequence specific for fibronectin. The hairpin structure inhibited the CHA reaction. On the other hand, the 3′-terminal region of HS2 was complementary to the RP DNA attached on AuNPs surface. In the absence of fibronectin, HS1 and HS2 remained intact since intramolecular hybridization inhibited CHA reaction, preventing the AuNPs aggregation. Upon addition of fibronectin, the CHA reaction could be initiated to form a HS1/fibronectin intermediate, which then triggered the formation of a stable hybridized duplex between HS1 and HS2, and the competitive releasing of fibronectin protein from HS1. The dissociated protein further initiated the next round of CHA reaction. After incubation with the RP-attached AuNPs solution, the formed dsDNA selectively hybridized with the RP DNA, resulting in the dissociation of RP from the AuNPs surface. Under high ionic-strength condition, RP-released AuNPs aggregated and showed purple–blue color change. Thus, fibronectin could be identified using only the naked eye or quantified sensitively by UV-vis spectrometry, with the LOD down to 2.3 pM.

2.3.2 Solution-Based LSPR for Protein Assays

Plasmon-enabled diagnostic assays are commonly performed in the solution phase. Thus, the stability of diagnostic systems must be considered. The colloidally stable plasmonic MNMs have good performance. The extremely large surface area due to the small size of the MNMs is particularly beneficial for protein sensing. And high diffusion rates in the samples allow for the increased speed and sensitivity compared to surface-based approaches. Those assays where the target molecules are directly captured on the MNMs surface tend to yield a relatively small shift in the LSPR peak. Therefore, the setup which can read the small shift in absorbance spectroscopy is necessary for protein detection. However, considering the cost, it is best that such equipment is the generic lab equipment, which enables to perform a complex analysis in clinical samples, such as a plate reader. A number of assays used dispersed plasmonic MNMs, especially AuNRs for direct protein detection (Wang et al., 2010; Tang et al., 2013; Li et al., 2016; Kim et al., 2018).

Wang et al. (2010) proposed an AuNRs-based biosensor for the detection of hepatitis B surface antigen (HBsAg) by using LSPR peak as signal output (Fig. 2.6). HBsAg is the most important marker for the laboratory diagnosis of hepatitis B. Its presence in serum indicates an acute and chronic hepatitis B virus infection and potential infectivity. To detect HBsAg, the surface of AuNRs was firstly modified with monoclonal hepatitis B surface antibody (HBsAb) via physical adsorption. Upon binding with HBsAg, the immunological reaction induced the change of local RI, resulting in the longitudinal wavelength shift of LSPR peak in the UV-Vis extinction spectrum. As a consequence, the binding of HBsAg to the molecular probe could be monitored by using AuNRs-based change of local RI. By this means, HBsAg was detected in Tris buffer in the range of 0.01–1 IU mL^{-1}, as well as in actual media of blood serum and plasma.

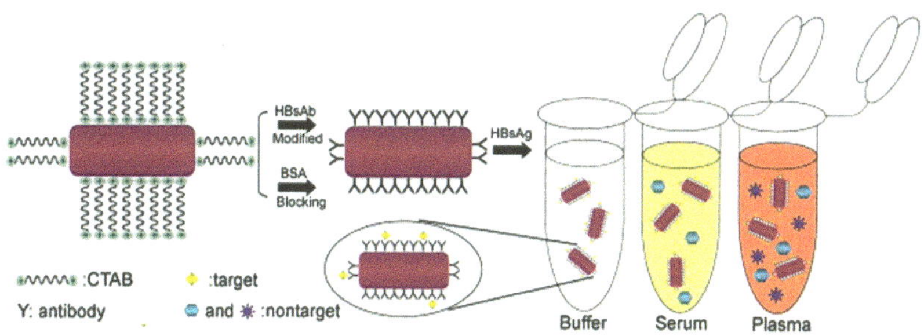

FIGURE 2.6 Schematic representation of the synthesis of antibody-functionalized AuNRs and the detection mechanism for the biosensor immunoassay in capturing targets in different matrices. *Reprinted from Wang, X.H., Li, Y., Wang, H.F., Fu, Q.X., Peng, J.C., Wang, Y.L., et al., 2010. Gold nanorod-based localized surface plasmon resonance biosensor for sensitive detection of hepatitis B virus in buffer, blood serum and plasma. Biosens. Bioelectron. 26, 404–410.*

2.3.3 Surface-Based LSPR for Protein Assays

Standard SPR assays work by measuring the target-induced change in reflectivity of a planar gold surface as a function of refractive index change. In general, commercially available SPR equipment consists of a microfluidic setup, an optical system, and an electronic system. To wash with buffers and sample solutions, the sensing surface was constructed on the microfluidic setup. The optical system was used to analyze surface biochemical activity, and the electronic system was used to process the obtained data. SPR-based systems have been designed for routine clinical and point-of-care (POC) applications. In fact, the SPR-based systems are typically more sensitive than current LSPR-based systems. However, it is still worth pursuing due to some inherent advantages of LSPR sensing. First of all, the absorbance or transmission as signal output in target assays simplifies the optics required compared to the reflectance as signal output. The advantage could enable the designed assay systems run high sensitivity assays with standard lab equipment and miniaturized optics. Secondly, the required minimum volume is further reduced in LSPR sensors. Therefore LSPR-based systems are less sensitive to the changes of bulk RI, but more specific in responsive to local RI changes. The introduction of MNMs into the LSPR-based system can create various protein detection strategies with good performance by integrating the advantages of LSPR and MNMs. For surface-based assays, the MNMs can be immobilized on a substrate to construct various nanobiosensors with the exciting advantages, e.g., washing steps can be performed with sequential washes over static sensor components (Polavarapu and Liz-Marzán, 2013).

By immobilizing MNMs on a surface, it is possible to integrate the advantageous plasmonic property of MNMs to obtain the benefits of surface-based protein sensing (Liu and Cheng, 2012; Guo and Kim, 2012; Jang et al., 2013; Inci et al., 2013; Kwon et al., 2012; Lane et al., 2015; Laing et al., 2016). Kwon et al. (2012) applied biofunctionalized gold nanomaterials with different shapes and sizes for the SPR detection of thrombin at attomolar level (Fig. 2.7). The shapes of the synthesized gold nanomaterials included cubic cages, rods, and quasi-spherical, whose size was at least one dimension in the 40–50 nm range. As part of a sandwich assay, these gold nanomaterials were respectively covalently conjugated with an antibody. The Au chip was modified with DNA aptamer probe specific to thrombin in advance. Thrombin was detected by the formation of aptamer–thrombin–antibody sandwich structure on Au SPR chip. Such a gold nanomaterials-based assay could detect thrombin at concentrations as low as 1 aM (quasi-spherical), 10 aM (rods), and 1 fM (cages) in optimized conditions, respectively. This was a significant improvement ($> 10^3$) compared to AuNPs-enhanced SPR system, which was attributed to the low nonspecific adsorption caused by the functionalization of both AuNMs and chip surfaces.

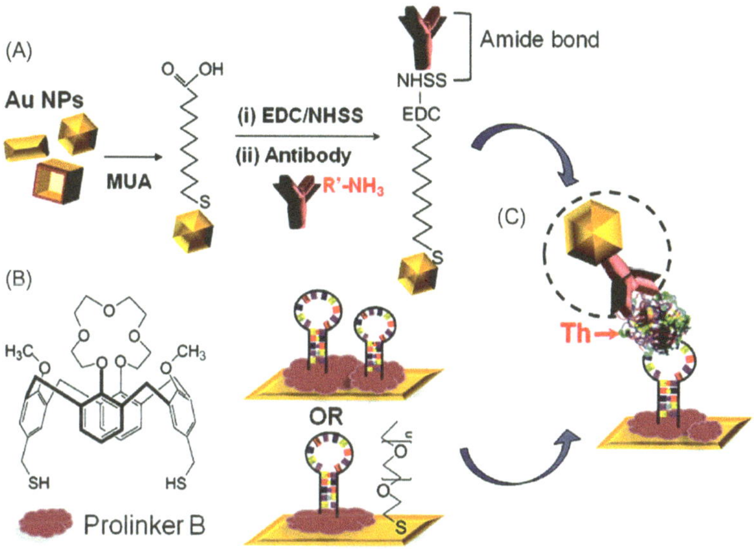

FIGURE 2.7 Schematic illustration of (A) the modification of antibody on gold nanomaterials; (B) the modification of aptamer on Au SPR chip; and (C) surface sandwich assay of thrombin. *Reprinted from Kwon, M.J., Lee, J., Wark, A.W., Lee, H.J., 2012. Nanoparticle-enhanced surface plasmon resonance detection of proteins at attomolar concentrations: comparing different nanoparticle shapes and sizes. Anal. Chem. 84, 1702–1707.*

In addition, upon adsorption on the Au chip, these gold nanomaterials could induce the increased density and large shifts in the local RI at the chip surface.

2.3.4 Refractive Index Shift-Based Direct Protein Assays

LSPR-induced field effects can decrease rapidly with distance from the particle surface, which gives rise to a nanoscale volume surrounding the particle. As a result, the perturbations in refractive index (RI) can be detected (Yang and Gao, 2015; Yu et al., 2017; McFarland and Van Duyne, 2003; Sönnichsen et al., 2005; Chen et al., 2008; Olson et al., 2015). If biological molecules are added into this volume, they can cause the change of the local RI and a shift in the LSPR of the particle can be measured. The change provides a possibility of developing biosensor systems, including those for sensitive protein detection. The general strategy benefits from being label-free, whereas in a large number of biosensors, the targets need to be labeled before they can be analyzed. The labeling process can interfere with the natural behavior and interactions of targets, and increase the difficulty of the assay. Although label-free methods can overcome these shortcomings, they traditionally suffered from the limitations of sensitivity due to the lack of signal amplification. Nevertheless, numerous advances in the field of

LSPR-based biosensors have given rise to very high sensitivities (Nath and Chilkoti, 2002; Cognet et al., 2003; Ament et al., 2012; Ma et al., 2015; McPeak et al., 2015).

Sensors generally feature two functional components: a recognition element and a transducer. The recognition element can selectively and specifically bind to the targets. And the transducer component provides the signaling of the capture event. The LSPR shift from MNMs can be used as the output signal. Various recognition interactions have been explored, including biotin–streptavidin, antibody–antigen, enzyme–peptide, aptamer–protein, toxin–receptor, and nucleic acid hybridization. Considering plasmonic nanoparticles can react to RI changes in their direct environment by a shift of the plasmon resonance, Ament et al. (2012) utilized single AuNR to detect single unlabeled protein with extremely high temporal resolution (Fig. 2.8). Rod-shaped AuNRs were immobilized in a glass capillary to introduce and change the liquid environment. The light scattering spectrum of a single AuNR was measured under illumination from a white light laser. The laser illuminated the particles in total internal reflection geometry, and the scattered light was picked up by the microscope and spectrally investigated. With this setup, single particle scattering spectra within a few milliseconds could be obtained. Moreover, this method allowed for monitoring of the dynamic process of a single protein binding event on a millisecond time scale. The technique even opened a new window into Brownian dynamics of unlabeled targets by resolving equilibrium coverage fluctuations. Therefore, the method showed the potential in studying protein folding dynamics, protein adsorption processes and kinetics, even the nonequilibrium soft matter dynamics on the single molecule level.

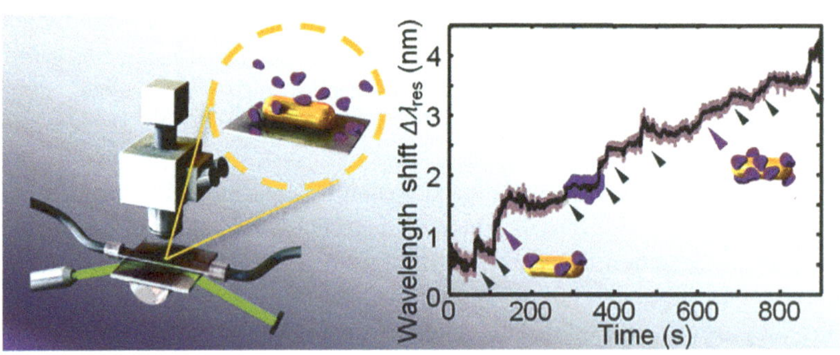

FIGURE 2.8 Experimental setup and time-resolved single protein attachment (left), and the change of resonance wavelength shift caused by single protein attachment events (right). *Reprinted from Ament, I., Prasad, J., Henkel, A., Schmachtel, S., Sönnichsen, C., 2012. Single unlabeled protein detection on individual plasmonic nanoparticles. Nano Lett. 12, 1092–1095.*

2.4 SURFACE-ENHANCED RAMAN SCATTERING-BASED PROTEIN ASSAYS

As a spectroscopic technique, surface-enhanced Raman scattering (SERS) provides possibility for the detection of protein at a single molecule level. The advantages of SERS-based protein assays can be summarized as follows. Firstly, SERS is especially suitable for rapid, label-free detection of biomarkers in biological matrices with high sensitivity, high structural specificity, and high flexibility. Secondly, the minimal background signals in aqueous biological samples due to the low interference from water makes it possible for ultrasensitive detection of protein. Thirdly, due to the intrinsic nature of the Raman effect, Raman measurements enable the detection of protein in a wide spectral range, with no need for a particular excitation/emission wavelength. Therefore, SERS-based protein assays can carry out multiplexed detection by Raman peaks. The narrow spectral widths of Raman peaks can minimize the overlap between multiple labels. Fourthly, SERS labels have high resistance to photobleaching by using a single excitation for multiple labels. Finally, the sample preparation is convenient and time-saving in SERS-based protein detection. The above- advantages of SERS make it a versatile technique for biosensing applications (Lane et al., 2015; Laing et al., 2016).

Proteins can be stably conjugated to substrates via the interaction with peptides or aptamers. Study on SERS of proteins has been performed for many years (Grabbe and Buck, 1989), much of which has focused on the method of attachment of the proteins to the metallic surface, for example, through a carboxyl group or an amine group. More recently, various methods have been developed by using different MNMs (Xu et al., 2016; Zhang et al., 2015a,b; Chuong et al., 2017). Xu et al. (2014) proposed an iodide-modified Ag nanoparticles (Ag IMNPs)-based label-free detection of proteins by SERS (Fig. 2.9). In this work, silver nanoparticles (AgNPs) could greatly enhance the Raman signal and increase the sensitivity. To avoid the interfering signal, a layer of iodide was modified on AgNPs as a barrier to prevent the direct interaction between the proteins and the AgNPs surface. The SERS signals of five typical proteins (lysozyme, avidin, bovine serum albumin, cytochrome c, and hemoglobin) were obtained with high reproducibility and high-quality by using this method. Without the chromophore, the measured SERS features of these proteins were almost the same as their respective normal Raman spectra. The results showed that the method could enable the qualitative detection of the five proteins simultaneously. Moreover, the conformational changes of proteins caused by hydrogen peroxide could be monitored in situ according to the changeable SERS signals. The method showed the promising future for high-throughput proteomics and studying the dynamics of proteins.

Xu et al. (2015) developed an approach for simultaneous and ultrasensitive detection of multiple disease biomarkers, including prostate specific

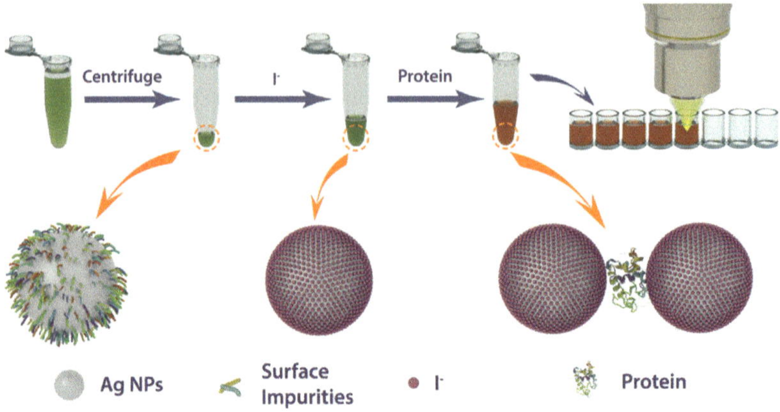

FIGURE 2.9 Schematic diagram of the method proposed for protein detection. *Reprinted from Xu, L.J., Zong, C., Zheng, X.S., Hu, P., Feng, J.M., Ren, B., 2014. Label-free detection of native proteins by surface-enhanced Raman spectroscopy using iodide-modified nanoparticles. Anal. Chem. 86, 2238–2245.*

antigen (PSA), thrombin, and mucin-1. The approach was constructed through self-assembly of AgNPs pyramids (denoted as Ag-pyramids), which was driven by disease biomarker-specific aptamers. In the presence of the designated disease biomarkers, the 3D spatial geometries Ag-pyramids could be altered to shorten the gap length due to the specific interaction between the biomarker and its aptamer, resulting in the enhancement of Raman signal (Fig. 2.10). The results showed that the designed Ag-pyramids were an excellent SERS platform for simultaneous detection of multiple protein targets. The intensity of the "signed" SERS had an excellent linear correlation with the corresponding biomarker concentration. The detection ranges were 1–500 aM, 0.1–50 fM, and 0.01–5 fM, with limits of detection of 0.96, 85, and 9.2 aM for PSA, thrombin, and mucin-1, respectively. The efficiency and selectivity of the Ag-pyramid sensor could provide accurate and distinct detection of disease biomarkers in clinical science.

2.5 ENZYME-CONTROLLED MNMS GROWTH FOR COLORIMETRIC DETECTION OF PROTEINS

As aforementioned, we discussed assays that use presynthesized MNMs and the change in LSPR arising from interactions of those MNMs with each other. However, dramatic shifts in the LSPR peak can also be observed when there is a physical change in the structure of MNMs, e.g., nucleation and growth of new particles or controlled growth of shell structures on the existing particles. In general, these shifts will be much larger than those observed in direct target capture. By integrating such changes in MNMs structure with the level of target proteins of interest, it is possible to create

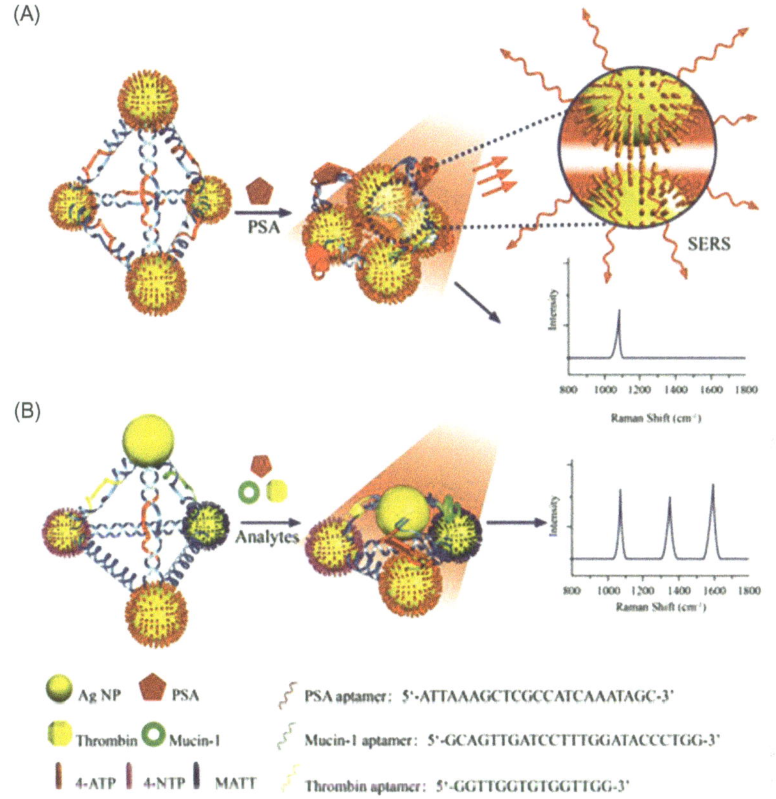

FIGURE 2.10 Scheme of Ag pyramids self-assembled by DNA frame for SERS analysis of biomarkers. (A) Ag-pyramid-mediated singlet SERS assay for PSA. (B) SERS encoded multiple Ag-pyramidal for the detection of PSA, thrombin, and mucin-1. *Reprinted from Xu, L.G., Yan, W.J., Ma, W., Kuang, H., Wu, X.L., Liu, L.Q., et al., 2015. SERS encoded silver pyramids for attomolar detection of multiplexed disease biomarkers. Adv. Mater. 27, 1706–1711.*

various approaches for protein assays. The examples discussed below will focus on the structural and material changes caused by different enzymes (Ma et al., 2016; Rodríguez-Lorenzo et al., 2012; Zhou et al., 2014; Liu et al., 2014a,b).

Liu et al. (2014a,b) developed a quantitative method for the detection of cancer biomarkers in clinical samples (Fig. 2.11). This method was dependent on the growth of small-sized AuNPs (5 nm in diameter) with the assistance of H_2O_2 and $AuCl_4^-$. In this work, glucose oxidase (GOx), which was conjugated with the detection antibody (Ab2) on the surfaces of magnetic beads (MBs), was employed to generate H_2O_2. Because the extinction coefficient of 5 nm AuNPs was much lower than those of larger-sized AuNPs, the diluted 5 nm AuNP solution at relatively low concentration (normally <10 nM) is colorless, while larger-sized AuNPs solution at the same

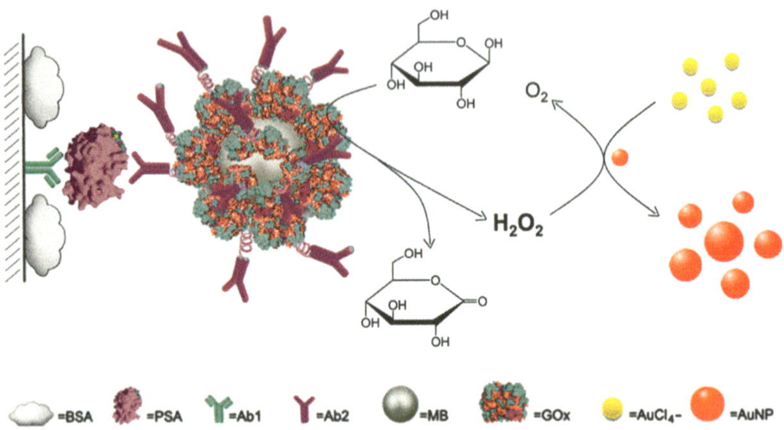

FIGURE 2.11 Schematic illustration of the quantitative immunoassay based on glucose oxidase (GOx)-catalyzed growth of AuNPs (5 nm). *Reprinted from Liu, D.B., Yang, J., Wang, H.F., Wang, Z.L., Huang, X.L., Wang, Z.T., et al., 2014a. Glucose oxidase-catalyzed growth of gold nanoparticles enables quantitative detection of attomolar cancer biomarkers. Anal. Chem. 86, 5800–5806.*

concentration is red. The concentration of GOx sandwiched complex via immunoreaction was proportional to the amount of the target prostate-specific antigen (PSA). PSA was first immobilized by the capture antibody (Ab1) modified on the substrate, and then Ab2 – GOx – MBs were conjugated by the captured target. GOx catalyzed the oxidation of glucose to generate H_2O_2, inducing the growth of 5-nm-sized AuNPs to the larger AuNPs. As a result, the colorless AuNPs solution turned red and an absorption band at around 530 nm could be observed, which in turn was utilized to quantify the cancer biomarker. Unlike previous AuNPs-based assays that depend on the red-to-blue (or purple) color change, the method was based on a colorless-to-red process. The generation of signal was similar to that of horseradish peroxidase (HRP)-based ELISA (from colorless to yellow), showing great promise of the assay for quantitative detection.

Similarly, Peng et al. (2015) adopted the AuNPs seeds-mediated growth to generate color changes based on plasmonic ELISA. The design provided a convenient and effective strategy for the detection of disease biomarkers (Fig. 2.12). In this method, alcohol dehydrogenase (ADH)-labeled secondary antibody was used as a probe for primary antibody recognition and signal amplification. To prepare such a probe, streptavidin (SA) was conjugated with ADH by a 24-unit ethylene glycol functionalized with succinimidyl and maleimido ends [SM(PEG)$_{24}$] linker. By introducing the biotinylated antimouse IgG (H + L) as the secondary antibody, the streptavidin-ADH was conjugated with biotinylated secondary antibodies easily and firmly through the biotin–streptavidin interaction, generating the biocatalytic cycle of the

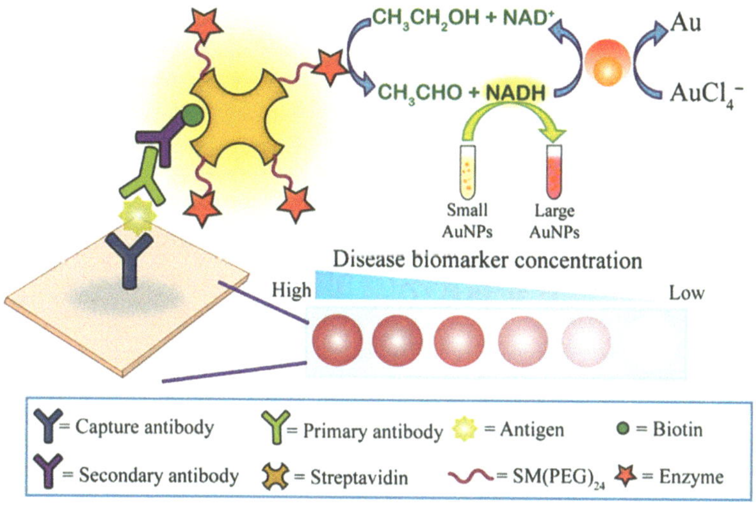

FIGURE 2.12 Schematic representation of the AuNPs growth-based sandwich ELISA and the principle of signal generation. *Reprinted from Peng, M.P., Ma, W., Long, Y.T., 2015. Alcohol dehydrogenase-catalyzed gold nanoparticle seed-mediated growth allows reliable detection of disease biomarkers with the naked eye. Anal. Chem. 87, 5891–5896.*

labeled enzyme in the ELISA assay. In the presence of biomarker antigens, ADH catalyzed the reaction between NAD^+ and ethanol to generate NADH and acetaldehyde. Then the generated NADH reduced $HAuCl_4$ and led to the enlargement of AuNPs seeds. As a result, the color of the solution changed from yellow to purple, which could be observed by the naked eye. By using this approach, two biomarkers, hepatitis B surface antigen (HBsAg) and α-fetoprotein (AFP) were detected with the LOD of naked eye detection down to 1.0×10^{-12} g mL^{-1}. Experiments with real serum samples from HBsAg infected patients were investigated, demonstrating the potential for clinical analysis. It was worth noting that the signal generation using AuNPs seeds-mediated growth in this experiment was much more reliable and specific than the aggregated colorimetric method.

2.6 NATURAL ENZYME ACTIVITY MIMICKING-BASED PROTEIN ASSAYS

Natural enzymes are extremely efficient to catalyze a variety of reactions. Their high substrate specificity, activities, and mild reaction conditions have attracted much attention in various fields, including biosensors, food industry, pharmaceutical process, and agrochemical production. However, the disadvantages of natural enzymes are still complicated, giving scientists headaches due to their low operational stability (denaturation and digestion), sensitivity of catalytic activity to environmental conditions, difficulties in recovery and recycling, and high costs in preparation and purification. These disadvantages greatly limit the applications of the natural enzymes (Garcia-Viloca et al., 2004; Jv et al., 2010; Long et al., 2011). To circumvent the aforementioned limitations, scientists place great hopes on artificial enzymes as alternatives to natural enzymes due to their low-cost and high stability (Kotov, 2010; Tao et al., 2015; Vázquez-González et al., 2017; Fan et al., 2017). Specially, with the emerging of nanotechnology, intensive research efforts have been ignited to design functional nanomaterials that exhibit various properties intrinsic to enzymes (Wei and Wang, 2013; Lin et al., 2014; Liu and Liu, 2015; Gao et al., 2017). A variety of nanoscale materials have been reported for their unique enzyme-mimicking catalytic activities, such as cerium oxide nanoparticles (Asati et al., 2009; Asati et al., 2011; Kim et al., 2014), magnetic nanoparticles (Wei and Wang, 2008; Dong et al., 2012; Liu et al., 2014a,b; Liang et al., 2012), AuNPs (Lien et al., 2013; He et al., 2013; Lien et al., 2014), V_2O_5 (André et al., 2011), $PtPd-Fe_3O_4$ (Sun et al., 2013), and graphene oxide (Song et al., 2010). Due to the excellent catalytic property of thiol monolayer protected AuNPs, Manea et al. (2004) called it "nanozyme" in analogy to the nomenclature of catalytic polymer (synzyme).

The catalytic activities of MNMs have been widely explored and used to mimic the function of natural enzymes, which provide the opportunities for the development of nanozymes with stable and high catalytic activity. (He et al., 2011; Zhou et al., 2009; Jiang et al., 2012; Zhang et al., 2011; Baruah et al., 2013). For example, historically, gold has been thought to be chemically inert, but the functionalized AuNPs with different surface modifications have been found to exhibit GOx- or peroxidase-like activity in recent years (Zhou et al., 2009; Comotti et al., 2004; Luo et al., 2010). MNMs with enzyme-mimicking activity have been applied in the detection of protein biomarkers in clinical samples (Lien et al., 2012; Zhu et al., 2014; Gao et al., 2015; Zhou et al., 2016). For example, Gao et al. (Gao et al., 2015) demonstrated that biofunctionalized AuNPs with a rationally designed peptide exhibited peroxidase-like activity (Fig. 2.13). The biofunctionalized AuNPs as a nanoprobe enabled the selective labeling and accurate determination of integrin GPIIb/IIIa on the human erythroleukemia cell line. After selective recognition and marking of integrin with the biofunctionalized AuNPs, the protein spatial distribution on cell membrane could be observed by the two-photon photoluminescence of the nanoprobes. More importantly, except for the qualitative observation of protein position, the expression level of integrin on human erythroleukemia cells could be detected quantitatively. In principle, the biofunctionalized AuNPs with intrinsic peroxidase-like activity could catalyze the oxidation of 3,3',5,5'-tetramethylbenzidine (TMB) in the presence of H_2O_2 after their conjugation on cells. Thus, the expression level of integrin was directly determined in an amplified and reliable colorimetric assay without cell lysis and protein extraction process. The method provides

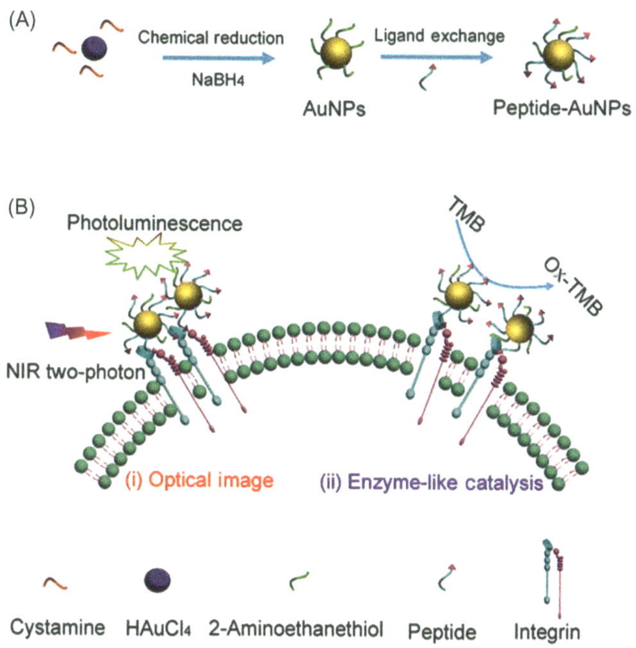

FIGURE 2.13 (A) Synthesis of peptide-conjugated AuNPs by chemical reduction and ligand exchange. (B) Nanoprobe integrates the signal generation and amplification for detection of integrin GPIIb/IIIa. *Reprinted from Gao, L., Liu, M.Q., Ma, G.F., Wang, Y.L., Zhao, L.N., Yuan, Q., et al., 2015. Peptideconjugatedgold nanoprobe: intrinsic nanozyme-linked immunsorbant assay of integrinexpression level on cell membrane. ACS Nano 9, 10979–10990.*

a simple and highly efficient strategy for protein detection, which holds a great promise for potential applications in universal quantification of other functional proteins on cell membrane directly.

Zhou et al. (2016) utilized AuNPs with glucose oxidase-like activity as a nanoprobe for the detection of DNA methyltransferase (MTase) activity (Fig. 2.14). Briefly, CdS quantum dots (QDs) were first conjugated on glassy carbon electrode (GCE), which could generate electrochemiluminescence (ECL) as output signal. The double-stranded DNA (dsDNA) containing the symmetric sequence of 5′-CCGG-3′ was then bonded to CdS QDs. Subsequently, the target M.SssI CpG MTase catalyzed the methylation of the specific CpG dinucleotides. Thereafter, M.SssI CpG MTase-treated electrode was incubated with a restriction endonuclease HpaII to recognize and cut off the 5′-CCGG-3′ sequence. The presence of the methylated CpG site in the 5′-CCGG-3′ sequence could block the recognition function of HpaII to protect the integrity of dsDNA. Later, AuNPs were combined with the end of the integrated dsDNA via Au-S bonds. According to the design, the higher M.SssI MTase activity could lead to more AuNPs–dsDNA conjugation. To obtain the enhanced ECL signal, glucose was added into the solution, which

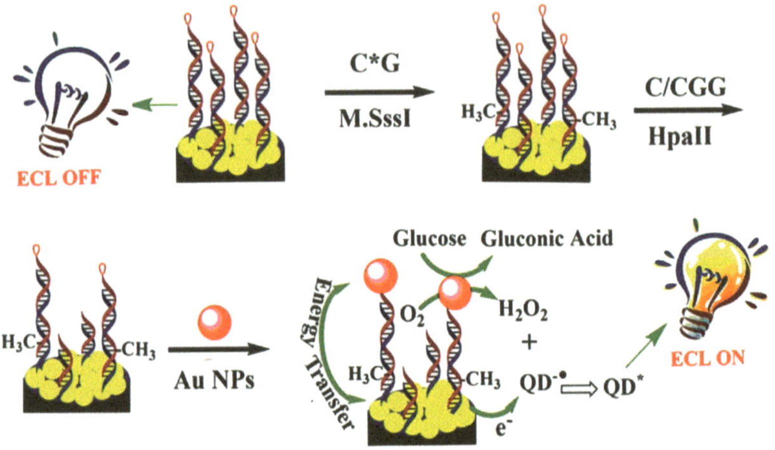

FIGURE 2.14 Schematic illustration of the AuNPs-based M.SssI MTase activity assay. *Reprinted from Zhou, H., Han, T.Q., Wei, Q., Zhang, S.S., 2016. Efficient enhancement of electrochemiluminescence from cadmium sulfide quantum dots by glucose oxidase mimicking gold nanoparticles for highly sensitive assay of methyltransferase activity. Anal. Chem. 88, 2976–2983.*

could be oxidized to produce gluconate and hydrogen peroxide (H_2O_2) by the catalysis of AuNPs. The produced H_2O_2 could serve as the ECL coreactant of CdS QDs and enhanced CdS QDs ECL via energy transfer (ET). Thus, the methylation event corresponding to the MTase activity could be monitored in the range from 1.0 to 120 U mL^{-1}, with the detection limit of 0.05 U mL^{-1}.

2.7 CONCLUSION

We have discussed a large variety of protein assays based on the excellent features of MNMs, including high surface to volume ratio, LSPR, SERS, and natural enzyme-mimicking activity. Through ingenious uses of synthetic chemistry and biochemistry, the intrinsic properties of MNMs have been exploited to develop all kinds of assays for various protein biomarkers related to a diverse set of diseases. Although a large amount of fundamental work about MNMs has been done over the last few decades, including various synthesis routes, all kinds of functionalization methods, and different protein biomarker assays, it is evident that there is still much to be done to achieve rapid, highly accurate, cost-effective, and point-of-care protein assays in a clinical setting. Looking to the future of such MNMs in relation to clinical applications, the continuous focus on the ideal attributes of clinically viable protein assays are still needed. The transfer from lab bench to clinic is a complex task, but the development of MNMs-based protein assays-enabled clinical applications is in a very good position to do just that.

REFERENCES

Aldewachi, H., Chalati, T., Woodroofe, M.N., Bricklebank, N., Sharrack, B., Gardiner, P., 2018. Gold nanoparticle-based colorimetric biosensors. Nanoscale 10, 18–33.

Ament, I., Prasad, J., Henkel, A., Schmachtel, S., Sönnichsen, C., 2012. Single unlabeled protein detection on individual plasmonic nanoparticles. Nano Lett. 12, 1092–1095.

André, R., Natálio, F., Humanes, M., Leppin, J., Heinze, K., Wever, R., et al., 2011. V_2O_5 nanowires with an intrinsic peroxidase-like activity. Adv. Funct. Mater. 21, 501–509.

Asati, A., Santra, S., Kaittanis, C., Nath, S., Perez, J.M., 2009. Oxidase-like activity of polymer-coated cerium oxide nanoparticles. Angew. Chem. Int. Ed. 121, 2344–2348.

Asati, A., Kaittanis, C., Santra, S., Perez, J.M., 2011. pH-tunable oxidase-like activity of cerium oxide nanoparticles achieving sensitive fluorigenic detection of cancer biomarkers at neutral pH. Anal. Chem. 83, 2547–2553.

Balasubramanian, S.K., Yang, L., Yung, L.Y.L., Ong, C.N., Ong, W.Y., Liya, E.Y., 2010. Characterization, purification, and stability of gold nanoparticles. Biomaterials 31, 9023–9030.

Baruah, B., Gabriel, G.J., Akbashev, M.J., Booher, M.E., 2013. Facile synthesis of silver nanoparticles stabilized by cationic polynorbornenes and their catalytic activity in 4-nitrophenol reduction. Langmuir 29, 4225–4234.

Busbee, B.D., Obare, S.O., Murphy, C.J., 2003. An improved synthesis of high-aspect-ratio gold nanorods. Adv. Mater. 15, 414–416.

Chang, C.C., Chen, C.P., Chen, C.Y., Lin, C.W., 2016. DNA base-stacking assay utilizing catalytic hairpin assembly-induced gold nanoparticle aggregation for colorimetric protein sensing. Chem. Commun. 52, 4167–4170.

Chen, H.J., Kou, X.S., Yang, Z., Ni, W.H., Wang, J.F., 2008. Shape and size dependent refractive index sensitivity of gold nanoparticles. Langmuir 24, 5233–5237.

Chen, J.B., Zuehlke, A., Deng, B., Peng, H.Y., Hou, X.D., Zhang, H.Q., 2017. A target-triggered DNAzyme motor enabling homogeneous, amplified detection of proteins. Anal. Chem. 89, 12888–12895.

Chen, J.Y., Wiley, B., McLellan, J., Xiong, Y.J., Li, Z.Y., Xia, Y.N., 2005. Optical properties of Pd-Ag and Pt-Ag nanoboxes synthesized via galvanic replacement reactions. Nano Lett. 5, 2058–2062.

Choi, W.K., Liew, T.H., Chew, H.G., Zheng, F., Thompson, C.V., Wang, Y., et al., 2008. A combined top-down and bottom-up approach for precise placement of metal nanoparticles on silicon. Small 4, 330–333.

Chuong, T.T., Pallaoro, A., Chaves, C.A., Li, Z., Lee, J., Eisenstein, M., et al., 2017. Dual-reporter SERS-based biomolecular assay with reduced false-positive signals. P. Natl. Acad. Sci. USA 114, 9056–9061.

Cognet, L., Tardin, C., Boyer, D., Choquet, D., Tamarat, P., Lounis, B., 2003. Single metallic nanoparticle imaging for protein detection in cells. P. Natl. Acad. Sci. USA 100, 11350–11355.

Comotti, M., Della Pina, C., Matarrese, R., Rossi, M., 2004. The catalytic activity of "naked" gold particles. Angew. Chem. Int. Ed. 116, 5936–5939.

Daniel, M.C., Astruc, D., 2004. Gold nanoparticles: assembly, supramolecular chemistry, quantum-size-related properties, and applications toward biology, catalysis, and nanotechnology. Chem. Rev. 104, 293–346.

Diao, J., Gall, K., Dunn, M.L., 2003. Surface-stress-induced phase transformation in metal nanowires. Nat. Mater. 2, 656–660.

Ding, C.F., Zhang, W., Wang, W., Chen, Y.Y., Li, X.Q., 2015. Amplification strategies using electrochemiluminescence biosensors for the detection of DNA, bioactive molecules and cancer biomarkers. TrAC-Trend Anal. Chem. 65, 137–150.

Dong, Y.L., Zhang, H.G., Rahman, Z.U., Su, L., Chen, X.J., Hu, J., et al., 2012. Graphene oxide–Fe_3O_4 magnetic nanocomposites with peroxidase-like activity for colorimetric detection of glucose. Nanoscale 4, 3969–3976.

Eustis, S., El-Sayed, M.A., 2006. Why gold nanoparticles are more precious than pretty gold: noble metal surface plasmon resonance and its enhancement of the radiative and nonradiative properties of nanocrystals of different shapes. Chem. Soc. Rev. 35, 209–217.

Fan, K.L., Wang, H., Xi, J.Q., Liu, Q., Meng, X.Q., Duan, D.M., et al., 2017. Optimization of Fe3O4 nanozyme activity via single amino acid modification mimicking an enzyme active site. Chem. Commun. 53, 424–427.

Fan, Z.X., Zhang, H., 2016. Crystal phase-controlled synthesis, properties and applications of noble metal nanomaterials. Chem.Soc. Rev. 45, 63–82.

Fenzl, C., Hirsch, T., Baeumner, A.J., 2016. Nanomaterials as versatile tools for signal amplification in (bio) analytical applications. TrAC-Trend Anal. Chem. 79, 306–316.

Gao, J., Huang, X.Y., Liu, H., Zan, F., Ren, J.C., 2012. Colloidal stability of gold nanoparticles modified with thiol compounds: bioconjugation and application in cancer cell imaging. Langmuir 28, 4464–4471.

Gao, L., Liu, M.Q., Ma, G.F., Wang, Y.L., Zhao, L.N., Yuan, Q., et al., 2015. Peptide-conjugated gold nanoprobe: intrinsic nanozyme-linked immunsorbant assay of integrin expression level on cell membrane. ACS Nano 9, 10979–10990.

Gao, L.Z., Fan, K.L., Yan, X.Y., 2017. Iron oxide nanozyme: a multifunctional enzyme mimetic for biomedical applications. Theranostics 7, 3207–3227.

García, I., Sánchez-Iglesias, A., Henriksen-Lacey, M., Grzelczak, M., Penadés, S., Liz-Marzán, L.M., 2015. Glycans as biofunctional ligands for gold nanorods: stability and targeting in protein-rich media. J. Am. Chem. Soc. 137, 3686–3692.

Garcia-Viloca, M., Gao, J., Karplus, M., Truhlar, D.G., 2004. How enzymes work: analysis by modern rate theory and computer simulations. Science 303, 186–195.

Grabbe, E.S., Buck, R.P., 1989. Surface-enhanced Raman spectroscopic investigation of human immunoglobulin G adsorbed on a silver electrode. J. Am. Chem. Soc. 111, 8362–8366.

Guo, L., Kim, D.H., 2012. LSPR biomolecular assay with high sensitivity induced by aptamer–antigen–antibody sandwich complex. Biosens. Bioelectron. 31, 567–570.

He, G., Luo, W.J., Li, P., Remmers, C., Netzer, W.J., Hendrick, J., et al., 2010. Gamma-secretase activating protein is a therapeutic target for Alzheimer's disease. Nature 467, 95–98.

He, X.R., Liu, H.B., Li, Y.L., Wang, S., Li, Y.J., Wang, N., et al., 2005. Gold nanoparticle-based fluorometric and colorimetric sensing of copper (II) ions. Adv. Mater. 17, 2811–2815.

He, W., Zhou, Y.T., Wamer, W.G., Hu, X., Wu, X., Zheng, Z., et al., 2013. Intrinsic catalytic activity of Au nanoparticles with respect to hydrogen peroxide decomposition and superoxide scavenging. Biomaterials 34, 765–773.

He, W.W., Liu, Y., Yuan, J.S., Yin, J.J., Wu, X.C., Hu, X.N., et al., 2011. Au@ Pt nanostructures as oxidase and peroxidase mimetics for use in immunoassays. Biomaterials 32, 1139–1147.

Heath, J.R., 1999. Nanoscale materials. Acc. Chem. Res. 32, 388.

Howes, P.D., Chandrawati, R., Stevens, M.M., 2014a. Colloidal nanoparticles as advanced biological sensors. Science 346, 1247390.

Howes, P.D., Rana, S., Stevens, M.M., 2014b. Plasmonic nanomaterials for biodiagnostics. Chem. Soc. Rev. 43, 3835–3853.

Hu, M., Chen, J.Y., Li, Z.Y., Au, L., Hartland, G.V., Li, X.D., et al., 2006. Gold nanostructures: engineering their plasmonic properties for biomedical applications. Chem. Soc. Rev. 35, 1084–1094.

Huang, X.Q., Tang, S.H., Liu, B.J., Ren, B., Zheng, N., 2011. Enhancing the photothermal stability of plasmonic metal nanoplates by a core-shell architecture. Adv. Mater. 23, 3420–3425.

Inci, F., Tokel, O., Wang, S., Gurkan, U.A., Tasoglu, S., Kuritzkes, D.R., et al., 2013. Nanoplasmonic quantitative detection of intact viruses from unprocessed whole blood. ACS Nano 7, 4733–4745.

Jang, H.R., Wark, A.W., Baek, S.H., Chung, B.H., Lee, H.J., 2013. Ultrasensitive and ultrawide range detection of a cardiac biomarker on a surface plasmon resonance platform. Anal. Chem. 86, 814–819.

Jiang, H., Chen, Z.H., Cao, H.Y., Huang, Y.M., 2012. Peroxidase-like activity of chitosan stabilized silver nanoparticles for visual and colorimetric detection of glucose. Analyst 137, 5560–5564.

Jiang, X.Y., Wang, H.J., Wang, H.J., Zhuo, Y., Yuan, R., Chai, Y.Q., 2017. Electrochemiluminescence biosensor based on 3-D DNA nanomachine signal probe powered by protein-aptamer binding complex for ultrasensitive mucin 1 detection. Anal. Chem. 89, 4280–4286.

Jung, H., Cha, H., Lee, D., Yoon, S., 2015. Bridging the nanogap with light: continuous tuning of plasmon coupling between gold nanoparticles. ACS Nano 9, 12292–12300.

Jv, Y., Li, B., Cao, R., 2010. Positively-charged gold nanoparticles as peroxidiase mimic and their application in hydrogen peroxide and glucose detection. Chem. Commun. 46, 8017–8019.

Kingsmore, S.F., 2006. Multiplexed protein measurement: technologies and applications of protein and antibody arrays. Nat. Rev. Drug Discov. 5, 310–320.

Kim, J., Oh, S.Y., Shukla, S., Hong, S.B., Heo, N.S., Bajpai, V.K., et al., 2018. Heteroassembled gold nanoparticles with sandwich-immunoassay LSPR chip format for rapid and sensitive detection of hepatitis B virus surface antigen (HBsAg). Biosens. Bioelectron. 107, 118–122.

Kim, M.I., Park, K.S., Park, H.G., 2014. Ultrafast colorimetric detection of nucleic acids based on the inhibition of the oxidase activity of cerium oxide nanoparticles. Chem. Commun. 50, 9577–9580.

Kotov, N.A., 2010. Inorganic nanoparticles as protein mimics. Science 330, 188–189.

Kumar, P.S., Pastoriza-Santos, I., Rodriguez-Gonzalez, B., De Abajo, F.J.G., Liz-Marzan, L.M., 2008. High-yield synthesis and optical response of gold nanostars. Nanotechnology 19, 015606.

Kwon, M.J., Lee, J., Wark, A.W., Lee, H.J., 2012. Nanoparticle-enhanced surface plasmon resonance detection of proteins at attomolar concentrations: comparing different nanoparticle shapes and sizes. Anal. Chem. 84, 1702–1707.

Laing, S., Gracie, K., Faulds, K., 2016. Multiplex in vitro detection using SERS. Chem. Soc. Rev. 45, 1901–1918.

Lane, L.A., Qian, X., Nie, S., 2015. SERS nanoparticles in medicine: from label-free detection to spectroscopic tagging. Chem. Rev. 115, 10489–10529.

Lee, A., Andrade, G.F.S., Ahmed, A., Souza, M.L., Coombs, N., Tumarkin, E., et al., 2011. Probing dynamic generation of hot-spots in self-assembled chains of gold nanorods by surface-enhanced Raman scattering. J. Am. Chem. Soc. 133, 7563–7570.

Lei, J.P., Ju, H.X., 2012. Signal amplification using functional nanomaterials for biosensing. Chem. Soc. Rev. 41, 2122–2134.

Li, C.L., Yamauchi, Y., 2013. Facile solution synthesis of Ag@ Pt core–shell nanoparticles with dendritic Pt shells. Phys. Chem. Chem. Phys. 15, 3490–3496.

Li, J., Zhao, T., Chen, T., Liu, Y., Ong, C.N., Xie, J., 2015. Engineering noble metal nanomaterials for environmental applications. Nanoscale 7, 7502–7519.

Li, W., Wang, L., Jiang, W., 2017. A catalytic assembled enzyme-free three-dimensional DNA walker and its sensing application. Chem. Commun. 53, 5527–5530.

Li, Y., Zhang, Y.L., Zhao, M., Zhou, Q.Q., Wang, L.L., Wang, H., et al., 2016. A simple aptamer-functionalized gold nanorods based biosensor for the sensitive detection of MCF-7 breast cancer cells. Chem. Commun. 52, 3959–3961.

Liang, J.J., Liu, H.W., Huang, C.H., Yao, C.Z., Fu, Q.Q., Li, X.Q., et al., 2015. Aggregated silver nanoparticles based surface-enhanced Raman scattering enzyme-linked immunosorbent assay for ultrasensitive detection of protein biomarkers and small molecules. Anal. Chem. 87, 5790–5796.

Liang, M.M., Fan, K.L., Pan, Y., Jiang, H., Wang, F., Yang, D.L., et al., 2012. Fe_3O_4 magnetic nanoparticle peroxidase mimetic-based colorimetric assay for the rapid detection of organophosphorus pesticide and nerve agent. Anal. Chem. 85, 308–312.

Lien, C.W., Huang, C.C., Chang, H.T., 2012. Peroxidase-mimic bismuth–gold nanoparticles for determining the activity of thrombin and drug screening. Chem. Commun. 48, 7952–7954.

Lien, C.W., Chen, Y.C., Chang, H.T., Huang, C.C., 2013. Logical regulation of the enzyme-like activity of gold nanoparticles by using heavy metal ions. Nanoscale 5, 8227–8234.

Lien, C.W., Tseng, Y.T., Huang, C.C., Chang, H.T., 2014. Logic control of enzyme-like gold nanoparticles for selective detection of lead and mercury ions. Anal. Chem. 86, 2065–2072.

Lin, Y., Ren, J., Qu, X., 2014. Catalytically active nanomaterials: a promising candidate for artificial enzymes. Acc. Chem. Res. 47, 1097–1105.

Liu, B.W., Liu, J.W., 2015. Accelerating peroxidase mimicking nanozymes using DNA. Nanoscale 7, 13831–13835.

Liu, Y., Cheng, Q., 2012. Detection of membrane-binding proteins by surface plasmon resonance with an all-aqueous amplification scheme. Anal. Chem. 84, 3179–3186.

Liu, D.B., Yang, J., Wang, H.F., Wang, Z.L., Huang, X.L., Wang, Z.T., et al., 2014a. Glucose oxidase-catalyzed growth of gold nanoparticles enables quantitative detection of attomolar cancer biomarkers. Anal. Chem. 86, 5800–5806.

Liu, Y., Yuan, M., Qiao, L.J., Guo, R., 2014b. An efficient colorimetric biosensor for glucose based on peroxidase-like protein-Fe_3O_4 and glucose oxidase nanocomposites. Biosens. Bioelectron. 52, 391–396.

Liz-Marzán, L.M., 2006. Tailoring surface plasmons through the morphology and assembly of metal nanoparticles. Langmuir 22, 32–41.

Long, Y.J., Li, Y.F., Liu, Y., Zheng, J.J., Tang, J., Huang, C.Z., 2011. Visual observation of the mercury-stimulated peroxidase mimetic activity of gold nanoparticles. Chem. Commun. 47, 11939–11941.

Luo, W.J., Zhu, C.F., Su, S., Li, D., He, Y., Huang, Q., et al., 2010. Self-catalyzed, self-limiting growth of glucose oxidase-mimicking gold nanoparticles. ACS Nano 4, 7451–7458.

Ma, X.M., Chen, Z.T., Kannan, P., Lin, Z.Y., Qiu, B., Guo, L.H., 2016. Gold nanorods as colorful chromogenic substrates for semiquantitative detection of nucleic acids, proteins, and small molecules with the naked eye. Anal. Chem. 88, 3227–3234.

Ma, X.Y., Truong, P.L., Anh, N.H., Sim, S.J., 2015. Single gold nanoplasmonic sensor for clinical cancer diagnosis based on specific interaction between nucleic acids and protein. Biosens. Bioelectron. 67, 59–65.

Manea, F., Houillon, F.B., Pasquato, L., Scrimin, P., 2004. Nanozymes: gold-nanoparticle-based transphosphorylation catalysts. Angew. Chem. Int. Ed. 43, 6165–6169.

Mayer, K.M., Lee, S., Liao, H., Rostro, B.C., Fuentes, A., Scully, P.T., et al., 2008. A label-free immunoassay based upon localized surface plasmon resonance of gold nanorods. ACS Nano 2, 687–692.

McFarland, A.D., Van Duyne, R.P., 2003. Single silver nanoparticles as real-time optical sensors with zeptomole sensitivity. Nano Lett. 3, 1057–1062.

McPeak, K.M., Van Engers, C.D., Bianchi, S., Rossinelli, A., Poulikakos, L.V., Bernard, L., et al., 2015. Ultraviolet plasmonic chirality from colloidal aluminum nanoparticles exhibiting charge-selective protein detection. Adv. Mater. 27, 6244–6250.

Mei, B.C., Susumu, K., Medintz, I.L., Mattoussi, H., 2009. Polyethylene glycol-based bidentate ligands to enhance quantum dot and gold nanoparticle stability in biological media. Nat. Protoc. 4, 412–423.

Nath, N., Chilkoti, A., 2002. A colorimetric gold nanoparticle sensor to interrogate biomolecular interactions in real time on a surface. Anal. Chem. 74, 504–509.

Nolan, G.P., 2007. What's wrong with drug screening today. Nat. Chem. Biol. 3, 187.

Nuraje, N., Mohammed, S., Yang, L.L., Matsui, H., 2009. Biomineralization nanolithography: combination of bottom-up and top-down fabrication to grow arrays of monodisperse gold nanoparticles along peptide lines. Angew. Chem. Int. Ed. 48, 2546–2548.

Olson, J., Dominguez-Medina, S., Hoggard, A., Wang, L.Y., Chang, W.S., Link, S., 2015. Optical characterization of single plasmonic nanoparticles. Chem.Soc. Rev. 44, 40–57.

Orendorff, C.J., Gearheart, L., Jana, N.R., Murphy, C.J., 2006. Aspect ratio dependence on surface enhanced Raman scattering using silver and gold nanorod substrates. Phys. Chem. Chem. Phys. 8, 165–170.

Peng, M.P., Ma, W., Long, Y.T., 2015. Alcohol dehydrogenase-catalyzed gold nanoparticle seed-mediated growth allows reliable detection of disease biomarkers with the naked eye. Anal. Chem. 87, 5891–5896.

Pissuwan, D., Valenzuela, S.M., Miller, C.M., Cortie, M.B., 2007. A golden bullet? Selective targeting of *Toxoplasma gondii* tachyzoites using antibody-functionalized gold nanorods. Nano Lett. 7, 3808–3812.

Polavarapu, L., Liz-Marzán, L.M., 2013. Towards low-cost flexible substrates for nanoplasmonic sensing. Phys. Chem. Chem. Phys. 15, 5288–5300.

Ralph, D.C., Black, C.T., Tinkham, M., 1997. Gate-voltage studies of discrete electronic states in aluminum nanoparticles. Phys. Rev. Lett. 78, 4087–4090.

Raveendran, P., Fu, J., Wallen, S.L., 2006. A simple and "green" method for the synthesis of Au, Ag, and Au–Ag alloy nanoparticles. Green Chem. 8, 34–38.

Rodríguez-Lorenzo, L., De La Rica, R., Álvarez-Puebla, R.A., Liz-Marzán, L.M., Stevens, M. M., 2012. Plasmonic nanosensors with inverse sensitivity by means of enzyme-guided crystal growth. Nat. Mater. 11, 604–607.

Rosi, N.L., Mirkin, C.A., 2005. Nanostructures in biodiagnostics. Chem. Rev. 105, 1547–1562.

Rusling, J.F., Bishop, G.W., Doan, N.M., Papadimitrakopoulos, F., 2014. Nanomaterials and biomaterials in electrochemical arrays for protein detection. J. Mater. Chem. B 2, 12–30.

Song, Y.J., Qu, K.G., Zhao, C., Ren, J.S., Qu, X.G., 2010. Graphene oxide: intrinsic peroxidase catalytic activity and its application to glucose detection. Adv. Mater. 22, 2206–2210.

Sönnichsen, C., Reinhard, B.M., Liphardt, J., Alivisatos, A.P., 2005. A molecular ruler based on plasmon coupling of single gold and silver nanoparticles. Nat. Biotechnol. 23, 741–745.

Sun, X.L., Guo, S.J., Chung, C.S., Zhu, W.L., Sun, S.H., 2013. A sensitive H_2O_2 assay based on dumbbell-like PtPd-Fe_3O_4 nanoparticles. Adv. Mater. 25, 132–136.

Tang, L., Casas, J., Venkataramasubramani, M., 2013. Magnetic nanoparticle mediated enhancement of localized surface plasmon resonance for ultrasensitive bioanalytical assay in human blood plasma. Anal. Chem. 85, 1431–1439.

Tao, Y., Ju, E., Ren, J., Qu, X., 2015. Bifunctionalized mesoporous silica-supported gold nanoparticles: intrinsic oxidase and peroxidase catalytic activities for antibacterial applications. Adv. Mater. 27, 1097–1104.

Tessler, L.A., Reifenberger, J.G., Mitra, R.D., 2009. Protein quantification in complex mixtures by solid phase single-molecule counting. Anal. Chem. 81, 7141–7148.

Tsai, M.F., Chang, S.H.G., Cheng, F.Y., Shanmugam, V., Cheng, Y.S., Su, C.H., et al., 2013. Au nanorod design as light-absorber in the first and second biological near-infrared windows for in vivo photothermal therapy. ACS Nano 7, 5330–5342.

Vázquez-González, M., Liao, W.C., Cazelles, R., Wang, S., Yu, X., Gutkin, V., et al., 2017. Mimicking horseradish peroxidase functions using Cu^{2+}-modified carbon nitride nanoparticles or Cu^{2+}-modified carbon dots as heterogeneous catalysts. ACS Nano 11, 3247–3253.

Wang, J., Wang, L.H., Liu, X.F., Liang, Z.Q., Song, S.P., Li, W.X., et al., 2007. A gold nanoparticle-based aptamer target binding readout for ATP assay. Adv. Mater. 19, 3943–3946.

Wang, S., Bi, S., Wang, Z.H., Xia, J.F., Zhang, F.F., Yang, M., et al., 2015a. A plasmonic aptasensor for ultrasensitive detection of thrombin via arrested rolling circle amplification. Chem. Commun. 51, 7927–7930.

Wang, W.J., Li, J.J., Rui, K., Gai, P.P., Zhang, J.R., Zhu, J.J., 2015b. Sensitive electrochemical detection of telomerase activity using spherical nucleic acids gold nanoparticles triggered mimic-hybridization chain reaction enzyme-free dual signal amplification. Anal. Chem. 87, 3019–3026.

Wang, X.H., Li, Y., Wang, H.F., Fu, Q.X., Peng, J.C., Wang, Y.L., et al., 2010. Gold nanorod-based localized surface plasmon resonance biosensor for sensitive detection of hepatitis B virus in buffer, blood serum and plasma. Biosens. Bioelectron. 26, 404–410.

Wang, Y.L., Xia, Y.N., 2004. Bottom-up and top-down approaches to the synthesis of monodispersed spherical colloids of low melting-point metals. Nano Lett. 4, 2047–2050.

Wei, H., Wang, E., 2008. Fe_3O_4 magnetic nanoparticles as peroxidase mimetics and their applications in H_2O_2 and glucose detection. Anal. Chem. 80, 2250–2254.

Wei, H., Wang, E., 2013. Nanomaterials with enzyme-like characteristics (nanozymes): next-generation artificial enzymes. Chem. Soc. Rev. 42, 6060–6093.

Wei, L.M., Wang, X.Y., Li, C., Li, X.X., Yin, Y.M., Li, G.X., 2015. Colorimetric assay for protein detection based on “nano-pumpkin” induced aggregation of peptide-decorated gold nanoparticles. Biosens. Bioelectron. 71, 348–352.

Willets, K.A., Van Duyne, R.P., 2007. Localized surface plasmon resonance spectroscopy and sensing. Annu. Rev. Phys. Chem. 58, 267–297.

Xia, Y., Xia, X., Wang, Y., Xie, S., 2013. Shape-controlled synthesis of metal nanocrystals. MRS Bull. 38, 335–344.

Xia, Y.N., Xiong, Y.J., Lim, B., Skrabalak, S.E., 2009. Shape-controlled synthesis of metal nanocrystals: Simple chemistry meets complex physics? Angew. Chem. Int. Ed. 48, 60–103.

Xu, L.G., Yan, W.J., Ma, W., Kuang, H., Wu, X.L., Liu, L.Q., et al., 2015. SERS encoded silver pyramids for attomolar detection of multiplexed disease biomarkers. Adv. Mater. 27, 1706–1711.

Xu, L.G., Zhao, S., Ma, W., Wu, X.L., Li, S., Kuang, H., et al., 2016. Multigaps embedded nanoassemblies enhance in situ Raman spectroscopy for intracellular telomerase activity sensing. Adv. Funct. Mater. 26, 1602–1608.

Xu, L.J., Zong, C., Zheng, X.S., Hu, P., Feng, J.M., Ren, B., 2014. Label-free detection of native proteins by surface-enhanced Raman spectroscopy using iodide-modified nanoparticles. Anal. Chem. 86, 2238–2245.

Yan, J., Song, S.P., Li, B., Zhang, Q., Huang, Q.A., Zhang, H., et al., 2010. An on-nanoparticle rolling-circle amplification platform for ultrasensitive protein detection in biological fluids. Small 6, 2520–2525.

Yang, X.J., Gao, Z.Q., 2015. Enzyme-catalysed deposition of ultrathin silver shells on gold nanorods: a universal and highly efficient signal amplification strategy for translating immunoassay into a litmus-type test. Chem. Commun. 51, 6928–6931.

Yu, W.B., Knauer, M., Kunas, C., Acaroz, U., Dietrich, R., Märtlbauer, E., 2017. A gold nanoparticles growth-based immunoassay for detection of antibiotic residues. Anal. Methods 9, 188–191.

Zhang, B., Liu, B.Q., Tang, D.P., Niessner, R., Chen, G.N., Knopp, D., 2012. DNA-based hybridization chain reaction for amplified bioelectronic signal and ultrasensitive detection of proteins. Anal. Chem. 84, 5392–5399.

Zhang, H., Ma, X.Y., Liu, Y., Duan, N., Wu, S.J., Wang, Z.P., et al., 2015a. Gold nanoparticles enhanced SERS aptasensor for the simultaneous detection of *Salmonella typhimurium* and *Staphylococcus aureus*. Biosens. Bioelectron. 74, 872–877.

Zhang, H.Q., Lai, M.D., Zuehlke, A., Peng, H.Y., Li, X.F., Le, X.C., 2015b. Binding-induced DNA nanomachines triggered by proteins and nucleic acids. Angew. Chem. Int. Ed. 54, 14326–14330.

Zhang, J.Z., 1997. Ultrafast studies of electron dynamics in semiconductor and metal colloidal nanoparticles: Effects of size and surface. Acc. Chem. Res. 30, 423–429.

Zhang, K., Hu, X.N., Liu, J.B., Yin, J.J., Hou, S., Wen, T., et al., 2011. Formation of PdPt alloy nanodots on gold nanorods: tuning oxidase-like activities via composition. Langmuir 27, 2796–2803.

Zheng, J., Li, N.X., Li, C.R., Wang, X.X., Liu, Y.C., Mao, G.B., et al., 2018. A nonenzymatic DNA nanomachine for biomolecular detection by target recycling of hairpin DNA cascade amplification. Biosens. Bioelectron. 107, 40–46.

Zheng, X.X., Liu, Q., Jing, C., Li, Y., Li, D., Luo, W.J., et al., 2011. Catalytic gold nanoparticles for nanoplasmonic detection of DNA hybridization. Angew. Chem. Int. Ed. 123, 12200–12204.

Zhou, C.H., Zhao, J.Y., Pang, D.W., Zhang, Z.L., 2014. Enzyme-induced metallization as a signal amplification strategy for highly sensitive colorimetric detection of avian influenza virus particles. Anal. Chem. 86, 2752–2759.

Zhou, H., Han, T.Q., Wei, Q., Zhang, S.S., 2016. Efficient enhancement of electrochemiluminescence from cadmium sulfide quantum dots by glucose oxidase mimicking gold nanoparticles for highly sensitive assay of methyltransferase activity. Anal. Chem. 88, 2976–2983.

Zhou, H., Liu, J., Xu, J.J., Zhang, S.S., Chen, H.Y., 2018. Optical nano-biosensing interface via nucleic acid amplification strategy: construction and application. Chem. Soc. Rev. 47, 1996–2019.

Zhou, X.C., Xu, W.L., Liu, G.K., Panda, D., Chen, P., 2009. Size-dependent catalytic activity and dynamics of gold nanoparticles at the single-molecule level. J. Am. Chem. Soc. 132, 138–146.

Zhu, Z., Guan, Z.C., Jia, S.S., Lei, Z.C., Lin, S.C., Zhang, H.M., et al., 2014. Au@Pt nanoparticle encapsulated target-responsive hydrogel with volumetric bar-chart chip readout for quantitative point-of-care testing. Angew. Chem. Int. Ed. 53, 12503–12507.

FURTHER READING

Liu, B.W., Liu, J.W., 2017. Surface modification of nanozymes. Nano Res. 10, 1125–1148.

Chapter 3

Quantum Dots and Nanoclusters

Jingjing Xu[1] and Ji Zheng[2]
[1]*Center for Molecular Recognition and Biosensing, School of Life Sciences, Shanghai University, Shanghai, P. R. China,* [2]*State Key Laboratory of Pharmaceutical Biotechnology, Department of Biochemistry, Nanjing University, Nanjing, P. R. China*

Chapter Outline

3.1 QUANTUM DOTS

3.1.1 Introduction of Quantum Dots

In 1981, Ekimov first discovered quantum dots in a glass matrix (Ekimov and Onushchenko, 1982). Then in 1985, Brus in Bell Labs found that quantum dots existed in colloidal materials as well (Brus, 1984). Soon after, Reed defined the concept of "quantum dot" (Reed et al., 1988), which was zero-dimensional nanosemiconductor materials. These spherical particles are usually prepared with atoms from groups II–VI, III–V, or IV–VI in the periodic table, and eventually become many different types of alloys (as summarized in Table 3.1). Nowadays, as far as we know, quantum dots are excellent luminescent materials with wide excitation range, and capable of giving narrow and symmetrical emission peaks after excitation (Service, 2000; Klostranec and Chan, 2006; Liu et al., 2010). Their unique physical

Nano-inspired Biosensors for Protein Assay with Clinical Applications.
DOI: https://doi.org/10.1016/B978-0-12-815053-5.00003-9

TABLE 3.1 Summary of Different Types of Quantum Dots

Type	Quantum Dots
II–VI	CdS, CdSe, CdTe, ZnS, ZnSe, ZnTe, HgS, HgSe, PbS, PbSe
III–V	GaAs, InGaAs, InP, InAs
IV	Si, Ge

properties and optical properties have made them attractive tools and vectors for research in molecular biology, material science, chemical analysis, etc. Quantum dots are becoming now one of the most exciting research objects in the 21st century.

3.1.2 Properties of Quantum Dots

Controllable emission light. The emission light given by quantum dots depends on the composition and dimension of the nanostructure. Most of the time, the emission spectrum, which mainly includes the intensity and emission wavelength, can be adjusted by modifying the reaction conditions or changing the composition or the size of the structure (Medintz et al., 2005). QDs in different sizes or of different compositions can be equally excited by a single light source, followed by distinctive emission spectrum with extremely little overlap, which thereby makes them particularly attractive for multiplexed cell imaging and analyzing (Liu et al., 2010). At present, the common QDs emit spectra distributed over the range of near-infrared–visible-ultraviolet (Han et al., 2001; Kairdolf et al., 2008).

Good light intensity and stability. Quantum dots show excellent light intensity and stability, which are 20 times or even 100 times higher than rhodamine 6G (Clapp et al., 2004; Bruchez and Hotz, 2007). Therefore, the QDs can be applied for observing the marked objects for a long time, as well as for studying the long-term interactions between biomolecules in living cells. With these advantages, QDs can be used for long-term tracking or real-time observation of target analytes. Hence, these QDs serve as a good tracer for clinical diagnosis (He et al., 2008).

A large Stokes shift and long fluorescence lifetime. Quantum dots have a large Stokes shift, which presents as a large gap between excitation and emission spectrum. Since each peak can be clearly identified, the detection of different fluorescent signals is no longer a problem (Dennis et al., 2012). In addition, the fluorescence lifetime of quantum dots can reach 50 ns, which is much longer than that of biological structures (<10 ns) (La Rosa et al., 2018; Mattoussi et al., 2000). Therefore, while most of the autofluorescence emitted by biological structures fades until it disappears, the fluorescent signal of QDs still exists and produces high intensity, which means the interference of background autofluorescence is negligible.

Active surface for functionalization and good biocompatibility. In addition to their excellent optical properties as mentioned above, quantum dots possess some other advantages, such as active surface for functionalization, good biocompatibility, etc. For the functionalization, various biomolecules like enzymes (Goldman et al., 2002), antibodies (Zhu et al., 2015), peptides (Qiu et al., 2010), DNA (Peng et al., 2007), are used, so as to facilitate the achievement of multipurpose analysis. Moreover, in the last two decades, a great progress has been made in overcoming some key limitations of QDs by improving the biocompatibility or reducing the cytotoxicity. Herein, the prospects of QDs in protein assays for clinical use are presented by discussing some representative examples.

3.1.3 Quantum Dots-Based Biosensors for Protein Assays

Currently, quantum dots are widely used to develop various biosensors due to the excellent optical properties. According to the sensing mechanism, quantum dots-based sensing systems can be divided into three categories as discussed below.

Biosensors based on ligand–target interactions. The first example is ligands-derivatized quantum dots that can send a corresponding signal upon binding to the target analytes. In this system, the quantity of biomolecules is reflected by the luminescent signal of QDs. In addition, according to the type of activation sources like external light, chemical reagents, and electricity, the QDs based luminescent sensors mainly include photoluminescent (PL) sensors (Silvi and Credi, 2015; Freeman and Willner, 2012), chemiluminescent (CL) sensors (Chen et al., 2014), and electrochemiluminescent (ECL) sensors (Wu et al., 2014).

Biosensors based on QDs–target interactions. The second example is biosensors based on the luminescent signal response of QDs upon directly binding to the target analytes. Specifically, under certain reaction conditions, the target analytes may interact with the surface of the quantum dots, by means of binding/dissociation, ligand/atomic exchange, and electron transfer (ET) (Wegner and Hildebrandt, 2015) or charge transfer (CT) (Algar et al., 2014). Thereafter, the attachment of target molecules or the replacement of natural ligands on QDs will affect the luminescence property of QDs, leading to a sensitive signal change. Generally speaking, the performance of this QDs-based biosensor relies on its surface sensitivity.

Biosensors based on fluorescence resonance energy transfer. Most of the time, quantum dots are capable of acting as an ideal energy transfer media, due to their outstanding optical properties, such as exceptional brightness, high quantum yield, large surface areas, broad absorption spectra, and narrow emission spectra (Nagy et al., 2013; Kim and Kim, 2012; Chen et al.,

2013; Clapp et al., 2010; Chen et al., 2012). Thus, the third example presented here (as shown in Fig. 3.1), is QDs-based resonance energy transfer sensing system. In this system, fluorescence resonance energy transfer (FRET) occurs when the QDs and the target fluorescent analytes become close enough. Specifically, when the emission spectrum of the fluorescence donors (referring to QDs) overlaps the absorption spectrum of the

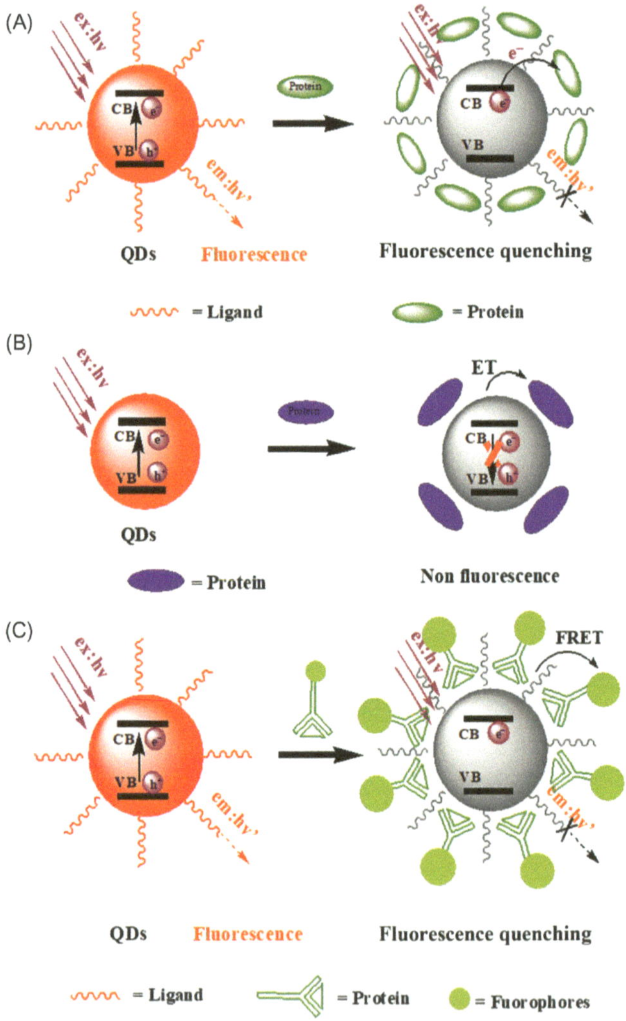

FIGURE 3.1 Three representative biosensors classified according to the sensing mechanism. (A) Biosensors based on ligand–target interactions. (B) Biosensors based on QDs–target interactions. (C) Biosensors based on fluorescence resonance energy transfer.

fluorescence receptor (referring to target analytes), and at least the distance between the two molecules is within 10 nm, FRET can be observed once an excitation light is given to the fluorescence donors (Chandan et al., 2018). The FRET phenomenon presents as a greatly enhanced emission intensity of the receptor, accompanied with a significantly decreased fluorescence intensity of the donors. As a powerful tool in life science, FRET plays an important role in the study of the distances between two molecules or two neighboring sites on a specific macromolecule. Thus, sensing systems based on FRET are considered as sensitive and reliable analytical tools, which are widely used for studying protein conformational changes (Krukenberg et al., 2011; Krusiński et al., 2010), protein–protein interactions, and enzyme activities (Boeneman et al., 2009), etc.

Recently, QDs have made tremendous progress in the application of protein analysis, which indicates their great value in clinical diagnosis.

3.1.3.1 Biosensors Based on Ligand–Target Interactions

In the system of biosensors based on ligand–target interactions, the identification and concentration of target proteins can be reflected by luminescent signal change of QDs (Xia et al., 2008; Claussen et al., 2015; Oh et al., 2015; Xiong et al., 2013; Zhang et al., 2013; Tang et al., 2013; Guo et al., 2013; Hu et al., 2010; Wang et al., 2015; Brazhnik et al., 2015). Recently, Chloe and coworkers (Kim and Searson, 2015) have established a magnetic beads–quantum dots (MBs-QDs) based sandwich assay to detect S100 calcium-binding protein B (S100B, a biomarker of traumatic brain injury). As shown in Fig. 3.2, the target proteins were captured by antibody-conjugated MBs. Afterwards, antibody-conjugated QDs were added, and then the QDs specifically bound to the target proteins were eluted from the MBs for optical measurements. By using this system, the limit of detection (LOD) of S100B in human serum was found to be as low as 10 pg mL^{-1}.

Moreover, in order to achieve simultaneous detection of multiple tumor markers, Ma and coworkers (Wang et al., 2015) have developed a novel immunochromatographic test strip (ICTS) based on multicolor quantum dots (QDs). As shown in Fig. 3.3, the system employed mouse anti-AFP monoclonal antibodies and mouse anti-CEA monoclonal antibodies as the detection line, and goat antimouse IgG antibodies as the control line. Anti-AFP monoclonal antibodies-conjugated QDs 546 and anti-CEA monoclonal antibodies-conjugated QDs 620 were mixed in a conjugate pad. Followed by adding the samples to be tested, the fluorescence intensities (at different wavelengths) of the detection line and the control line were detected, and the calculated intensity ratio reflects the concentration of corresponding protein. In the meantime, a simultaneous quantitative detection of multiple tumor markers was achieved. Under optimal conditions, AFP and CEA were quickly identified within 15 minutes, with a LOD of 3 ng/mL and 2 ng/mL,

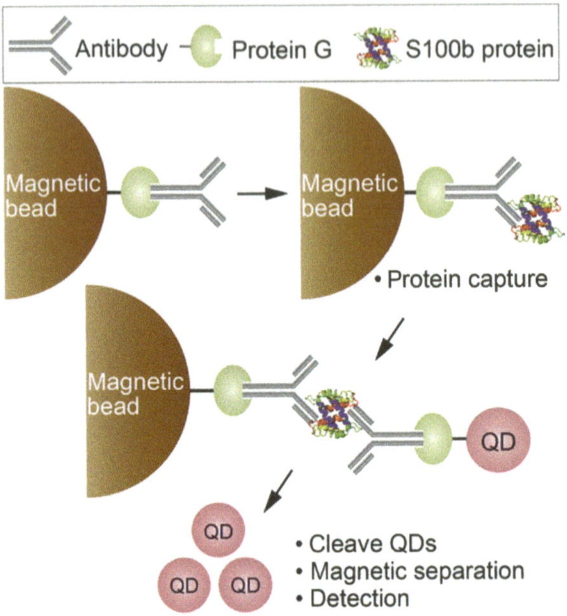

FIGURE 3.2 Schematic illustration of the MBs-QDs based sandwich protein assay. *Reprinted from Kim, C., Searson, P.C., 2015. Magnetic bead-quantum dot assay for detection of a biomarker for traumatic brain injury. Nanoscale 7, 17820–17826.*

respectively. This example demonstrated the great potential of multicolor QDs-based ICTS for the simultaneous quantitative detection of multiple tumor markers.

3.1.3.2 Biosensors Based on QDs–Target Interactions

Nowadays, the switch sensor based on quantum dots is a popular research object. The "on" and "off" signals exhibited by QDs are usually induced by three types of reactions, such as electron transfer, energy transfer, and ligand exchange (Ma and Su, 2011). Zhang's group (Chang et al., 2017) have constructed a fluorescence switch sensor based on boronic acid-capped Mn-doped ZnS quantum dots (QDs@MPS@AAPBA) to detect glycoproteins, and transferrin (TRF) and horseradish peroxidase (HRP) are selective as models of glycoproteins to evaluate sensitivity and selectivity of the probe. As shown in Fig. 3.4, in the absence of glycoproteins, the relatively weaker fluorescence intensity of QDs@MPS@AAPBA induced by effective electron transfer was observed. When the glycoproteins were covalently bound onto the boronic acid functionalized QDs, leading to the blockage of electron transfer, a relatively higher signal would be emitted, thus realizing the sensitive detection of glycoproteins in serum. The LOD of TRF and HRP was found to be 1.44×10^{-10} M and 3.36×10^{-10} M, respectively.

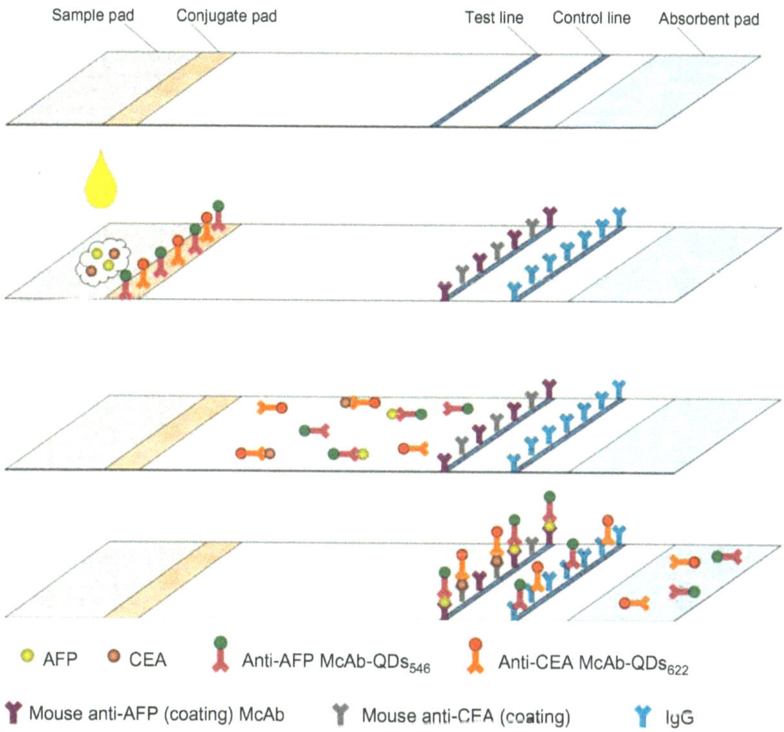

FIGURE 3.3 Schematic illustration of simultaneous detection of AFP and CEA using multi-color QDs-based immunochromatographic test strip. *Reprinted from Wang, C.Y., Hou, F., Ma, Y.C., 2015. Simultaneous quantitative detection of multiple tumor markers with a rapid and sensitive multi-color quantum dots based immunochromatographic test strip. Biosens. Bioelectron. 68, 156–162.*

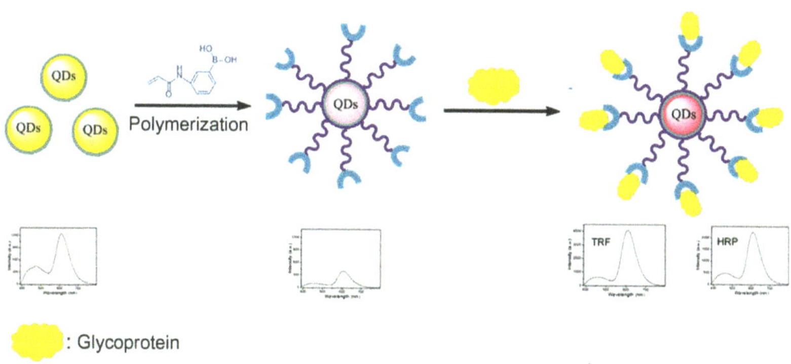

FIGURE 3.4 Schematic illustration of fluorescence turn-on sensor based on boronic acid functionalized Mn-doped ZnS QDs for the detection of glycoproteins. *Reprinted from Chang, L.F., Wu, H.C., He, X.W., Chen, L.X., Zhang, Y.K., 2017. A highly sensitive fluorescent turn-on biosensor for glycoproteins based on boronic acid functional polymer capped Mn-doped ZnS quantum dots. Anal. Chim. Acta, 91–98.*

Moreover, Wu and colleagues (Wu et al., 2011) developed a QDs-based phosphorescence biosensor for simultaneous discrimination of multiple proteins, by utilizing the triple-channel optical properties (fluorescence (FL), phosphorescence (Ph), light scattering (LS)) of Mn-doped ZnS (Mn-ZnS) QDs (as shown in Fig. 3.5). Mn-ZnS QDs are the most studied doped quantum dots with both defect- and dopant-dependent emission (Zou et al., 2011). Defect-related emission can produce fluorescence in the wavelength range of 360–520 nm, which originates from the bandgap transition of ZnS with ns-scale lifetime, and the doping associated emission Ph is exhibited in the wavelength range of 500–700 nm, due to the triplet transition of $Mn^{2+}(^4T_1\text{-}^6A_1)$ doped in the host lattice of ZnS, and the decay time is much longer than FL. In addition, Mn-ZnS QDs can also scatter incident light, and the aggregation of quantum dot nanoparticles can greatly increase the light scattering intensity. During the detection, a variety of optical detection techniques and statistical models were used to achieve the identification and differentiation of various proteins in the same time, such as cytochrome c (Cyt c), hemoglobin (Hb), human serum albumin (HSA), lysozyme (Lys), myoglobin (Mb), papain (Pap), transferrin (Tf), and ovalbumin (Ob). According to the spectral response patterns obtained from the FL and Ph of Mn-ZnS QDs the eight target proteins can be classified into three types. The first group contains Cyt c, Mb, and Hb, all of which quench the FL and Ph of Mn-ZnS QDs. All three metalloproteins contain hemin, an effective electron transfer quencher for CdSe@ZnS QDs (Sharon et al., 2010). In the presence of these proteins, the FL and Ph lifetimes of Mn-ZnS QDs are greatly shortened, thus making it possible to quench mechanisms through electron transfer, and the difference in quenching depends on the structure diversity of these proteins (Ma et al., 2005). HSA, TF, and Ob, which are classified into the second type, enhance intensity of Ph and FL, as well as the lifetime of FL and Ph of Mn-ZnS QDs. Studies have shown that HSA or Ob can enhance the fluorescence of CdTe QDs due to the reduction of surface defects (Gerhards et al., 2008). Therefore, they speculate that the interaction of HSA, TF, and Ob with Mn-ZnS QDs can also lead to a reduction in surface defects, thereby enhancing the strength of FL and Ph. Lys and papain Pap fall into the third category, which effectively increases the Ph of Mn-ZnS QDs, but leads to minimal variations of the FL. The changes of FL are minimized, especially in the presence of Lys and Pap, the LS signal of Mn-ZnS QDs is obviously larger than that of other proteins. It has been reported that highly aggregated Mn-ZnS QDs give an improved Ph signal (He et al., 2009). Therefore, in the presence of Lys and pap, the Ph-increment of mn-ZnS QDs exceeds that of FL. Further PCA analysis of three-channel response data at different protein concentrations shows that the eight proteins can be distinguished when the concentration is not less than 0.5 μM. The performance of this three-channel sensing device based on phosphorescent quantum dots further proves the identification of proteins in human urine.

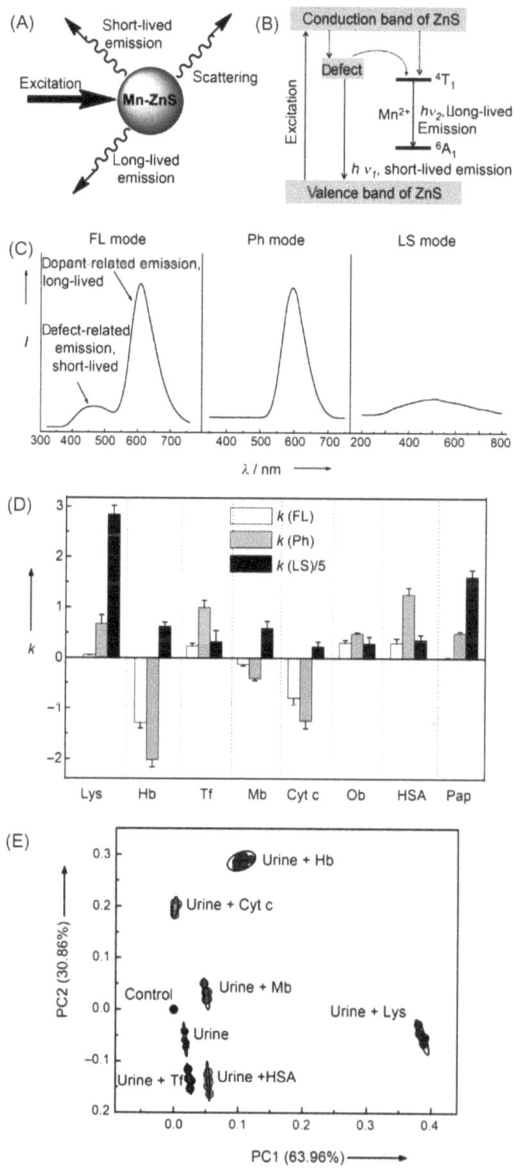

FIGURE 3.5 (A–C) Schematic illustration of triple-channel phosphorescence biosensor based on Mn-ZnS QDs; (D) Fingerprints of eight selected proteins based on the patterns of corresponding k values obtained from the FL, Ph, and LS of Mn–ZnS QDs; and (E) PCA plot for the discrimination of eight proteins (0.5 μM) based on the triple-channel optical properties of Mn–ZnS QDs. *Reprinted from Wu, P., Miao, L.N., Wang, H.F., Shao, X., Yan, X.G., Yan, X.P., 2011. A multidimensional sensing device for the discrimination of proteins based on manganese-doped ZnS quantum dots. Angew. Chem. Int. Ed. 50, 8118–8121.*

3.1.3.3 *Biosensors Based on Fluorescence Resonance Energy Transfer*

Recently, a great number of biosensors have been developed based on FRET properties of QDs. For instance, Chi et al. (2011) established a QDs-based biosensing system derivatized with stem-loop structures formed by single-strand aptamers of thrombin (as shown in Fig. 3.6). Since BOBO-3(B) (1,1′-(4,4,7,7-tetramethyl-4,7-diazaundecamethylene)-bis-4-[3-methyl-2,3-dihydro-(benzo-1,3 thiazole)-2-methylidene]-pyridinium tetraiodide) can specifically insert into the double-strand region of the stem-loop structures, leading to FRET (fluorescence quenching) between QDs and BOBO-3(B). However, the stem-loop structure would be unfolded upon the binding of thrombin, bound to the DNA, leading to the release of BOBO-3 (B) from the surface of the QDs, thus recovering the fluorescence signal of QDs, and then enabling the analysis of thrombin. The LOD of thrombin was 1 nM with a nM to μM dynamic range and thrombin detection in the presence of 15% diluted serum is also demonstrated.

Li et al. (2015) designed a dual FRET-based fluorescence probe, which consisted of a quencher–fluorophore pair and two specific linkers, so as to achieve the sequential detection of MMP-2 and caspase-3 (as shown in Fig. 3.7). In the beginning, the fluorophore (FAM) in the probe was quenched by Dabcyl. When incubated with MMP-2 or caspase-3, the FRET ended as a result of the breakage of corresponding peptides. At the same time, the fluorescence signal of FAM was recovered, which reflected the concentration of MMP-2 or caspase-3. Taking advantage of the dual system, the probe exhibited high sensitivity and selectivity toward MMP-2 and caspase-3 at the same time.

3.1.4 Conclusion for Quantum Dots-Based Biosensors

For the development of QDs-based biosensors, the main challenge is the specific detection of target proteins in complex samples. In recent years, a great

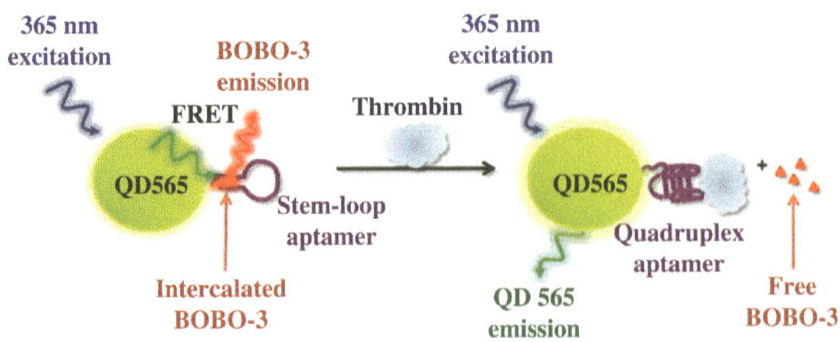

FIGURE 3.6 Label-free detection of thrombin with a novel QD-apt:B beacon. The binding of thrombin induces the conformational change of the aptamer on QDs, and then causes the dissociation of the stained BOBO-3 from QD-apt, resulting in the quantification of thrombin. *Reprinted from Chi, C.W., Lao, Y.H., Li, Y.S., Chen, L.C., 2011. A quantum dot-aptamer beacon using a dnaintercalating dye as the fret reporter: application to label-free thrombin detection. Biosens. Bioelectron. 26, 3346–3352.*

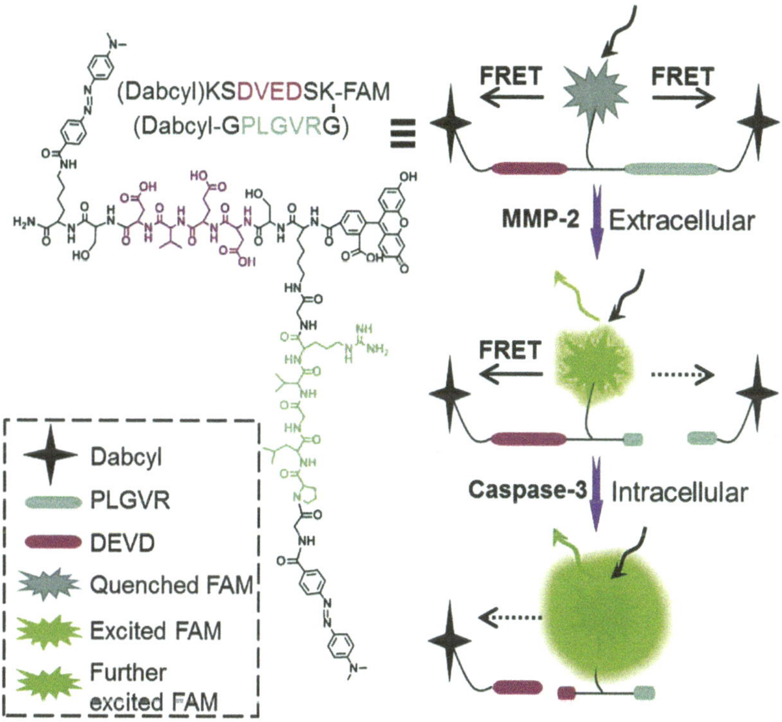

FIGURE 3.7 The chemical structure, and the proposed mechanism of the dual FRET-based fluorescence probe for the detection of MMP-2 and caspase-3. *Reprinted from Li, S.Y., Liu, L.H., Cheng, H., Li, B., Qiu, W.X., Zhang, X.Z., 2015. A dual-fret-based fluorescence probe for the sequential detection of MMP-2 and caspase-3. Chem. Commun., 51, 14520–14523.*

amount of effort has been devoted to overcome the challenge. And some breakthroughs have also been made by taking advantage of aptamers, peptides, etc. Moreover, detection efficacy has been highly optimized by combining the superior luminescence property of QDs with some recognition elements, like molecularly imprinted polymers. Generally speaking, the recent advances in material science have brought great benefits to the medical arena, by providing accurate detection of biomarker proteins. We predict, that in the near future, quantum dots-based biosensors will establish a complete pathway from research to industrialization and solve more practical problems.

3.2 NANOCLUSTERS

3.2.1 Introduction of Nanoclusters

Usually, those metal nanomaterials (MNMs), who belong to zero-dimensional nanostructures, are generally spherical nanomaterials, such as gold nanoparticles, metal nanoclusters (MNCs), etc. Recently, MNCs have attracted considerable attention, due to their high sensitivity.

The metal nanoclusters are molecular aggregates composed of several to several dozens of metal atoms, which present a diameter of less than 2 nm. As shown in Fig. 3.8, the size of MNCs approaches the Fermi wavelength of the electrons in the conduction band, thus the electron energy level easily changes from the quasi-continuous state to the discrete state, leading to the formation of cleavage energy (Díez and Ras, 2011). The interference factors in other bulk metallic materials, such as plasmon resonance conductivity, thermal conductivity, and light reflection, all disappear in the case of MNCs, due to their unique discrete energy levels and quantum size effects. Therefore, compared to other larger metal nanomaterials, MNCs exhibit significantly different optical, electrical and chemical properties, such as strong fluorescence, excellent light stability, and good biocompatibility (Zheng et al., 2007).

At present, the widely reported MNCs are mainly AuNCs, AgNCs, CuNCs, as well as alloy nanoclusters composed of two kinds of metal atoms (Zhang and Wang, 2014). In this report, we focus on the advantages of these MNCs, such as their ultrasmall particle size, strong fluorescence, and low toxicity (Shang et al., 2011). Some representative examples of their application in protein assay for clinical use are described here, and their current prospects and limits are discussed.

3.2.2 Properties of Metal Nanoclusters

The physical and chemical properties of metal nanomaterials depend mainly on their particle sizes. Thus, MNCs exhibit outstanding advantages in many aspects, such as fluorescence, catalytic properties, and biocompatibility, due to their ultrasmall crystal structures. In particular, the fluorescence of MNCs has better light stability, greater Stokes shift, and adjustable fluorescence emission.

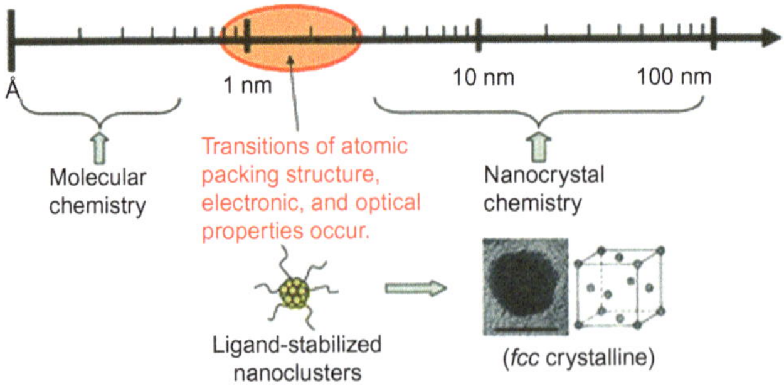

FIGURE 3.8 The position of metal nanoclusters that connects metal atoms and metal sodium particles on the scale. *Reprinted from Jin, R.C., 2010. Quantum sized, thiolate-protected gold nanoclusters. Nanoscale 2, 343–362.*

The main optical properties of MNCs are their unique fluorescence emission mechanism. However, the mechanism of fluorescence emission mechanism performed by MNCs has not been completely explained until now. Because the composition of the MNCs, the synthesis approach and environment may cause the change of emission mechanism, leading to a change of fluorescence emission spectrum (Wu and Jin, 2010). This unique optical property fills the gap between metal atoms and relatively larger nanomaterials.

Essentially, the fluorescence signal emitted by MNCs is a kind of photoluminescence excited by external light. Specifically, the absorption of photoelectrons makes MNCs energetic, which facilitates the electrons to transition from active state to the low-energy stable state, leading to fluorescence emission. In special cases, MNCs can be excited by two-photon light. Two-photon excitation is particularly suitable for deep tissues and intracellular imaging (Bianchini et al., 2012) due to its strong penetrating power in near-infrared regions. For example, it was found that some AgNCs activated by two-photon light could emit completely different signals at 660, 680, and 710 nm (Helmchen and Denk, 2005), thus they can be used as an effective tool for multiple bioimaging.

In addition, the composition and synthesis approach of MNCs also have important influences on the emission spectrum of MNCs. Taking AuNCs as an example (shown in Fig. 3.9), using different ligands as stabilizers, AuNCs emitted fluorescence signal of different wavelengths (Jin, 2010). This example shows the terrific potential of MNCs for the detection of multiple proteins due to its signal tunable property.

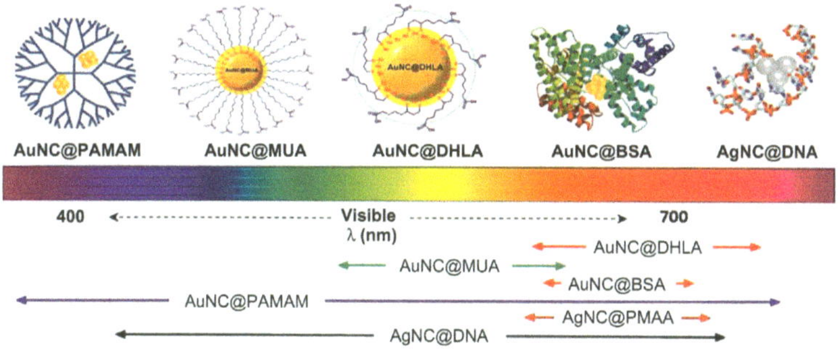

FIGURE 3.9 Schematic representation of signal tunable AuNCs functionalized with different ligands. *Reprinted from Lin, C.J., Lee, C.H., Hsieh, J.T., Wang, H.H., Li, J.K., Shen, J.L., et al., 2009. Review: synthesis of fluorescent metallic nanoclusters toward biomedical application: recent progress and present challenges. J. Med. Biol. Eng. 29, 276–283.*

3.2.3 Metal Nanoclusters-Based Biosensors for Proteins Assays

3.2.3.1 Biosensors Based on AuNCs

Among all the metal nanoclusters, gold nanoclusters have the best stability, thus are widely used in the detection of proteins for clinical use.

As shown in Fig. 3.10, Triulzi and coworkers developed a biosensing system based on polyamidoamine(PAMAM)-dendrimer-wrapped AuNCs, which were derivatized with anti-IgG that enabled quantitative detection of human IgG (Triulzi et al., 2006). Because of the high quantum yield and strong membrane permeability of PAMAM dendrimer package, when combined with AuNCs, the complex could be used as an immunofluorescence sensor for antibody detection. The construction of quantitative immunofluorescence assays and biosensors is realizable due to the linear relationship in both micromolar and nanomolar ranges.

In addition, Hu et al. developed a biosensing system based on BSA-stabilized AuNCs (shown in Fig. 3.11). In the system, the presence of trypsin

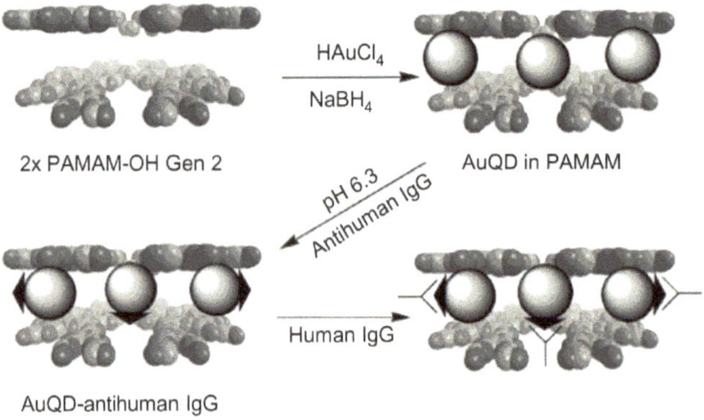

FIGURE 3.10 Schematic illustration of polyamidoamine-dendrimer-wrapped AuNCs for IgG detection. *Reprinted from Triulzi, R.C., Micic, M., Giordani, S., Serry, M., Chiou, W.A., Leblanc, R.M., 2006. Immunoasssay based on the antibody-conjugated pamam-dendrimer-gold quantum dot complex. Chem. Commun. 48, 5068–5070.*

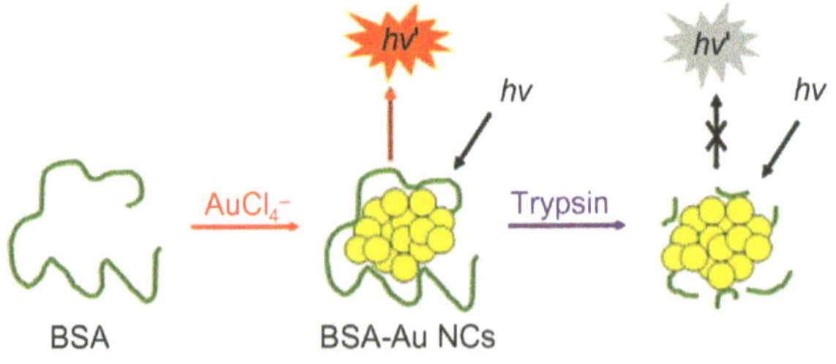

FIGURE 3.11 Schematic illustration of BSA-stabilized AuNCs for trypsin detection. *Reprinted from Hu, L.Z., Han, S., Parveen, S., Yuan, Y.L., Zhang, L., Xu, G.B., 2012. Highly sensitive fluorescent detection of trypsin based on bsa-stabilized gold nanoclusters. Biosens. Bioelectron. 32, 297–299.*

would reduce the fluorescence intensity by digesting BSA, producing signal-off process. To be more specific, trypsin cleaves other proteins from the side of positively charged lysine and arginine, leading to BSA hydrolyzation and BSA-AuNCs decomposition, resulting in a significant decrease in the fluorescence intensity. Finally, the detection range was found to be 0.01–100 μg/mL in urine samples (Hu et al., 2012).

3.2.3.2 Biosensors Based on AgNCs

Compared with gold nanoclusters, silver nanoclusters are cheaper and more stable. In addition, AgNCs can be simply synthesized with designable functions. Sharma et al. (2011) developed a biosensing system based on aptamer-functionalized AgNCs for the detection of thrombin. As shown in Fig. 3.12, in the presence of thrombin, the fluorescence was quenched when an aliquot of the agglomerated AgNCs was added, and the concentration of thrombin was reflected by the changes in fluorescence intensity.

Because fluorescein-rich DNA fragments can enhance the fluorescence intensity of silver nanoclusters, a hairpin-structured DNA-AgNCs biosensor was designed. As shown in Fig. 3.13 (Li et al., 2017), since transcription factors (TFs) can specifically bind to double strands of DNA, DNA duplexes that bind TFs were not opened by exonuclease III (ExoIII) digestion, but DNA duplexes without TFs-protection would be digested by ExoIII, and the open single strand could be complementary to the hairpin probe

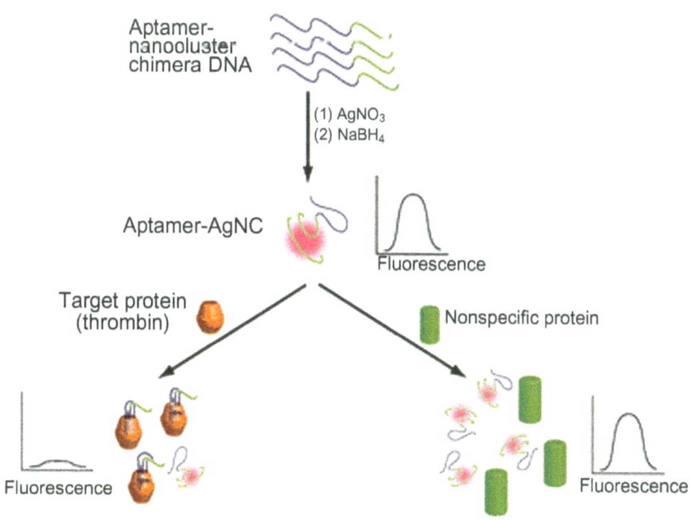

FIGURE 3.12 Schematic illustration of Aptamer functionalized AgNCs for the detection of thrombin. *Reprinted from Sharma, J., Yeh, H.C., Yoo, H., Werner, J.H., Martinez, J.S., 2011. Silver nanocluster aptamers: in situ generation of intrinsically fluorescent recognition ligands for protein detection. Chem. Commun. 47, 2294–2296.*

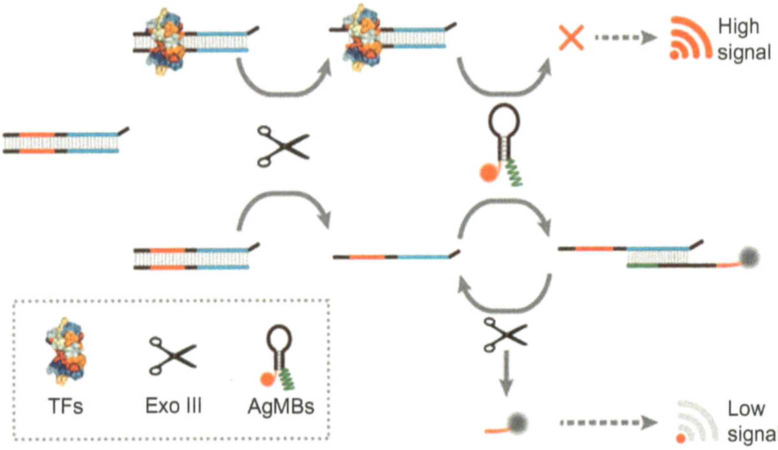

FIGURE 3.13 The detection of TFs by using AgNCs. *Reprinted from Li, B.Z., Xu, L., Chen, Y., Zhu, W.Y., Shen, X., Zhu, C.H., et al., 2017. Sensitive and label-free fluorescent detection of transcription factors based on dna-ag nanoclusters molecular beacons and exonuclease iii-assisted signal amplification. Anal. Chem. 89, 7316–7323.*

DNA-AgNCs, thus reducing the fluorescence intensity. This method is extremely sensitive to TFs and achieves a detection limit of 10 pM, and a linear detection range was found to be from 30 pM to 1.5 nM.

3.2.3.3 Biosensors Based on CuNCs

Usually, copper ions (Cu^{2+}) can interact with DNA to form copper nanoclusters, and its fluorescence properties are stable. Since copper is cheaper than gold and silver, copper nanoclusters can be synthesized with less cost, thus providing an important nanomaterial for the detection of proteins.

Copper nanoclusters (CuNCs) are novel fluorescent probes that can be protected by bovine serum albumin (BSA) to form BSA-CuNCs. As shown in Fig. 3.14 (Li et al., 2016), copper ions were protected by BSA under the catalysis of hydrazine hydrate ($N_2H_4 \cdot H_2O$) to form BSA-CuNCs, which emitted fluorescence, followed by further functionalization with 3-aminophenylboronic acid (APBA) to selectively discriminate sugars. Proteins, when specifically bound to glycoproteins, induced fluorescence quenching of BSA-CuNCs. This method is highly sensitive for the detection of glycoproteins, with a detection range of 5–220 nM and a quantitative spike recovery rate from 95% to 104%.

Double-stranded DNA-templated copper nanoparticles are also called copper nanoclusters, which exhibit fluorescence properties when CuNCs are formed at low concentrations, so as to detect proteins. As shown in Fig. 3.15 (Zhou et al., 2011), the sensor was composed of two nucleic acid strands, one was an ATP aptamer strand, named "ABA," and the other one was

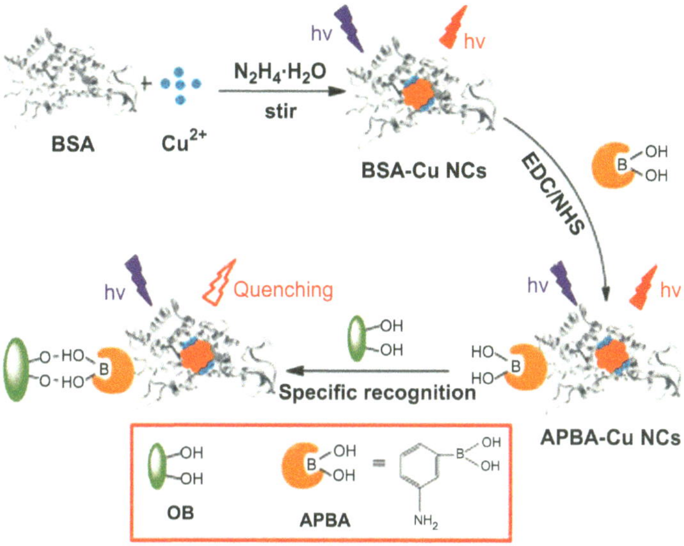

FIGURE 3.14 Synthesis of 3-aminophenylboronic acid-functionalized CuNCs (APBA-Cu NCs) and application for the selective recognition of glycoproteins. *Reprinted from Li, X.G., Zhang, F., Gao, Y., Zhou, Q.M., Zhao, Y., Li, Y., et al., 2016. Facile synthesis of red emitting 3-aminophenylboronic acid functionalized copper nanoclusters for rapid, selective and highly sensitive detection of glycoproteins. Biosens. Bioelectron. 86, 270–276.*

"c-ABA" strand which is complementary to the ABA strand. Cu^+ was reduced by ascorbic acid in buffer solution at room temperature to form dsDNA-CuNCs with high fluorescence. However, if ATP specifically bound to ABA chains, no CuNCs with high fluroscene will be formed. The linear detection range of this system was found to be 0.05–500 μM with a LOD of 28 nM.

3.2.3.4 Biosensors Based on Bimetallic Nanoclusters

At present, since different nanoclusters can emit different fluorescence signals, gold nanocluster materials with two or more kinds of metal materials which could produce different fluorescence signals simultaneously are being studied. For example, gold/silver nanoclusters, silver/copper nanoclusters, etc., have good fluorescence characteristics, and can be widely applied to the detection of proteins.

Incorporation of silver into gold nanoclusters can result in significantly enhanced fluorescence, so gold and silver bimetallic nanoclusters were designed to detect proteins. As shown in Fig. 3.16 (Zhou et al., 2016), gold and silver bimetallic nanoclusters were synthesized using bovine serum albumin (BSA) as a protective and reducing agent, and pyrophosphatase (PPase) as catalyst of the conversion from single-molecule inorganic pyrophosphate (PPi, $P_2O_7^{4-}$) to orthophosphoric acid, so as to detect pyrophosphatase.

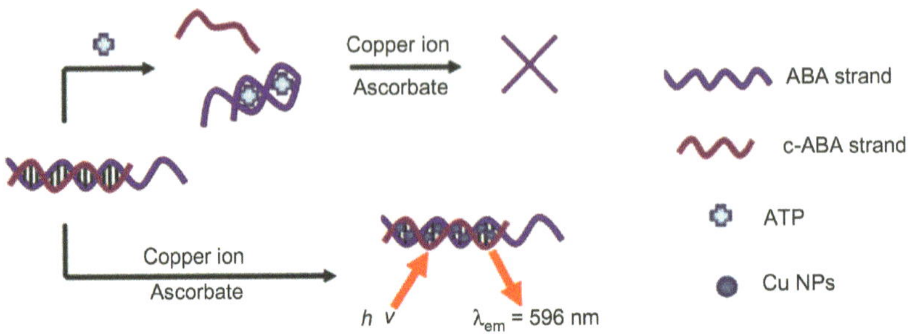

FIGURE 3.15 Representation of the sensing procedure for analysis of ATP based on the dsDNA-Cu NPs. *Reprinted from Zhou, Z.X., Du, Y., Dong, S.J., 2011. Double-strand dna-templated formation of copper nanoparticles as fluorescent probe for label-free aptamer sensor. Anal. Chem. 83, 5122–5127.*

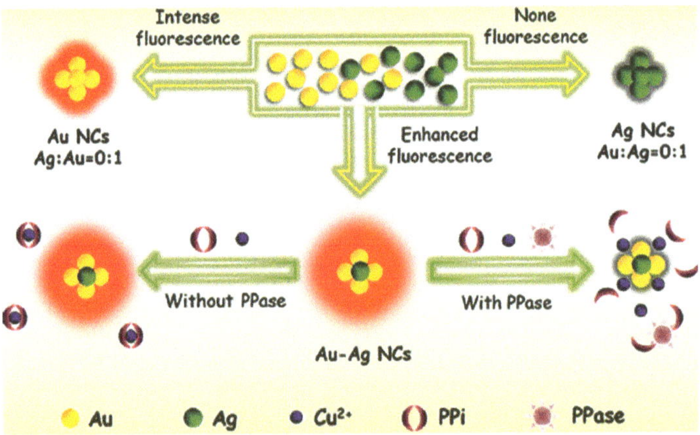

FIGURE 3.16 Gold – silver bimetallic nanoclusters (Au – Ag NCs)-based fluorescent probes for pyrophosphatase (PPase) activity. *Reprinted from Zhou, Q., Lin, Y., Xu, M., Gao, Z., Yang, H.H., Tang, D., 2016. Facile synthesis of enhanced fluorescent gold-silver bimetallic nanocluster and its application for highly sensitive detection of inorganic pyrophosphatase activity. Anal. Chem. 88, 8886–8892.*

When copper ions were added, the coordination bonds between Cu^{2+} and PPi were formed, which helped to protect the fluorescence of gold–silver nanoclusters. In the presence of pyrophosphorylase, Cu^{2+} attaching to the gold–silver nanoclusters would quench the fluorescence. The detection limit of this method was found to be 0.1–30 mU/mL under the optimal conditions, and the minimum detection concentration is 0.03 mU/mL.

Copper–silver nanoclusters are biosensors with specific fluorescence detection proteins synthesized using DNA as a template. As shown in Fig. 3.17 (Li et al., 2013), acetylcholinesterase could hydrolyze acetylcholine into sulfhydryl thiocholine, then DNA-Cu/AgNCs were bound to quench

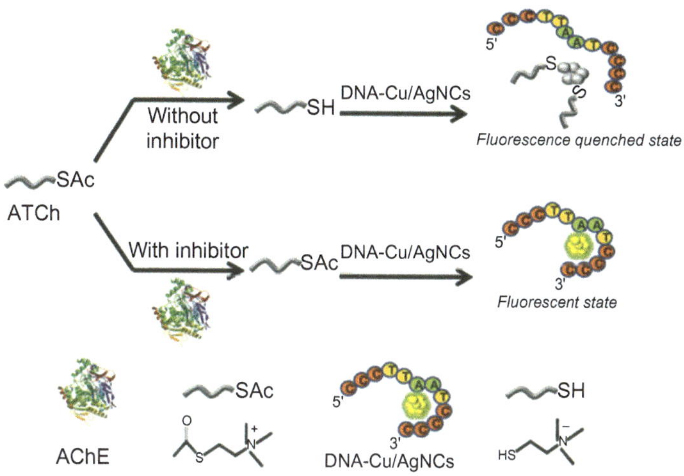

FIGURE 3.17 Mechanism of the DNA-Cu/AgNCs-based fluorometric assay for AChE activity and inhibitor detection. *Reprinted from Li, W.H., Li, W., Hu, Y.F., Xia, Y.L., Shen, Q.P., Nie, Z., et al., 2013. A fluorometric assay for acetylcholinesterase activity and inhibitor detection based on DNA-templated copper/silver nanoclusters. Biosens. Bioelectron. 47, 345–349.*

fluorescence. However, when the acetylcholinesterase inhibitor existed, acetylcholine hydrolysis didn't occur, and the fluorescence of DNA-Cu/AgNCs were not influenced. The method determined the minimum AChE activity of 0.05 mU/mL, and the linear range from 0.05 to 2.0 mU/mL, realizing a label-free and rapid detection with high sensitivity.

The fluorescence characteristics of metal nanoclusters can be applied to a variety of experimental methods for the detection of various biomolecules. In addition to the detection of proteins, the specific binding of aptamers and membrane proteins can also be used for cell visualization and biomarker cells for cancers (Li et al., 2014). Detection of biological macromolecules except proteins and nucleic acids can also be achieved (Chen et al., 2016). In addition, metal ions can also be detected (Ding et al., 2014).

3.3 CONCLUSIONS

In addition to the physicochemical properties of nanomaterials, quantum dots and nanoclusters also possess some unique properties. For example, the quantum dots with a particle size of 1–100 nm have a high quantum yield, which emits fluorescence with high intensity to ensure sensitivity. The fluorescence emission spectrum is narrow, the cross-covering area of adjacent emission spectrum is small, the resolution is high, the light is stable, the photobleaching resistance is stable and can withstand multiple excitations and emissions, and the color can be adjusted arbitrarily in the visible light range. The same material will emit different wavelengths of light with different sizes, manifesting as a single excitation light source, multiple emission wavelengths, and can be modified and connected. It is easy to disperse in solution and be operated experimentally. However, its disadvantage lies in the inherent toxicity of components (such as Pb and Cd in II–VI groups) and its unclear biodegradation mechanism. Compared with quantum dots, nanoclusters are smaller in size, about 1–2 nm, and have strong fluorescence emission, good photostability, and high conductivity. Combining the properties of many metals, bimetallic or trimetallic nanocomposites with high bioactivity and high utilization ratio can be developed. In addition, gold nanoparticles exhibit good electrochemical properties due to their unique surface plasmon resonance properties and can also help enhance Raman scattering signals. Of course, each material has its own imperfections, and metal nanoclusters are composed of only a few atoms, in order to reduce the surface energy, so it is easy to have a violent reaction with each other and to polymerize irreversibly. The reduction of metal ions in an aqueous solution tends to result in large nanoparticles rather than small nanoclusters, so it is necessary to introduce a stable material or ligand as a protective agent in the preparation process. In addition, ligands coated on the surface of particles can significantly affect the fluorescence emission properties of particles, especially silver nanoclusters, besides the inherent instability resulting from their size. They are also very easily oxidized by oxygen in the air or electron receptors in the air, and their stable lifetime in aqueous solution is very short (Zhang and Wang, 2014). Their biocompatibility is better than quantum dots, but the quantum yields are generally lower (usually less than 5%), and their preparation costs are higher. Therefore, we can consider the choice of materials based on different purposes to achieve better overall performance.

REFERENCES

Algar, W.R., Stewart, M.H., Scott, A.M., Moon, W.J., Medintz, I.L., 2014. Quantum dots as platforms for charge transfer-based biosensing: challenges and opportunities. J. Mater. Chem. B 2, 7816–7827.

Bianchini, P., Harke, B., Galiani, S., Vicidomini, G., Diaspro, A., 2012. Single-wavelength two-photon excitation stimulated emission depletion (SW2PE-STED) superresolution imaging. Proc. Natl. Acad. Sci. U.S.A. 109, 6390–6393.

Boeneman, K., Mei, B.C., Dennis, A.M., Bao, G., Deschamps, J.R., Mattoussi, H., et al., 2009. Sensing caspase 3 activity with quantum dot-fluorescent protein assemblies. J. Am. Chem. Soc. 131, 3828–3829.

Brazhnik, K., Sokolova, Z., Baryshnikova, M., Bilan, R., Efimov, A., Nabiev, I., 2015. Quantum dot-based lab-on-a-bead system for multiplexed detection of free and total prostate-specific antigens in clinical human serum samples. Nanomed. Nanotechnol. Biol. Med. 11, 1065–1075.

Bruchez, M.P., Hotz, C.Z. (Eds.), 2007. Quantum Dots: Applications in Biology (Vol. 374). Springer Science & Business Media.

Brus, L.E., 1984. Electron–electron and electron-hole interactions in small semiconductor crystallites: the size dependence of the lowest excited electronic state. J. Chem. Phys. 80, 4403–4409.

Chandan, H.R., Schiffman, J.D., Balakrishna, R.G., 2018. Quantum dots as fluorescent probes: synthesis, surface chemistry, energy transfer mechanisms, and applications. Sens. Actuators B 258, 1191–1214.

Chang, L.F., Wu, H.C., He, X.W., Chen, L.X., Zhang, Y.K., 2017. A highly sensitive fluorescent turn-on biosensor for glycoproteins based on boronic acid functional polymer capped Mn-doped ZnS quantum dots. Anal. Chim. Acta 91–98.

Chen, A.Y., Ma, S.Y., Zhuo, Y., Chai, Y.Q., Yuan, R., 2016. In situ electrochemical generation of electrochemiluminescent silver naonoclusters on target-cycling synchronized rolling circle amplification platform for microrna detection. Anal. Chem. 88, 3203–3210.

Chen, G., Song, F., Xiong, X., Peng, X., 2013. Fluorescent nanosensors based on fluorescence resonance energy transfer (FRET). Ind. Eng. Chem. Res. 52, 11228–11245.

Chen, H., Lin, L., Li, H.H., Lin, J.M., 2014. Quantum dots-enhanced chemiluminescence: mechanism and application. Coordin. Chem. Rev. 263-264, 86–100.

Chen, N.T., Cheng, S.H., Liu, C.P., Souris, J.S., Chen, C.T., Mou, C.Y., et al., 2012. Recent advances in nanoparticle-based förster resonance energy transfer for biosensing, molecular imaging and drug release profiling. Int. J. Mol. Sci. 13, 16598–16623.

Chi, C.W., Lao, Y.H., Li, Y.S., Chen, L.C., 2011. A quantum dot-aptamer beacon using a dna intercalating dye as the fret reporter: application to label-free thrombin detection. Biosens. Bioelectron. 26, 3346–3352.

Clapp, A.R., Medintz, I.L., Mauro, J.M., Fisher, B.R., Bawendi, M.G., Mattoussi, H., 2004. Fluorescence resonance energy transfer between quantum dot donors and dye-labeled protein acceptors. J. Am. Chem. Soc. 126, 301–310.

Clapp, A.R., Medintz, I.L., Mattoussi, H., 2010. Forster resonance energy transfer investigations using quantum-dot fluorophores. Chemphyschem 7, 47–57.

Claussen, J.C., Malanoski, A., Breger, J.C., Oh, E., Walper, S.A., Susumu, K., et al., 2015. Probing the enzymatic activity of alkaline phosphatase within quantum dot bioconjugates. J. Chem. Phys. Chem. C 119, 2208–2221.

Dennis, A.M., Rhee, W.J., Sotto, D., Dublin, S.N., Bao, G., 2012. Quantum dot-fluorescent protein fret probes for sensing intracellular ph. ACS Nano 6, 2917–2924.

Díez, I., Ras, R.H., 2011. Fluorescent silver nanoclusters. Nanoscale 3, 1963–1970.

Ding, H., Liang, C., Sun, K., Wang, H., Hiltunen, J.K., Chen, Z., et al., 2014. Dithiothreitol-capped fluorescent gold nanoclusters: an efficient probe for detection of copper(II) ions in aqueous solution. Biosens. Bioelectron. 59, 216–220.

Ekimov, A.I., Onushchenko, A.A., 1982. Quantum size effect in the optical-spectra of semiconductor micro-crystals. Sov. Phys. Semiconduct.-USSR 16, 775–778.

Freeman, R., Willner, I., 2012. Optical molecular sensing with semiconductor quantum dots (QDs). Chem. Soc. Rev. 41, 4067–4085.

Gerhards, C., Schulz-Drost, C., Sgobba, V., Guldi, D.M., 2008. Conjugating luminescent CdTe quantum dots with biomolecules. J. Chem. Phys. Chem. B 112, 14482–14491.

Goldman, E.R., Anderson, G.P., Tran, P.T., Mattoussi, H., Charles, P.T., Mauro, J.M., 2002. Conjugation of luminescent quantum dots with antibodies using an engineered adaptor protein to provide new reagents for fluoroimmunoassays. Anal. Chem. 74, 841–847.

Guo, Z.Y., Hao, T.T., Du, S.P., Chen, B.B., Wang, Z.B., Li, X., et al., 2013. Multiplex electrochemiluminescence immunoassay of two tumor markers using multicolor quantum dots as labels and graphene as conducting bridge. Biosens. Bioelectron. 44, 101–107.

Han, M.Y., Gao, X.H., Su, J.Z., Nie, S., 2001. Quantum-dot-tagged microbeads for multiplexed optical coding of biomolecules. Nat. Biotechnol. 19, 631–635.

He, Y., Lu, H.T., Sai, L.M., Su, Y.Y., Hu, M., Fan, C.H., et al., 2008. Microwave synthesis of water-dispersed CdTe/CdS/ZnS core-shell-shell quantum dots with excellent photostability and biocompatibility. Adv. Mater. 20, 3416–3421.

He, Y., Wang, H.F., Yan, X.P., 2009. Self-assembly of Mn-doped ZnS quantum dots/octa (3-aminopropyl) octasilsequioxane octahydrochloride nanohybrids for optosensing DNA. Chem.-Eur. J 15, 5436–5440.

Helmchen, F., Denk, W., 2005. Deep tissue two-photon microscopy. Nat. Methods 2, 932–940.

Hu, L.Z., Han, S., Parveen, S., Yuan, Y.L., Zhang, L., Xu, G.B., 2012. Highly sensitive fluorescent detection of trypsin based on bsa-stabilized gold nanoclusters. Biosens. Bioelectron. 32, 297–299.

Hu, M., Yan, J., He, Y., Lu, H.T., Weng, L.X., Song, S.P., et al., 2010. Ultrasensitive, multiplexed detection of cancer biomarkers directly in serum by using a quantum dot-based microfluidic protein chip. ACS Nano 4, 488–494.

Jin, R.C., 2010. Quantum sized, thiolate-protected gold nanoclusters. Nanoscale 2, 343–362.

Kairdolf, B.A., Smith, A.M., Nie, S., 2008. One-pot synthesis, encapsulation, and solubilization of size-tuned quantum dots with amphiphilic multidentate ligands. J. Am. Chem. Soc. 130, 12866–12867.

Kim, C., Searson, P.C., 2015. Magnetic bead-quantum dot assay for detection of a biomarker for traumatic brain injury. Nanoscale 7, 17820–17826.

Kim, G.B., Kim, Y.P., 2012. Analysis of protease activity using quantum dots and resonance energy transfer. Theranostics 2, 127–138.

Klostranec, J.M., Chan, W.C., 2006. Quantum dots in biological and biomedical research: recent progress and present challenges. Adv. Mater. 18, 1953–1964.

Krukenberg, K.A., Street, T.O., Lavery, L.A., Agard, D.A., 2011. Conformational dynamics of the molecular chaperone HSP 90. Q. Rev. Biophys. 44, 229–255.

Krusiński, T., Ozyhar, A., Dobryszycki, P., 2010. Dual FRET assay for detecting receptor protein interaction with DNA. Nucleic Acids Res. 38, e108-e108.

La Rosa, M., Denisov, S.A., Jonusauskas, G., Mcclenaghan, N., Credi, A., 2018. Designed long lived emission from CdSe quantum dots by reversible electronic energy transfer with a surface bound chromophore. Angew. Chem. Int. Ed 130 (12), 3158–3161.

Li, B.Z., Xu, L., Chen, Y., Zhu, W.Y., Shen, X., Zhu, C.H., et al., 2017. Sensitive and label-free fluorescent detection of transcription factors based on dna-ag nanoclusters molecular beacons and exonuclease iii-assisted signal amplification. Anal. Chem. 89, 7316–7323.

Li, J.J., You, J., Zhuang, Y.P., Han, C.P., Hu, J.F., Wang, A., et al., 2014. A "light-up" and "spectrum-shift" response of aptamer-functionalized silver nanoclusters for intracellular mRNA imaging. Chem. Commun. 50, 7107–7110.

Li, S.Y., Liu, L.H., Cheng, H., Li, B., Qiu, W.X., Zhang, X.Z., 2015. A dual-fret-based fluorescence probe for the sequential detection of MMP-2 and caspase-3. Chem. Commun. 51, 14520–14523.

Li, W.H., Li, W., Hu, Y.F., Xia, Y.L., Shen, Q.P., Nie, Z., et al., 2013. A fluorometric assay for acetylcholinesterase activity and inhibitor detection based on DNA-templated copper/silver nanoclusters. Biosens. Bioelectron. 47, 345–349.

Li, X.G., Zhang, F., Gao, Y., Zhou, Q.M., Zhao, Y., Li, Y., et al., 2016. Facile synthesis of red emitting 3-aminophenylboronic acid functionalized copper nanoclusters for rapid, selective and highly sensitive detection of glycoproteins. Biosens. Bioelectron. 86, 270–276.

Lin, C.J., Lee, C.H., Hsieh, J.T., Wang, H.H., Li, J.K., Shen, J.L., et al., 2009. Review: synthesis of fluorescent metallic nanoclusters toward biomedical application: recent progress and present challenges. J. Med. Biol. Eng. 29, 276–283.

Liu, J., Lau, S.K., Varma, V.A., Moffitt, R.A., Caldwell, M., Liu, T., et al., 2010. Molecular mapping of tumor heterogeneity on clinical tissue specimens with multiplexed quantum dots. ACS nano 4, 2755–2765.

Ma, Q., Su, X.G., 2011. Recent advances and applications in QDs-based sensors. Analyst 136, 4883.

Ma, Y., Bai, H., Yang, C., Yang, X., 2005. A sensitive method for the detection of proteins by high-efficiency fluorescence quenching. Analyst 130, 283–285.

Mattoussi, H., Mauro, J.M., Goldman, E.R., Anderson, Gg. P., Sunder, V.C., Mikulec, F.V., et al., 2000. Self-assembly of CdSe – ZnS quantum dot bioconjugates using an engineered recombinant protein. J. Am. Chem. Soc. 122, 12142–12150.

Medintz, I.L., Uyeda, H.T., Goldman, E.R., Mattoussi, H., 2005. Quantum dot bioconjugates for imaging, labelling and sensing. Nature Mater. 4, 435–446.

Nagy, A., Gemmill, K.B., Delehanty, J.B., Medintz, I.L., Sapsford, K.E., 2013. Peptide-functionalized quantum dot biosensors. IEEE J. Sel. Top. Quant. 20, 115–126. IEEE J SEL TOP QUANT.

Oh, S.D., Duong, H.D., Rhee, J.I., 2015. Simple and sensitive progesterone detection in human serum using a CdSe/ZnS quantum dot-based direct binding assay. Anal. Biochem. 483, 54–61.

Peng, H., Zhang, L.J., Kjällman, T.H., Soeller, C., Travas-Sejdic, J., 2007. DNA hybridization detection with blue luminescent quantum dots and dye-labeled single-stranded DNA. J. Am. Chem. Soc. 129, 3048–3049.

Qiu, T., Zhao, D., Zhou, G.H., Liang, Y., He, Z.K., Liu, Z.H., et al., 2010. A positively charged QDs-based fret probe for micrococcal nuclease detection. Analyst 135, 2394–2399.

Reed, M.A., Randall, J.N., Aggarwal, R.J., Matyi, R.J., Moore, T.M., Wetsel, A.E., 1988. Observation of discrete electronic states in a zero-dimensional semiconductor nanostructure. Phys. Rev. Lett. 60, 535–537.

Service, R.F., 2000. Atom-scale research gets real. Science 290, 1524–1531.

Shang, L., Dong, S.J., Nienhaus, G.U., 2011. Ultra-small fluorescent metal nanoclusters: synthesis and biological applications. Nano. Today 6, 401–418.

Sharma, J., Yeh, H.C., Yoo, H., Werner, J.H., Martinez, J.S., 2011. Silver nanocluster aptamers: in situ generation of intrinsically fluorescent recognition ligands for protein detection. Chem. Commun. 47, 2294–2296.

Sharon, E., Freeman, R., Willner, I., 2010. CdSe/ZnS quantum dots-G-quadruplex/hemin hybrids as optical DNA sensors and aptasensors. Anal. Chem. 82, 7073–7077.

Silvi, S., Credi, A., 2015. Luminescent sensors based on quantum dot-molecule conjugates. Chem. Soc. Rev. 44, 4275–4289.

Tang, D.P., Hou, L., Niessner, R., Xu, M.D., Gao, Z.Q., Knopp, D., 2013. Multiplexed electrochemical immunoassay of biomarkers using metal sulfide quantum dot nanolabels and trifunctionalized magnetic beads. Biosens. Bioelectron. 46, 37–43.

Triulzi, R.C., Micic, M., Giordani, S., Serry, M., Chiou, W.A., Leblanc, R.M., 2006. Immunoasssay based on the antibody-conjugated pamam-dendrimer-gold quantum dot complex. Chem. Commun. 48, 5068–5070.

Wang, C.Y., Hou, F., Ma, Y.C., 2015. Simultaneous quantitative detection of multiple tumor markers with a rapid and sensitive multicolor quantum dots based immunochromatographic test strip. Biosens. Bioelectron. 68, 156–162.

Wegner, K.D., Hildebrandt, N., 2015. Quantum dots: bright and versatile in vitro and in vivo fluorescence imaging biosensors. Chem. Soc. Rev. 44, 4792–4834.

Wu, P., Miao, L.N., Wang, H.F., Shao, X., Yan, X.G., Yan, X.P., 2011. A multidimensional sensing device for the discrimination of proteins based on manganese-doped ZnS quantum dots. Angew. Chem. Int. Ed. 50, 8118–8121.

Wu, Z.K., Jin, R.C., 2010. On the ligand's role in the fluorescence of gold nanoclusters. Nano. Lett. 10, 2568–2573.

Xia, Z., Xing, Y., So, M.K., Ai, L.K., Sinclair, R., Rao, J., 2008. Multiplex detection of protease activity with quantum dot nanosensors prepared by intein-mediated specific bioconjugation. Anal. Chem. 80, 8649–8655.

Xiong, W.W., Yang, G.H., Wu, X.C., Zhu, J.J., 2013. Aqueous synthesis of color-tunable $CuInS_2$/ZnS nanocrystals for the detection of human interleukin 6. ACS Appl. Mater. Interfaces. 5, 8210–8216.

Zhang, L., Wang, E., 2014. Metal nanoclusters: new fluorescent probes for sensors and bioimaging. Nano. Today 9, 132–157.

Zhang, R.Q., Liu, S.L., Zhao, W., Zhang, W.P., Yu, X., Li, Y., et al., 2013. A simple point-of-care microfluidic immunomagnetic fluorescence assay for pathogens. Anal. Chem. 85, 2645–2651.

Zheng, J., Nicovich, P.R., Dickson, R.M., 2007. Highly fluorescent noble metal quantum dots. Annu. Rev. Phys. Chem. 58, 409–431.

Zhou, Q., Lin, Y., Xu, M., Gao, Z., Yang, H.H., Tang, D., 2016. Facile synthesis of enhanced fluorescent gold-silver bimetallic nanocluster and its application for highly sensitive detection of inorganic pyrophosphatase activity. Anal. Chem. 88, 8886–8892.

Zhou, Z.X., Du, Y., Dong, S.J., 2011. Double-strand dna-templated formation of copper nanoparticles as fluorescent probe for label-free aptamer sensor. Anal. Chem. 83, 5122–5127.

Zhu, X.L., Liu, J.W., Peng, H.Y., Jiang, J.H., Yu, R.Q., 2015. A novel fluorescence assay for inorganic pyrophosphatase based on modulated aggregation of graphene quantum dots. Analyst 141, 251–255.

Zou, W.S., Sheng, D., Ge, X., Qiao, J.Q., Lian, H.Z., 2011. Room-temperature phosphorescence chemosensor and Rayleigh scattering chemodosimeter dual-recognition probe for 2,4,6-trinitrotoluene based on manganese-doped ZnS quantum dots. Anal. Chem. 83, 30–37.

FURTHER READING

Wu, P., Hou, X., Xu, J.J., Chen, H.Y., 2014. Electrochemically generated versus photoexcited luminescence from semiconductor nanomaterials: bridging the valley between two worlds. Chem. Rev. 114, 11027–11059.

Chapter 4

Other Nanomaterials

Jingjing Xu[1] and Shuai Wu[2]

[1]Center for Molecular Recognition and Biosensing, School of Life Sciences, Shanghai University, Shanghai, P. R. China, [2]State Key Laboratory of Pharmaceutical Biotechnology, Department of Biochemistry, Nanjing University, Nanjing, P. R. China

Chapter Outline

4.1 INTRODUCTION

With the rapid development of materials science (Farka et al., 2017a), biosensors based on silicon nanomaterials (Peng et al., 2014), upconversion nanomaterials (Chen et al., 2014), liposomes (Liu and Boyd, 2013), and virus nanoparticles (Cao et al., 2016) etc., have been successfully developed. In addition, these nanomaterials-based biosensors have attracted much attention due to their excellent performance in protein assays for clinical use. Some representative examples are highlighted here, and their future prospects are discussed.

Nano-inspired Biosensors for Protein Assay with Clinical Applications.
DOI: https://doi.org/10.1016/B978-0-12-815053-5.00004-0

4.2 SILICON NANOMATERIALS

Given the natural abundance of silicon on the earth, silicon-based nanomaterials have received extensive attention, due to their special optical and electrical properties, excellent biocompatibility, and high surface-to-volume ratio. Currently, various types of silicon nanostructures, such as small luminescent silicon nanoparticles, one-dimensional silicon nanowires (SiNWs) or nanotubes, and large mesoporous silicon nanoparticles (SiNPs), have been successfully prepared due to the enormous progress made in chemical synthesis field (He et al., 2010). Differences in size and shape bring about differences in properties. Furthermore, the nature of silane makes silicon nanomaterials easy to be covalently conjugated with biomolecules (antibodies, aptamers, and peptides), so as to achieve the goal of target recognition. As a result, they have been widely accepted as potential building blocks to fabricate high-performance optical and electrochemical biosensors for sensitive detection of proteins for clinical applications (Wang et al., 2006a).

4.2.1 Luminescent Silicon Nanoparticles as Signaling Probes for Protein Assays

Owing to indirect band gap, the bulk silicon widely favored in the microelectronics industry exhibits poor optical properties. Interestingly, great changes take place when the size is reduced to the nanoscale, strong and various visible luminescence could be emitted by silicon nanoparticles. It has been confirmed that these size- and surface-dependent optical properties are produced and controlled by quantum confinement effects (Cheng et al., 2014). Luminescent silicon nanoparticles (SiNPs), which are zero-dimensional silicon-based nanostructures, possess unique luminescence features (e.g., fluorescence or electrochemiluminescence) for high quantum efficiency, strong luminescence, size-dependent, low toxicity, and ultrahigh photostability. Recently, with the great development in the rational design of luminescent SiNPs, the observed luminescence, controllable features, and surface group modification facilitate the construction of a variety of functionalized biosensors for monitoring protein levels with high sensitivity (Gonzalez and Veinot, 2016).

Typically, fluorescent SiNPs are representative luminescent SiNPs, which include two types of nanoparticles, such as silicon Quantum dots (SiQDs) and dye-doped fluorescent SiNPs. SiQDs having ultrasmall size (<10 nm) but strong fluorescence are mainly used for bioimaging (McVey and Tilley, 2014). On the other hand, a large amount of dye molecules could be encapsulated into silicon nanoparticles to form the dye-doped fluorescent SiNPs with improved strong fluorescence (Yan et al., 2007). In most cases (as shown in Fig. 4.1), SiNPs are modified with recognition molecules and directly used as detectable signaling markers to form a classic sandwich-type hybridization for protein detection (Wu et al., 2008, 2009). In the early

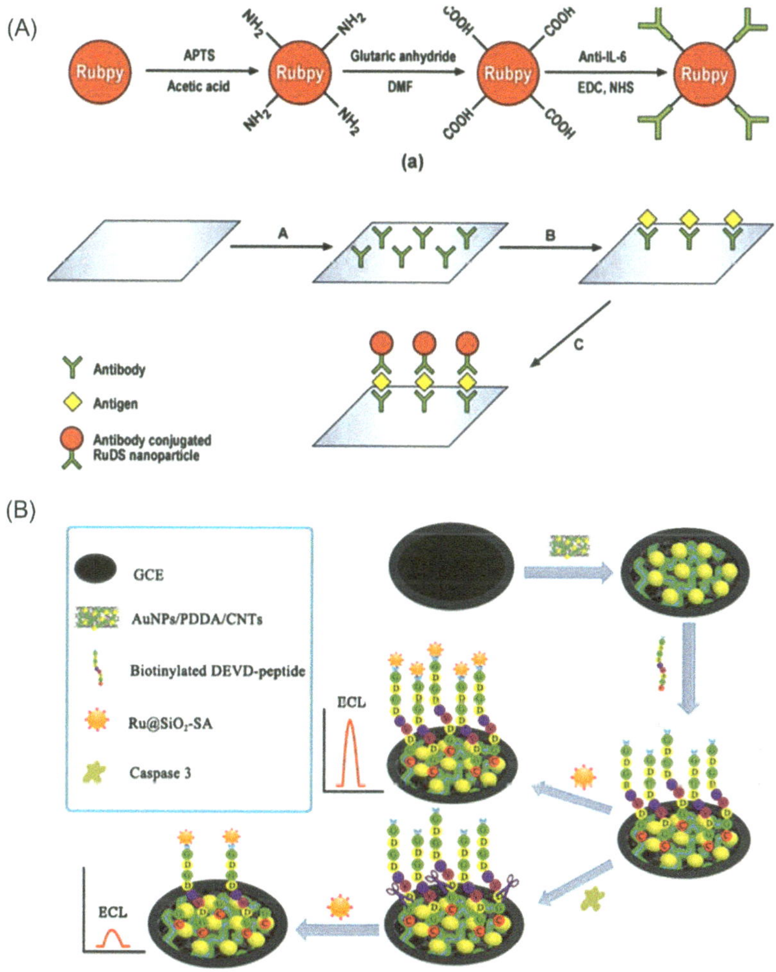

FIGURE 4.1 (A) Schematic representation of the procedure for preparing antibody-conjugated RuDS nanoparticles and the fluorescent immunoassay of IL-6 based on RuDS-labeled protein microarray format; (B) Schematic representation of the detection of caspase-3 Activity. *Reprinted from Wu, H., Huo, Q.S., Varnum, S., Wang, J., Liu, G.G., Nie, Z.M., et al., 2008. Dye-doped silica nanoparticle labels/protein microarray for detection of protein biomarkers. Analyst 133, 1550–1555; Dong, Y.-P., Chen, G., Zhou, Y., Zhu, J.-J., 2016. Electrochemiluminescent sensing for caspase-3 activity based on Ru(bpy)32 + -doped silica nanoprobe. Anal. Chem. 88, 1922–1929.*

stages, Tan's group made great contributions to the development of bioconjugated SiNPs to recognize proteins (Santra et al., 2001; Wang et al., 2005). Specifically, they prepared monodisperse luminophore-doped SiNPs by water-in-oil microemulsion approach. Mouse antihuman CD10 antibody was immobilized onto the silanized surface, thus antibody-conjugated nanoparticles showed excellent performance, and have been applied effectively for target detection. Later, in order to improve the diagnostic accuracy, they also developed dual-luminophore-doped SiNPs for parallel and high-throughput signaling. By integrating a different ratio of the two dyes in a single NP, a series of fluorescent probes were successfully created for multiple proteins detection. Moreover, Yu et al. reported a kind of novel core/shell near-IR fluorescent nanoparticle for alpha fetoprotein (AFP) detection in whole-blood samples, so as to expand the utility of this probe. The detection range was found to be 1.9–51.9 ng mL^{-1} (Deng et al., 2006).

In addition, SiNPs-based electrochemiluminescence (ECL) probes also become an ideal signaling transmitter in protein detection. Zhu's group (Dong et al., 2016) combined novel streptavidin-modified $Ru(bpy)_3^{2+}$-doped silica nanoparticles with DEVD-peptide sequence immobilized on the interface. The system achieved the measurement of caspase-3 activity during cell apoptosis. Using similar principles and methods, various luminophore and specific functional molecules could be incorporated into the inner space or attach on the surface of SiNPs. The resulting SiNPs have been used to recognize disease markers, such as thrombin (Wang et al., 2005), prostate specific antigen (PSA) (Xu et al., 2017), and tumor necrosis factor-alpha (TNF-α) (Wang et al., 2006b), etc.

Intriguingly, Yang and coworkers (Sun et al., 2016) recently made an attempt to improve the conventional enzyme-linked immunosorbent assay (ELISA) system using maturely available alkaline phosphatase (ALP) labeled antibody. As shown in Fig. 4.2, in the presence of antigen, the substrate (ascorbic acid 2-phosphate) could be transformed into strong reducing agent (ascorbic acid). Then it promoted the transformation of amine-containing silane molecules to fluorescent silicon nanoparticles, which could be

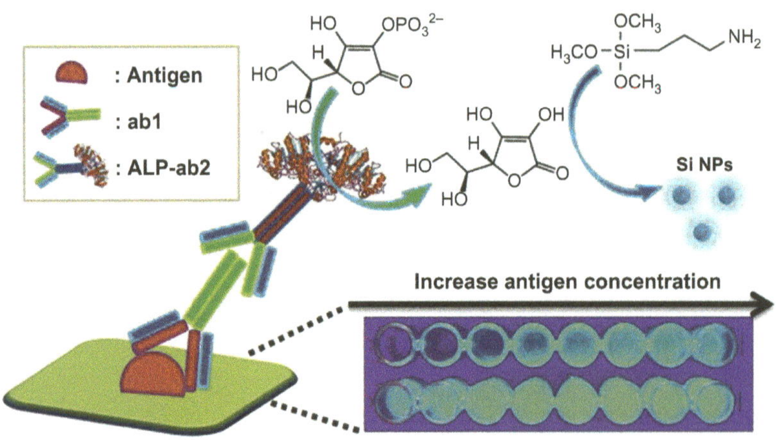

FIGURE 4.2 Schematic representation of the fluorescence immunoassay platform via enzyme-enabled in situ synthesis of fluorescent silicon nanoparticles. *Reprinted from Sun, J., Hu, T., Chen, C., Zhao, D., Yang, F., Yang, X., 2016. Fluorescence immunoassay system via enzyme-enabled in situ synthesis of fluorescent silicon nanoparticles. Anal. Chem. 88, 9789–9795.*

observed by the naked eye under UV lamp. AFP, as the model protein target, could be detected effectively as low as 1 ng mL^{-1}. This signal generation mechanism, which made the resultant fluorescent SiNPs, has become a powerful biosensor for disease diagnosis.

4.2.2 Silicon Nanowires-Based Field-Effect Transistor for Protein Assays

Semiconducting silicon nanowires (SiNWs), which are representative one-dimensional silicon nanomaterials, have shown their strong advantages for designing field-effect transistor (FET)-based electrical devices, which enable the great selectivity and sensitivity for rapid label-free detection of various proteins with high signal-to-noise ratio. As early as 2001, Lieber's group (Cui et al., 2001) was the first to assemble a highly sensitive FET detector with antigen-modified silicon nanowires, so as to detect protein molecules using biotin–streptavidin as model system, in which the detection limit was found to be as low as 10 pM. This successful example laid the foundation for the development of FET-based silicon nanowires for the early determination of cancer over the past two decades. Subsequently, they employed this well-established sensing platform to build arrays of silicon-nanowire field-effect devices for multiplexed electrical detection of cancer markers, including prostate specific antigen (PSA), carcinoembryonic antigen (CEA), and mucin-1 (as shown in Fig. 4.3) (Zheng et al., 2005). The limit of detection

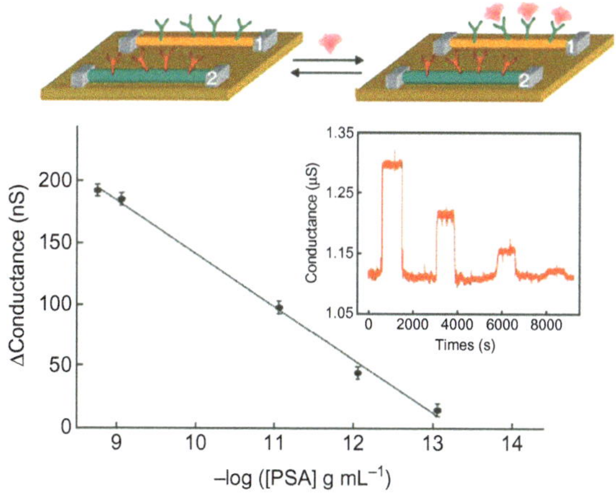

FIGURE 4.3 Schematic representation of the two nanowire devices and the change in conductance versus PSA concentration of a p-type silicon nanowire modified with PSA-Ab1 receptor. Inset: Conductance-versus-time information recorded after alternate transfer of PSA and pure buffer solutions. *Reprinted from Zheng, G., Patolsky, F., Cui, Y., Wang, W.U., Lieber, C.M., 2005. Multiplexed electrical detection of cancer markers with nanowire sensor arrays. Nature Biotechnol. 23, 1294–1301.*

could attain a level of 0.9 ng mL^{-1} in undiluted serum samples, which demonstrates the potentiality of SiNWs-based FET for practical diagnosis with clinical significance.

In order to overcome the problem that material and device would degrade during the process of traditional "top-down" fabrication methods, the group of Mark A. Reed (Stern et al., 2007) presented a progress that integrated complementary metal oxide semiconductor (CMOS) technology with SiNWs FET sensors for label-free immunodetection with the limit below 100 fM, and the stability and practical value of the sensor had been greatly improved for further molecular-level diagnosis. Following the trend, similar researches were reported where different targets such as human cardiac troponin T or I, protein markers of acute myocardial infarction, could be respectively or simultaneously detected. In addition, different recognition molecules like nucleic acid aptamers could be used as functionalization elements on SiNWs. For instance, Jo and coworkers (Lee et al., 2009) modified the silicon nanowires with nucleic acid aptamers that specifically recognized vascular epidermal growth factor (VEGF, a marker supposed to control neovascularization promoting the tumor growth) with high sensitivity.

However, the biggest obstacle for clinical cancer diagnosis is achieving the purpose of point-of-care diagnostics while the tumor biomarkers are in complex samples such as whole blood samples, undiluted urine samples, etc. To solve this problem, researchers, such as Fahmy's group (Stern et al., 2010), used primary antibody to pretreat microfluidic chips based on silicon nanoribbons (as shown in Fig. 4.4). The treated samples were injected into the detector to perform preseparation of the target molecules and their FET signal was detected. Using this two-step system, two different cancer antigens (PSA and carbohydrate antigen 15.3) could be specifically detected in a 10 mL whole-blood sample, and the detection limits were 2.5 ng mL^{-1} and 30 U mL^{-1}, respectively. This work demonstrated the great portability and versatility of FET nanosensor to test the levels of protein in physiological solution, which may accelerate the speed of clinical diagnosis in the future.

4.2.3 Mesoporous Silica Nanoparticles-Based Biosensors

Mesoporous silica nanoparticles are well known as an important class of porous materials with huge surface-to-volume ratios and highly ordered pore size distribution, which provide an attractive platform for encapsulating all kinds of molecules, such as fluorescent dyes or other signal molecules that can be measured by the reading device (Lee et al., 2011). All of these characteristics plus the inherent properties and advantages of silicon materials assure MSNs with great prospects for developing biosensors inspired by biochannels and biogates for protein detection. In this field, MSNs are usually designed to be smart nanodevices containing switchable "molecular gatekeepers" for advanced controlled-release or delivery of drugs. The gated materials are

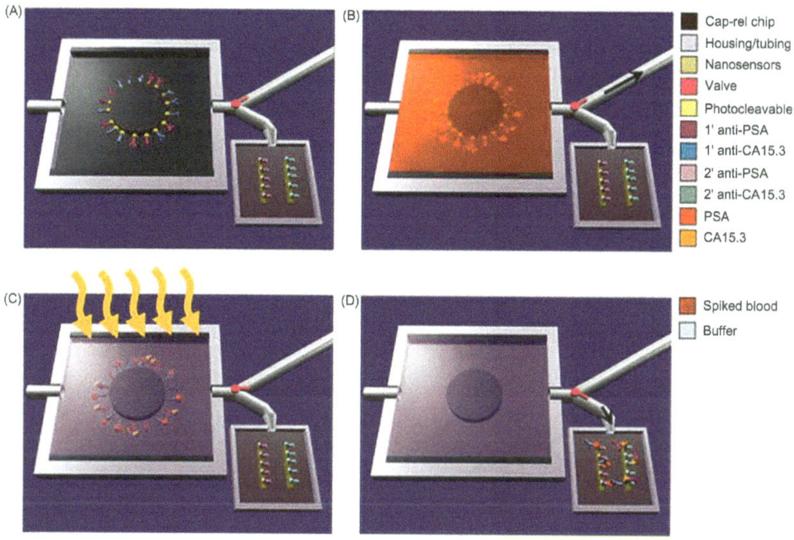

FIGURE 4.4 Schematic of MPC (microfluidic purification chip) operation. (A) Primary antibodies were bound to the MPC with a photocleavable crosslinker for the recognition of PSA and carbonhydrate antigen 15.3 (CA15.3). (B) Whole blood was injected onto the chip. The biomarkers in the sample bound with their antibodies. (C) Wash the chip volume (5 μl) with sensing buffer before UV irradiation (orange arrows). During UV exposure, the photolabile crosslinker were cleaved to release the antibody-antigen complexes into solution. (D) Switch the valve on the nanosensor reservoir to transfer the 5 μl volume, thus realize the label-free sensing of specific biomarkers. *Reprinted from Stern, E., Vacic, A., Rajan, N.K., Criscione, J.M., Park, J., Ilic, B.R., et al., 2010. Label-free biomarker detection from whole blood. Nat. Nanotechnol. 5, 138–142.*

mainly composed of diversiform stimuli-responsive groups like nucleic acid or peptide that can work as caps for protein events. The changes of enzyme activity often reflect the occurrence of many diseases and are often used as effective biomarkers for the diagnosis of certain types of cancer. Thus, by designing the enzyme substrate as a gated switch, great numbers of on-demand biosensors were developed. For instance, Perez' group (Bernardos et al., 2009) showed evidence that a lactose derivative could be covalently anchored onto inorganic MSNs loaded with $[Ru(bipy)_3]^-$, the activity of β-D-galactosidase could be evaluated by monitoring release process of the entrapped guests, which were triggered by enzyme controlled uncapping. In the next work (Coll et al., 2011), enzyme-mediated controlled release systems were achieved by immobilizing long modular protease-sensitive peptide sequences on mesoporous silica scaffoldings with zero release. Proteases are considered to be extremely important signaling molecules for diseases, including cardiovascular diseases, cancers, etc. Thus, it is particularly important to detect various protease activities in vitro and in vivo.

Recently, Chen and coworkers (Huang et al., 2012) put forward a new idea of embedding a broad-spectrum nanoquencher within MSNs instead of signal guests reversely, which allowed a series of specific substrate peptides labeled with different dyes to conjugate onto the surface of MSN for multiplexed imaging of various caspases activities in living cells (as shown in Fig. 4.5). The presence of related proteases would induce the cleavage of peptides, then the fluorescence of report probe section would be recovered. This will not only further broaden the application of MSNs for intracellular detection, but also help to promote the research on complex protease signaling mechanisms. Certainly, DNA could also act as a stimuli response element to design telomerase sensors (Qian et al., 2013).

Similarly, as shown in Fig. 4.6, Ren and coworkers (Ren et al., 2014) proposed a homogeneous electrochemical immunoassay integrated with

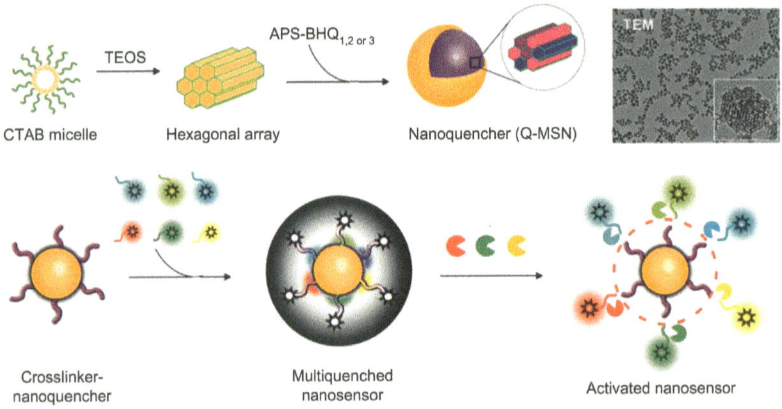

FIGURE 4.5 Schematic representation of the design of nonfluorescent and broad-spectrum nanoquencher and nanosensor. A graph showing Q-MSN synthesis and the TEM image of Q-MSN. *Reprinted from Huang, X., Swierczewska, M., Choi, K.Y., Zhu, L., Bhirde, A., Park, J., et al., 2012. Multiplex imaging of an intracellular proteolytic cascade by using a broad-spectrum nanoquencher. Angew. Chem., Int. Ed. 51, 1625–1630.*

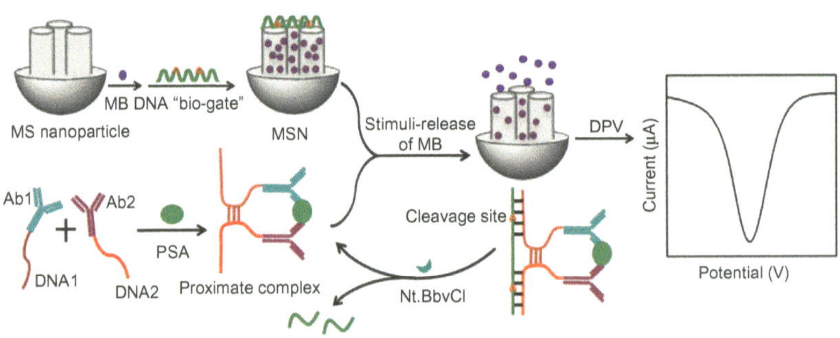

FIGURE 4.6 Schematic representation of the homogeneous electrochemical immunoassay using proximity hybridization-responsive mesoporous silica nanoprobe. *Reprinted from Ren, K., Wu, J., Zhang, Y., Yan, F., Ju, H., 2014. Proximity hybridization regulated DNA biogate for sensitive electrochemical immunoassay. Anal. Chem. 86, 7494–7499.*

DNA-gated MSN probe. The presence of target resulted in the form of a stable double-stranded structure and thus the biogate was opened. Therefore, prostate-specific antigen could be detected with a limit of 1.3 pg mL^{-1}.

4.3 UPCONVERSION NANOMATERIALS

Photon upconversion (UC) is an anti-Stokes process, in which the sequential absorption of multiple low-energy photons leads to high-energy emission. In fact, this phenomenon has been known for over 50 years (Auzel, 2004). Early studies focused on how to improve the synthesis method and made great breakthroughs to increase the quantum yield. Upconversion nanomaterials, also named luminescent upconverting nanoparticles (UCNPs) (Haase and Schafer, 2011), are particles approximately of 100 nm in diameter possessing the unique photophysical characteristic that converts the long-wavelength radiation, usually near-infrared radiation (NIR), to short-wavelength radiation, such as visible light. Currently, UCNPs mainly constitute crystalline host materials doped with various rare ions, especially lanthanide elements. Doped rare ions are divided into activator and sensitizer according to the function. Erbium (Er^{3+}) or thulium (Tm^{3+}), having multiple energy levels ions and longer lifetime of metastable energy levels, are widely used as luminescence centers of UCNPs. This ensures the preconditions of efficient upconversion. Sensitizers like ytterbium (Yb^{3+}) should exhibit strong absorption of excitation light (e.g., 980 nm) and deliver the energy to activator ions. The host material is a support that holds doped rare ions with good light transmission and lower phonon energy such as halogenide and fluoride, etc. Compared to organic fluorophores, UCNPs show exceptional properties, including large anti-Stokes shifts, sharp emission bandwidths and high photochemical stability. In addition, since the peak emission wavelength of the UCNPs is not related to the size, multicolor emission can be easily achieved by controlling the ratio between the two components. These nanomaterials avoid the problem of signal interference, like autofluorescence arising from complex biological system. Moreover, excellent surface chemistry and low toxicity make them suitable for the field of bioapplications (Duan et al., 2018). It is noteworthy that UCNPs have been used in the design of heterogeneous and homogeneous immunoassays for the detection of various proteins since the first report about the potential of UCNPs in diagnostics in 1999 (Zijlmans et al., 1999).

4.3.1 Upconverting Nanoparticles as Signaling Probes for Protein Assays

As one class of luminescent material, these nanoparticles can be directly used as reporter probes for the detection of proteins after the modification by biomolecules. Chen and coworkers (Huang et al., 2014) have reported that the fluoride $LiLuF_4$ as the host material and doped lanthanide ions Ln^{3+} could be assembled by layer-by-layer (LBL) method to form novel core/shell UCNPs with high quantum yields, which could convert 980 nm NIR laser irradiation to the intense light of green (Yb/Er) or blue (Yb/Tm). As shown in Fig. 4.7, by employing their superior upconversion luminescence (UCL), they designed a sensitive bioprobe for the for the detection of the β-subunit of human chorionic gonadotropin (β-hCG) by modifying avidin onto the surface of nanoparticles with a detection limit of 3.8 ng mL^{-1}. Using the similar UCL-based immunoassay, the biomarker of myocardial infarction, cardiac troponin I, could also be sensitively detected with the limit of detection of 3.14 ng L^{-1} (Kale et al., 2016). Specific doped lanthanide-ions enable the UCNPs multicolor emission under single-wavelength excitation, which meets the requirements of multiple testing in diagnosis. The group of Soukka carried out a series of works (Pakkila et al., 2012). They established a platform within the antibody array targeting prostate specific antigen, thyroid stimulating hormone, and luteinizing hormone. This system eliminated interference from autofluorescence, and quantitative multianalysis immunoassays could be achieved with good performance by utilizing luminescence upconversion of single-crystal nanoparticles as signal markers.

In another interesting study, Soukka's group developed (Kale et al., 2016) a dual-mode multiplexed assay for the detection of proteins in real samples by coupling two types of UCNPs as reporters. The targets could be simultaneous differentiated according to the color and the position of nanoparticles, and the obtained system is readily suitable for further wide application in diagnosis. Detecting proteins at single-molecule levels plays a great role to capture and understand heterogeneity in real samples for clinical use. As shown in Fig. 4.8, Gorris and coworkers (Farka et al., 2017b) developed an upconversion-linked immunosorbent assay using the classical sandwich model for counting diagnostic biomarkers at single-molecule level under a wide-field epifluorescence microscope, 980 nm as the excitation wavelength.

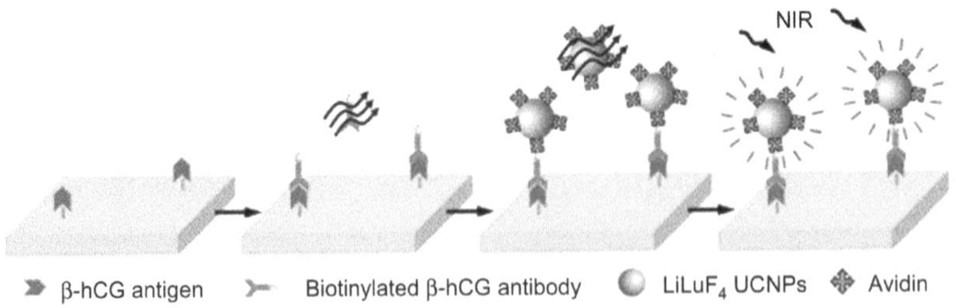

FIGURE 4.7 Schematic representation of the detecting system for the measurement of β−hCG level based on avidin-modified luminescent upconversion nanoparticles. *Reprinted from Huang, P., Zheng, W., Zhou, S., Tu, D., Chen, Z., Zhu, H., et al., 2014. Lanthanide-doped LiLuF4 upconversion nanoprobes for the detection of disease biomarkers. Angew. Chem. Int. Ed. 53, 1252−1257.*

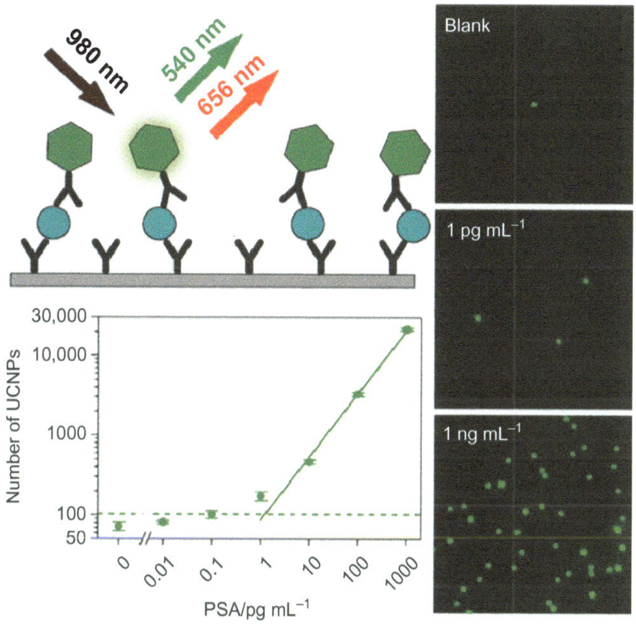

FIGURE 4.8 Schematic representation of the scheme of single molecule upconversion-linked immunosorbent assay with extended dynamic range for the sensitive detection of diagnostic biomarkers. *Reprinted from Farka, Z., Juřík, T., Kovář, D., Trnková, L., Skládal, P., 2017a. Nanoparticle-based immunochemical biosensors and assays: recent advances and challenges. Chem. Rev. 117, 9973–10042.*

The limit of detection of PSA achieved a very low concentration of 1.2 pg mL^{-1} in 25% blood serum, which shows much improved sensitivity compared with conventional ELISA assays. What is more, it is urgent to develop the sensors for in situ analysis of multiple biomarkers to provide more accurate information for cancer diagnosis.

As shown in Fig. 4.9, Zhang and coworkers (Zhou et al., 2015) realized that single-band upconversion nanoparticles with different colors could be synthesized by coating programmable dye-doped SiNPs on the surface to assimilate the unwanted emission bands. Thus, three types of antibody, which target significant breast cancer biomarkers, estrogen receptors (ERs), progesterone receptors (PRs), and human epithelial growth factor receptor-2 (HER2), were conjugated on single-band UCNPs to form nanoprobes that allowed quantitative and simultaneous bioanalysis of multiple biomarkers at low levels in breast cancer cells and tissues. Compared to traditional immunohistochemical technology, it showed more advantages and capacity for further applications in the life sciences.

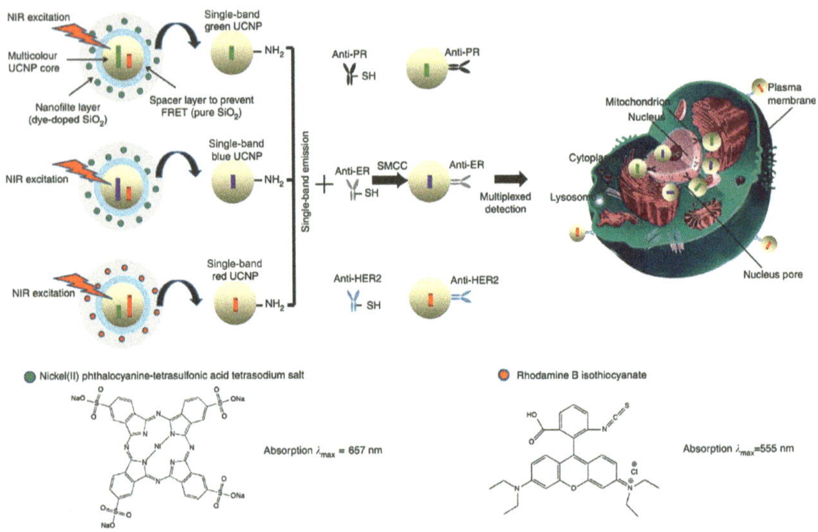

FIGURE 4.9 Schematic representation of the synthesis principle and process of single-band upconversion nanoparticles for multiple detection of cancer biomarker. *Reprinted from Zhou, L., Wang, R., Yao, C., Li, X., Wang, C., Zhang, X., et al., 2015. Single-band upconversion nanoprobes for multiplexed simultaneous in situ molecular mapping of cancer biomarkers. Nat. Commun. 6, 6938.*

4.3.2 Energy Resonance Transfer-Based Upconversion Nanoplatforms for Protein Assays

Energy resonance transfer (ERT) refers to a phenomenon of energy transfer between two closely (usually <10 nm) luminescent molecules, which are an excited donor molecule and an acceptor molecule. It is deemed to a be reliable and powerful analytical technique and has been widely used under physiological conditions. Apparently, all aforementioned merits of lanthanide-doped UCNPs are uniquely positioned to mediate energy transfer to an acceptor by absorbing low-energy light (Su et al., 2017). Currently, investigators have designed some UCNPs-based RET nanoforms that show a response to specific changes in the protein biosensors.

By using different acceptors of nanomaterials, such as gold nanoparticles (Lin et al., 2013) and carbon nanomaterials (Zhang et al., 2011), or organic dyes like rhodamine B (Liu et al., 2018), or various recognition elements such as antibody (Kuningas et al., 2007), aptamers (Wang et al., 2011), and peptides (Wang et al., 2012), plenty of significant proteins such as carcinoembryonic antigen (CEA)(Wu et al., 2015), platelet-derived growth factor-BB (PDGF-BB) (Lin et al., 2013) and proteases (Rantanen et al., 2008) have been detected by using FRET in upconversion nanoplatforms. As shown in Fig. 4.10, aptamer assembled on the surface of UCNPs binds to the carbon nanoparticles via $\pi-\pi$ stacking interaction, FRET occurs, and the luminesce

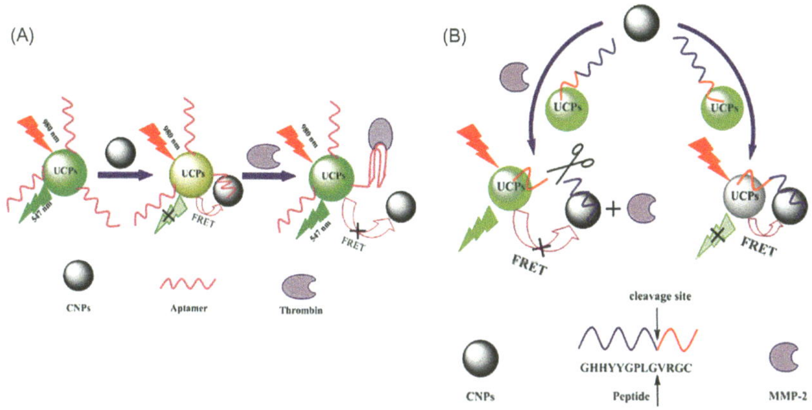

FIGURE 4.10 Schematic representation of the FRET biosensor based on aptamer (A)/peptide (B)-linked upconversion nanoparticles for biomolecular detection. *Reprinted from Wang, Y., Bao, L., Liu, Z., Pang, D.W., 2011. Aptamer biosensor based on fluorescence resonance energy transfer from upconverting phosphors to carbon nanoparticles for thrombin detection in human plasma. Anal. Chem. 83, 8130–8137; Wang, Y., Shen, P., Li, C., Wang, Y., Liu, Z., 2012. Upconversion fluorescence resonance energy transfer based biosensor for ultrasensitive detection of matrix metalloproteinase-2 in blood. Anal. Chem. 84, 1466–1473.*

of UCNPs is quenched (Wang et al., 2011). The strong interaction between aptamer and thrombin would block the process of FRET, resulting in a detectable signal. Analogously, the same group employed a polypeptide chain constituted of the specific substrate reactive site sequence and a π-rich motif bound to carbon nanoparticles for a homogenous detection of matrix metalloproteinase-2 (MMP-2, an extremely crucial biomarker in blood) (Wang et al., 2012). When the proteinase reacted at the cleavage site of the peptide, the donor–receptor pairs separated and the concentration of MMP-2 was measured within the range from 10–500 pg mL^{-1}. The group also combined the upconversion nanosystem with the microfluidic paper-based analytical device for a point-of-care test, advancing the development of UCNPs for portable and rapid clinical assays (He and Liu, 2013).

4.4 LIPOSOME

Liposomes have been widely used as nanodevices, cell membrane mimetics for molecular carriers, and cell membrane mimetics for drug delivery and gene delivery. In the manufacture of biosensors, liposomes have particular advantages in signal amplification because of their excellent carrier properties, enabling the encapsulation of signal-labeling compounds in a variety of sensing modes (Jesorka and Orwar, 2008). In addition, the surface of liposomes could be modified to realize the recognition of various types of analytes, which is an important application in biological analysis.

4.4.1 Liposome-Based Colorimetric Assays

Colorimetry is a simple method that has been widely used for biochemical analysis and the construction of biosensors. Compared with other methods, the colorimetric method has some obvious advantages, such as low cost, simple instruments (or, in the case of naked eye detection, no instruments), and can be qualitatively or semiqualitatively identified by the naked eye. However, colorimetry is generally less sensitive. To solve this problem, dye-encapsulated liposomes have been used as signal amplifiers to increase the sensitivity of colorimetric biosensors.

Strip-based liposome sandwich immunoassays have been developed for the detection of antigens, antibodies, proteins, and large biological macromolecules. In a typical liposome-strip sandwich immunoassay device, the antianalyte antibody is immobilized on the AC region of the band, and the capillary action causes the analyte and the antibody-labeled dye-encapsulated liposome to sequentially or together pass through the AC area. When a sample is added, the analyte forms an immune complex with the immobilized antibody and liposome. The optical density in the AC region is proportional to the concentration of the analyte in the sample (Liu and Boyd, 2013). For example, Sanku Mallik and coworkers (Banerjee et al., 2010) have developed a new liposome-based method using an ELISA-like signal amplification strategy for MMP-9 detection, which did not require antibodies. The system could detect recombinant MMP-9 of 10 nM and higher concentrations in complex protein mixtures. Furthermore, Steven's group (Aili et al., 2011) reported a liposome-based strategy for the detection of phospholipase (PL). Liposome-loaded polypeptides could be hetero-associated with a second polypeptide immobilized on gold nanoparticles by rational design, in which the enzymatic substrate was separated from the surface of the nanoparticles. When PL was present, it could trigger release of the polypeptide from the liposome, inducing folding-dependent nanoparticle bridging aggregation. The colorimetric response caused by aggregation allowed direct and continuous detection of PL in the picomolar range. The advantages of this strategy, including its high specificity, rapidity, and flexibility, make it suitable for a series of applications from point-of-care diagnostics to high-throughput drug screening.

4.4.2 Liposome-Based Electrochemical Assays

As a sensitive, fast and convenient analytical technique, electrochemistry has a wide range of applications in the field of biosensors. Electronic signals are stable and sensitive, and can be quickly and easily detected. Besides, the electrochemical device is easily miniaturized to develop a portable sensor without requiring a complicated large-scale instrument. Owing to these advantages, sensors where liposomes were adapted to encapsulated electrochemical labels as signal amplifiers have generated great interest in biochemical analysis.

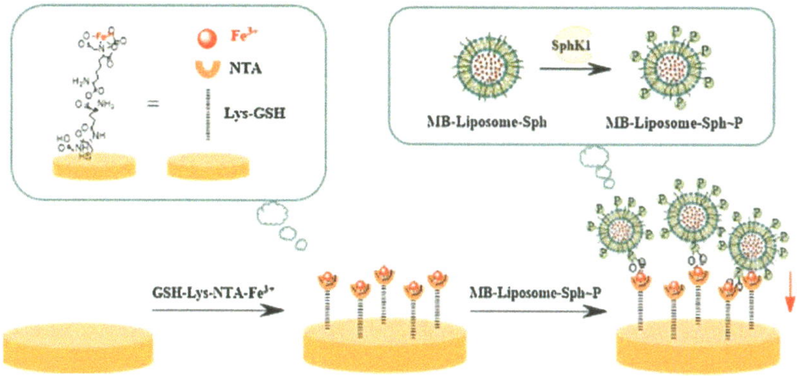

FIGURE 4.11 Schematic representation of the electrochemical assay of lipid kinase activity based on liposomes. *Reprinted from Gao, T., Gu, S., Mu, C., Zhang, M., Yang, J., Liu, P., et al., 2017. Electrochemical assay of lipid kinase activity facilitated by liposomes. Electrochim. Acta 252, 362–367.*

Li's group (Gao et al., 2017) used liposomes to provide a biomimetic membrane environment, creating a simple, sensitive, and efficient electrochemical method for lipid kinase detection (as shown in Fig. 4.11). In this work, liposomes provide a favorable catalytic environment for the lipid kinases. At the same time, liposomes act as a carrier for a large number of signaling molecules to enhance the electrochemical signal. Therefore, the problems involved in the lipid kinase assay can be solved, and the high sensitivity of detection is ensured due to the enrichment of signal molecules (methylene blue, MB). Compared to the methods currently reported, this new method eliminates the complex process of processing lipid substrates/products, and therefore, the assay procedure is simpler. An important lipid kinase sphingosine kinase 1 (SphK1) was selected as the target of the assay. With the promotion of liposome-based electrochemical signal amplification, the relatively broad lower limit of detection (LOD) drops to 2.33 pmol min^{-1} mg^{-1}. In addition, Yang and coworkers (Lin et al., 2016) developed an enzyme-encapsulated liposome-linked immunosorbent assay for sensitive detection of proteins and used a personal glucose meter (PGM) to facilitate portable quantitative readings. The recognition elements were coated onto the surface of liposomes encapsulating a large amount of amyloglucosidase or invertase, such as aptamers or antibodies, for target recognition. By converting the target recognition signal into the amount of glucose with the encapsulated enzyme, disease biomarkers such as thrombin or creactive protein (CRP) can be quantified by PGM with a detection limit of 1.8 or 0.30 nM, respectively.

4.4.3 Liposome-Based Fluorescent Assays

In the construction of liposome-based biosensors, fluorescent dyes are widely used as encapsulants for signal amplification. Due to the convenience (no separation or wash steps are required) and potential commercial value, it is desirable to construct a homogeneous liposomal immunoassay system. The homogeneous liposome immunoassay can be achieved by fluorescence enhancement of fluorescence resonance energy transfer (FRET) and self-quenching dye. Liposomes were typically destabilized and lysed using cell lysing reagents to release self-quenching dye encapsulated in liposomes, thus enhancing fluorescence intensity for homogeneous liposome immunoassay.

Based on DNA-encapsulated liposomes, (Wen et al., 2005) proposed a novel liposome-rolling circle amplification (RCA) immunoassay for ultrasensitive detection of proteins. This technique utilized antibodies to modify liposomes encapsulated with DNA primer probes as detection reagents. In the presence of target, the DNA primer probe was released from the liposome and the linear RCA reaction was then initiated, resulting in long tandem repeat sequences that can be sensitively detected using bead-based fluorometric assays. A highly selective biomarker associated with prostate cancer, prostate specific antigen (PSA), demonstrated this technology. The results showed that the technology had a dynamic response to PSA over the concentration range of 6-decades from 0.1 fg mL^{-1} to 0.1 ng mL^{-1}, with a detection limit as low as 0.08 fg mL^{-1} and a high dose-response sensitivity. Since primary amplification of large amounts of DNA primers in liposomes further induced secondary amplification of RCA, this technique exhibits very high sensitivity. In addition, (Chen et al., 2013) designed a novel, simple, and convenient platform to prepare 11-MUA − Au ND/Lip hybrids that can be used to detect PLC. PLC selectively hydrolyzed its specific phosphatidylcholine from the liposome to produce DAG, which further minimized the fluorescence quenching of the 11-MUA − Au ND/Lip hybrids under oxygen. Under the optimal conditions, 11-MUA − Au ND/Lip hybrids showed high sensitivity (LOD: 0.21 nM) and high selectivity for the detection of PLC.

4.5 VIRUS NANOPARTICLES

Recently, viruses have been shown to be highly selective and sensitive in detecting target analytes, such as explosives, proteins, bacteria, viruses, spores, and toxins (Mao et al., 2009). Bacteriophages (often shortened to phages) specifically infect bacteria and are currently the most studied viruses. Target-specific nonbacteriophages (and the peptides and proteins that they carry) can be identified by well-established phage display technology. By lysing the phage, the bacteria can be specifically destroyed to release cell-specific marker molecules, such as enzymes that can be assayed. In addition, since phages have good chemical and thermal stability, they can be conjugated to nanomaterials and immobilized on the surface of biosensor devices. Therefore, as a novel nanomaterial, phages have great potential in construction of biosensing platforms.

To improve the diagnostic accuracy of cancer patients infected with *Candida albicans*. As shown in Fig. 4.12, Mao and coworkers

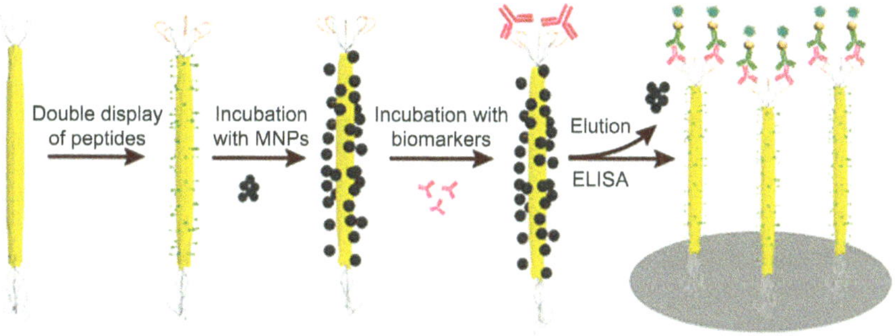

FIGURE 4.12 Schematic representation of the detection of anti-Sap2-IgG from human serum by using ASIT-MNP-phage. *Reprinted from Wang, Y., Ju, Z., Cao, B., Gao, X., Zhu, Y., Qiu, P., et al., 2015. Ultrasensitive rapid detection of human serum antibody biomarkers by biomarker-capturing viral nanofibers. ACS Nano, 9, 4475–4483.*

(Wang et al., 2015) used monofilament fd bacteriophage, a human-safe bacterial-specific virus, to genetically display two functional peptides, one of which recognized markers of infection (antiaspartyl protease 2 IgG antibody) in the serum of cancer patients and another was responsible for binding magnetic nanoparticles (MNPs). The resulting phage was first modified with MNPs and the biomarkers in the serum were captured with the modified phage. Then, the phage-bound biomarkers were separated by magnetic enrichment and biochemically detected. This method greatly enhanced the sensitivity and specificity for detecting biomarkers. The average time to detect each serum sample was only about 6 hours, which was significantly shorter than the clinically method that consumes 1 week. The detection limit of this nanobiotechnology method was about 1.1 pg mL^{-1}, which was about two orders of magnitude lower than that of the traditional antigen method, creating a new approach for virus-based disease diagnosis. The same group further confirmed that phage nanofibers which recognized biomarkers were also effective probes in early cancer diagnosis that could detect biomarkers (e.g., p53 antibody) of cancer patients (Pan et al., 2015).

Biosensors based on viruses other than phages for protein have not been extensively studied, but can be imagined if the viruses are chemically conjugated with functional molecules or inorganic nanomaterials that provide unique physical properties (optical, electronic, and magnetic). Besides, such nanocomposites which offer target specificity by displaying a target-specific element on the capsid can be used for building the novel sensing platform.

4.6 CONCLUSIONS

Thanks to the rapid development of nanomaterial science, new nanomaterials including silicon nanomaterials, upconverting nanomaterials, liposomes, and virus nanoparticles have been explored to build ultrasensitive biosensing platforms for protein detection. Additionally, nanocomposite systems have been widely used in the development of biosensors. However, there are still many problems to be solved in the realization of the standard synthesis, long-lasting stability, easy availability, and economic efficiency of nanomaterials. Besides, in the sensing systems using nanomaterials for protein detection, the interaction mechanisms between nanomaterials and proteins have not been fully clarified. Therefore, reducing the nonspecific adsorption in the detection process and improving the sensitivity are still worth investigating. Furthermore, the miniaturization of biosensors for point-of-care testing and the construction of sensing systems for high-throughput detection of proteins still need to be explored.

REFERENCES

Aili, D., Mager, M., Roche, D., Stevens, M.M., 2011. Hybrid nanoparticle-liposome detection of phospholipase activity. Nano Lett. 11, 1401–1405.

Auzel, F., 2004. Upconversion and Anti-Stokes Processes with f and d Ions in Solids. Chem. Rev. 104, 139–174.

Banerjee, J., Hanson, A.J., Nyren-Erickson, E.K., Ganguli, B., Wagh, A., Muhonen, W.W., et al., 2010. Liposome-mediated amplified detection of cell-secreted matrix metalloproteinase-9. Chem. Commun. 46, 3209–3211.

Bernardos, A., Aznar, E., Marcos, M.D., Martinez-Manez, R., Sancenon, F., Soto, J., et al., 2009. Enzyme-responsive controlled release using mesoporous silica supports capped with lactose. Angew. Chem., Int. Ed. 48, 5884–5887.

Cao, B., Yang, M., Mao, C., 2016. Phage as a genetically modifiable supramacromolecule in chemistry, materials and medicine. Acc. Chem. Res. 49, 1111–1120.

Chen, G., Qiu, H., Prasad, P.N., Chen, X., 2014. Upconversion nanoparticles: design, nanochemistry, and applications in theranostics. Chem. Rev. 114, 5161–5214.

Chen, W.Y., Chen, L.Y., Ou, C.M., Huang, C.C., Wei, S.C., Chang, H.T., 2013. Synthesis of fluorescent gold nanodot-liposome hybrids for detection of phospholipase C and its inhibitor. Anal. Chem. 85, 8834–8840.

Cheng, X., Lowe, S.B., Reece, P.J., Gooding, J.J., 2014. Colloidal silicon quantum dots: from preparation to the modification of self-assembled monolayers (SAMs) for bio-applications. Chem. Soc. Rev. 43, 2680–2700.

Coll, C., Mondragon, L., Martinez-Manez, R., Sancenon, F., Marcos, M.D., Soto, J., et al., 2011. Enzyme-mediated controlled release systems by anchoring peptide sequences on mesoporous silica supports. Angew. Chem., Int. Ed. 50, 2138–2140.

Cui, Y., Wei, Q., Park, H., Lieber, C.M., 2001. Nanowire nanosensors for highly sensitive and selective detection of biological and chemical species. Science 293, 1289–1292.

Deng, T., Li, J.S., Jiang, J.H., Shen, G.L., Yu, R.Q., 2006. Preparation of near-IR fluorescent nanoparticles for fluorescence-anisotropy-based immunoagglutination assay in whole blood. Adv. Funct. Mater. 16, 2147–2155.

Dong, Y.-P., Chen, G., Zhou, Y., Zhu, J.-J., 2016. Electrochemiluminescent sensing for caspase-3 activity based on Ru(bpy)32 + -doped silica nanoprobe. Anal. Chem. 88, 1922–1929.

Duan, C., Liang, L., Li, L., Zhang, R., Xu, Z.P., 2018. Recent progress in upconversion luminescence nanomaterials for biomedical applications. J. Mater. Chem. B 6, 192–209.

Farka, Z., Juřík, T., Kovář, D., Trnková, L., Skládal, P., 2017a. Nanoparticle-based immunochemical biosensors and assays: recent advances and challenges. Chem. Rev. 117, 9973–10042.

Farka, Z., Mickert, M.J., Hlavacek, A., Skladal, P., Gorris, H.H., 2017b. Single molecule upconversion-linked immunosorbent assay with extended dynamic range for the sensitive detection of diagnostic biomarkers. Anal. Chem. 89, 11825–11830.

Gao, T., Gu, S., Mu, C., Zhang, M., Yang, J., Liu, P., et al., 2017. Electrochemical assay of lipid kinase activity facilitated by liposomes. Electrochim. Acta 252, 362–367.

Gonzalez, C.M., Veinot, J.G.C., 2016. Silicon nanocrystals for the development of sensing platforms. J. Mater. Chem. C 4, 4836–4846.

Haase, M., Schafer, H., 2011. Upconverting nanoparticles. Angew. Chem., Int. Ed. 50, 5808–5829.

He, M., Liu, Z., 2013. Paper-based microfluidic device with upconversion fluorescence assay. Anal. Chem. 85, 11691–11694.

He, Y., Fan, C., Lee, S.-T., 2010. Silicon nanostructures for bioapplications. Nano Today 5, 282–295.

Huang, P., Zheng, W., Zhou, S., Tu, D., Chen, Z., Zhu, H., et al., 2014. Lanthanide-doped $LiLuF_4$ upconversion nanoprobes for the detection of disease biomarkers. Angew. Chem. Int. Ed. 53, 1252–1257.

Huang, X., Swierczewska, M., Choi, K.Y., Zhu, L., Bhirde, A., Park, J., et al., 2012. Multiplex imaging of an intracellular proteolytic cascade by using a broad-spectrum nanoquencher. Angew. Chem., Int. Ed. 51, 1625–1630.

Jesorka, A., Orwar, O., 2008. Liposomes: technologies and analytical applications. Annu. Rev. Anal. Chem. 1, 801–832.

Kale, V., Pakkila, H., Vainio, J., Ahomaa, A., Sirkka, N., Lyytikainen, A., et al., 2016. Spectrally and spatially multiplexed serological array-in-well assay utilizing two-color upconversion luminescence imaging. Anal. Chem. 88, 4470–4477.

Kuningas, K., Pakkila, H., Ukonaho, T., Rantanen, T., Lovgren, T., Soukka, T., 2007. Upconversion fluorescence enables homogeneous immunoassay in whole blood. Clin. Chem. 53, 145–146.

Lee, H.S., Kim, K.S., Kim, C.J., Hahn, S.K., Jo, M.H., 2009. Electrical detection of VEGFs for cancer diagnoses using anti-vascular endotherial growth factor aptamer-modified Si nanowire FETs. Biosens. Bioelectron. 24, 1801–1805.

Lee, J.E., Lee, N., Kim, T., Kim, J., Hyeon, T., 2011. Multifunctional mesoporous silica nanocomposite nanoparticles for theranostic applications. Acc. Chem. Res. 44, 893–902.

Lin, B., Liu, D., Yan, J., Qiao, Z., Zhong, Y., Yan, J., et al., 2016. Enzyme-encapsulated liposome-linked immunosorbent assay enabling sensitive personal glucose meter readout for portable detection of disease biomarkers. ACS Appl. Mater. Interfaces 8, 6890–6897.

Lin, F.B., Yin, B.D., Li, C.Z., Deng, J.H., Fan, X.Y., Yi, Y.H., et al., 2013. Fluorescence resonance energy transfer aptasensor for platelet-derived growth factor detection based on upconversion nanoparticles in 30% blood serum. Anal. Methods 5, 699–704.

Liu, L., Zhang, H., Song, D., Wang, Z., 2018. An upconversion nanoparticle-based fluorescence resonance energy transfer system for effectively sensing caspase-3 activity. Analyst 143, 761–767.

Liu, Q., Boyd, B.J., 2013. Liposomes in biosensors. Analyst 138, 391–409.

Mao, C., Liu, A., Cao, B., 2009. Virus-based chemical and biological sensing. Angew. Chem., Int. Ed. 48, 6790–6810.

McVey, B.F.P., Tilley, R.D., 2014. Solution synthesis, optical properties, and bioimaging applications of silicon nanocrystals. Acc. Chem. Res. 47, 3045–3051.

Pakkila, H., Yliharsila, M., Lahtinen, S., Hattara, L., Salminen, N., Arppe, R., et al., 2012. Quantitative multianalyte microarray immunoassay utilizing upconverting phosphor technology. Anal. Chem. 84, 8628–8634.

Pan, P., Wang, Y., Zhu, Y., Gao, X., Ju, Z., Qiu, P., et al., 2015. Nontoxic virus nanofibers improve the detection sensitivity for the anti-p53 antibody, a biomarker in cancer patients. Nano Res. 8, 3562–3570.

Peng, F., Su, Y., Zhong, Y., Fan, C., Lee, S.-T., He, Y., 2014. Silicon nanomaterials platform for bioimaging, biosensing, and cancer therapy. Acc. Chem. Res. 47, 612–623.

Qian, R., Ding, L., Ju, H., 2013. Switchable fluorescent imaging of intracellular telomerase activity using telomerase-responsive mesoporous silica nanoparticle. J. Am. Chem. Soc. 135, 13282–13285.

Rantanen, T., Järvenpää, M.-L., Vuojola, J., Kuningas, K., Soukka, T., 2008. Fluorescence-quenching-based enzyme-activity assay by using photon upconversion. Angew. Chem., Int. Ed. 120, 3871–3873.

Ren, K., Wu, J., Zhang, Y., Yan, F., Ju, H., 2014. Proximity hybridization regulated DNA biogate for sensitive electrochemical immunoassay. Anal. Chem. 86, 7494–7499.

Santra, S., Zhang, P., Wang, K., Tapec, R., Tan, W., 2001. Conjugation of biomolecules with luminophore-doped silica nanoparticles for photostable biomarkers. Anal. Chem. 73, 4988–4993.

Stern, E., Klemic, J.F., Routenberg, D.A., Wyrembak, P.N., Turner-Evans, D.B., Hamilton, A. D., et al., 2007. Label-free immunodetection with CMOS-compatible semiconducting nanowires. Nature 445, 519–522.

Stern, E., Vacic, A., Rajan, N.K., Criscione, J.M., Park, J., Ilic, B.R., et al., 2010. Label-free biomarker detection from whole blood. Nat. Nanotechnol. 5, 138–142.

Su, Q., Feng, W., Yang, D., Li, F., 2017. Resonance energy transfer in upconversion nanoplatforms for selective biodetection. Acc. Chem. Res. 50, 32–40.

Sun, J., Hu, T., Chen, C., Zhao, D., Yang, F., Yang, X., 2016. Fluorescence immunoassay system via enzyme-enabled in situ synthesis of fluorescent silicon nanoparticles. Anal. Chem. 88, 9789–9795.

Wang, G.X., Zhang, B.L., Yu, Z.L., Qu, M.Z., 2005. Manganese oxide/MWNTs composite electrodes for supercapacitors. Solid State Ionics 176, 1169–1174.

Wang, J., Liu, G., Engelhard, M.H., Lin, Y., 2006a. Sensitive immunoassay of a biomarker tumor necrosis factor-α based on poly(guanine)-functionalized silica nanoparticle label. Anal. Chem. 78, 6974–6979.

Wang, L., Wang, K., Santra, S., Zhao, X., Hilliard, L.R., Smith, J.E., et al., 2006b. Watching silica nanoparticles glow in the biological world. Anal. Chem. 78, 646–654.

Wang, Y., Bao, L., Liu, Z., Pang, D.W., 2011. Aptamer biosensor based on fluorescence resonance energy transfer from upconverting phosphors to carbon nanoparticles for thrombin detection in human plasma. Anal. Chem. 83, 8130–8137.

Wang, Y., Shen, P., Li, C., Wang, Y., Liu, Z., 2012. Upconversion fluorescence resonance energy transfer based biosensor for ultrasensitive detection of matrix metalloproteinase-2 in blood. Anal. Chem. 84, 1466–1473.

Wang, Y., Ju, Z., Cao, B., Gao, X., Zhu, Y., Qiu, P., et al., 2015. Ultrasensitive rapid detection of human serum antibody biomarkers by biomarker-capturing viral nanofibers. ACS Nano 9, 4475–4483.

Wen, H.W., Borejsza-Wysocki, W., De Cory, T.R., Durst, R.A., 2005. Development of a competitive liposome-based lateral flow assay for the rapid detection of the allergenic peanut protein Ara h1. Anal. Bioanal. Chem. 382, 1217–1226.

Wu, H., Huo, Q.S., Varnum, S., Wang, J., Liu, G.G., Nie, Z.M., et al., 2008. Dye-doped silica nanoparticle labels/protein microarray for detection of protein biomarkers. Analyst 133, 1550–1555.

Wu, Y.F., Chen, C.L., Liu, S.Q., 2009. Enzyme-functionalized silica nanoparticles as sensitive labels in biosensing. Anal. Chem. 81, 1600–1607.

Wu, Z., Li, H., Liu, Z., 2015. An aptasensor for carcinoembryonic antigen based on upconversion fluorescence resonance energy transfer. Sens. Actuators, B 206, 531–537.

Xu, D.-D., Deng, Y.-L., Li, C.-Y., Lin, Y., Tang, H.-W., 2017. Metal-enhanced fluorescent dye-doped silica nanoparticles and magnetic separation: a sensitive platform for one-step fluorescence detection of prostate specific antigen. Biosens. Bioelectron. 87, 881–887.

Yan, J., Estévez, M.C., Smith, J.E., Wang, K., He, X., Wang, L., et al., 2007. Dye-doped nanoparticles for bioanalysis. Nano Today 2, 44–50.

Zhang, C., Yuan, Y., Zhang, S., Wang, Y., Liu, Z., 2011. Biosensing platform based on fluorescence resonance energy transfer from upconverting nanocrystals to graphene oxide. Angew. Chem., Int. Ed. Engl. 50, 6851–6854.

Zheng, G., Patolsky, F., Cui, Y., Wang, W.U., Lieber, C.M., 2005. Multiplexed electrical detection of cancer markers with nanowire sensor arrays. Nature Biotechnol. 23, 1294–1301.

Zhou, L., Wang, R., Yao, C., Li, X., Wang, C., Zhang, X., et al., 2015. Single-band upconversion nanoprobes for multiplexed simultaneous in situ molecular mapping of cancer biomarkers. Nat. Commun. 6, 6938.

Zijlmans, H.J.M.A.A., Bonnet, J., Burton, J., Kardos, K., Vail, T., Niedbala, R.S., et al., 1999. Detection of cell and tissue surface antigens using up-converting phosphors: a new reporter technology. Anal. Biochem. 267, 30–36.

FURTHER READING

Ali, M.A., Mondal, K., Jiao, Y., Oren, S., Xu, Z., Sharma, A., et al., 2016. Microfluidic immuno-biochip for detection of breast cancer biomarkers using hierarchical composite of porous graphene and titanium dioxide nanofibers. ACS Appl. Mater. Interfaces 8, 20570–20582.

Yang, K., Feng, L.Z., Shi, X.Z., Liu, Z., 2013. Nano-graphene in biomedicine: theranostic applications. Chem. Soc. Rev. 42, 530–547.

Part II

Molecular Recognition in Protein Assay

A biosensor, consisting of a biorecognition element, a biotransducer component, and a signal output system, employs the biomolecules such as antibodies, nucleic acids, peptides, and small molecules for the detection of analytes by converting a biological response into an electrical or optical signal. One of the essential components in the fabrication of a biosensor for protein analysis is the molecular recognition part that is capable of identifying the protein of interest in a given sample and ensuring a high degree of selectivity. So, the establishment of the molecular recognition part is the first step for the fabrication of nano-inspired biosensor, which plays a crucial role in protein assay.

Molecular recognition refers to the specific interaction between two or more molecules through noncovalent bonding such as hydrogen bonding, metal coordination, hydrophobic forces, van der Waals forces, $\pi-\pi$ interactions, halogen bonding, electrostatic, and/or electromagnetic effects. It can occur in between receptor–ligand, antigen–antibody, DNA–protein, peptide–protein, small molecule–protein/peptide, etc. Among them, antigen–antibody interaction is a specific chemical interaction between antibodies produced by B cells of the white blood cells and antigens during immune reaction. The antibodies specifically bind to antigens to form an antigen–antibody complex through weak and noncovalent interactions such as electrostatic interactions, hydrogen bonds, Van der Waals forces, and hydrophobic interactions. Each antibody is capable of binding to only one specific antigen and the specificity of the binding is due to the specific chemical constitution of each antibody. For DNA–protein interaction, it occurs when a protein binds to a molecule of DNA to regulate the biological function of DNA. Generally, the proteins can specifically or nonspecifically bind to DNA in the major groove. The specific interaction between DNA and protein can be utilized for the detection of protein. Up to now, a type of single-stranded oligonucleotide (called aptamer), screened from a large random DNA or RNA library, represents a promising alternative for antibodies in the field of molecule recognition, in view of the advantages including the comparable binding affinity and selectivity towards protein of interest, cost-effective chemical synthesis, high stability and remarkable flexibility, and convenience in design and modification. Apart from antigen or antibody and aptamer, protein can also be specifically recognized by peptides. For example, Peptide T, a short peptide derived from the HIV envelope protein gp120, can bind to the CCR5 receptor so as to block binding and infection of viral strains. In addition, small molecule ligands with high specificity and affinity can interact with the protein/peptide. For instance, the antibiotic vancomycin selectively binds to the peptides with terminal D-alanyl-D-alanine in bacterial cells through five hydrogen bonds.

In this part, we aim to provide an overview of major advances in the field of nano-inspired biosensors for detecting disease protein biomarkers with a special focus on the recognition elements. According to the molecular recognition component and mechanism, the biosensors can be classified as immune-biosensor, aptasensors, peptide-based biosensors, and biosensor-based on protein–small molecule interaction, and so on.

Chapter 5

Immuno-Biosensor

Juan Zhang and Jing Zhao

Center for Molecular Recognition and Biosensing, School of Life Sciences, Shanghai University, Shanghai, P. R. China

Chapter Outline

Immuno-biosensor is a type of biosensor to detect the formation of an immunocomplex using an antibody or antigen as a bioreceptor. In the 1920s, Michael Heidelberger and Oswald Avery first discovered the binding of an antigen to an antibody to form a precipitation. Antibody–antigen interaction is the most well-known biological affinity system in living organisms. An antibody, a part of the immunoglobulin superfamily, specially recognizes and binds to an epitope at the surface of a given antigen through spatial complementary, which is like a key fitting into a lock. Although antibody–antigen interaction is driven by noncovalent weak molecular forces, including electrostatic interaction, hydrogen bond, hydrophobic interaction, and van der Waals force, the immunocomplex is highly stable and selective for biological recognition. In the 1950s, Rosalyn Sussman Yalow and Solomon Berson first reported biological assays based on antibody–antigen interaction, which was called immunoassay. Today, the most widely used immunoassay is the enzyme-linked immunosorbent assay (ELISA). A sandwich structure is used for both molecular recognition and signal output using antibody as a recognition element and a redox enzyme as a signal label. The original immunosensor is fabricated based on a principle similar to ELISA.

Nano-inspired Biosensors for Protein Assay with Clinical Applications.
DOI: https://doi.org/10.1016/B978-0-12-815053-5.00005-2

5.1 CANCER DIAGNOSTICS AND MONITORING

Cancer is a leading cause of death worldwide, which seriously threatens human health today. Cancer is a group of diseases initiated by abnormal cell growth, and has the potential to spread from one part of the body to another. Early detection, followed by precise treatment is believed to be the most effective method to improve survival rate. However, most patients can only be diagnosed at the middle stage or even late stage. Tumor markers are substances that may reveal the occurrence and development of cancers. Since tumor markers are critical to early detection of tumors, they are the most extensively studied target molecules in the development of biosensors.

Elevated prostate-specific antigen (PSA) level in blood is usually found in men with prostate cancer, which is approved by the FDA to monitor the occurrence and progression of prostate cancer. PSA is also a popular tumor target in the development of biosensors. In order to improve the sensitivity, different strategies are designed to enhance the signal response. Li et al. (2016) proposed an ultrasensitive dynamic light scattering (DLS)-based immunosenosr for the detection of tumor marker protein, taking PSA as an example (Fig. 5.1). A DLS technique was combined with ELISA as a novel DLS-ELISA. Manganese dioxide (MnO_2) nanosheet-modified gold nanoparticles (GNPs) were used to label antibody, and the labeled antibody was immobilized onto substrate for the formation of an immunocomplex. Wrapped GNPs were released to the solution with the decomposition of

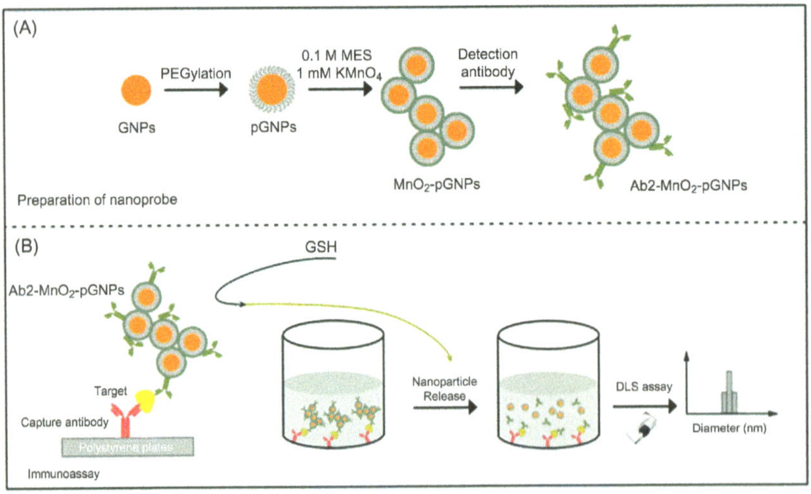

FIGURE 5.1 Representation of the dynamic light scattering-linked immunosorbent assay for protein detection performed in one of the 96-well polystyrene plates. *Recreated from Li, C., Ma, J., Fan, Q., Tao, Y., Li, G., 2016. Dynamic light scattering (DLS)-based immunoassay for ultra-sensitive detection of tumor marker protein. Chem. Commun. 52, 7850–7853, with kind permission of The Royal Society of Chemistry.*

MnO_2, and thus characterized by DLS measurement. The detection limit was proven to be at fM level, which is ultrasensitive compared to the commercial ELISA test. Xu et al. (2011) fabricated a cathodic electrogenerated chemiluminescence (ECL) immunosensor for the detection of PSA. Increased amount of glucose oxidase (GOx) and secondary antibody (Ab_2) were loaded onto a gold nanorod for recognition of target protein and amplification of the ECL signal, while functionalized graphene provided an increased surface for immobilization of the primary antibody with improved electron transfer. The immunosensor displayed high sensitivity and specificity toward PSA detection with a detection limit of 8 pg mL^{-1}. Liu et al. (2014) reported a colorimetric immunoassay for ultrasensitive and quantitative detection of tumor marker PSA. Glucose oxidase (GOx)-labeled Ab_2 was firstly immobilized on the surface of magnetic beads (MBs), which were attached onto the substrate through formation of a sandwich structure upon target recognition. GOx catalyzed the oxidation of glucose to generate H_2O_2, and thus promoted the growth of gold nanoparticles for colorimetric detection. The method demonstrated a four orders of magnitude lower detection limit than a commercial ELISA test. Liu et al. (2013) synthesized a gold nanoparticle (AuNP)-based fluorescence-activatable probe for detection of PSA at an ultralow level. A fluorescent dye rhodamine B isothiocyanate (RBITC) was immobilized on AuNP surface by means of an ITC group, where AuNP acted as a quencher. Negatively charged groups on Ab_2 bound to RBITC through positively charged quaternary ammonium groups. After immobilizing onto a substrate through immunoreaction, RBITC was released from the gold surface with the addition of cysteamine, thereby generating fluorescent signal for determination. The detection limit of PSA was estimated to be as low as 0.032 pg mL^{-1}. Zhu et al. (2016) utilized a PSA aptamer to replace Ab_2 for ultrasensitive electrochemical detection of the prostate cancer biomarker. Both PSA aptamer and primer of rolling circle amplification (RCA) were functionalized on the surface of gold nanoparticles, and immobilized on the electrode surface through immunoreaction. Subsequently, repeated poly(thymine) sequences were produced as a result of RCA, and worked as a template for the formation of copper nanoparticles. By tracing electrochemical responses of copper ions released from copper nanoparticles, the method demonstrated a remarkable detection limit of 0.020 ± 0.001 fg mL^{-1} PSA.

Carcinoembryonic antigen (CEA) is a type of broad-spectrum tumor marker, which is widely found in many cancers. Pang et al. (2015) developed a highly sensitive ECL immunosensor for detection of CEA using GO/MWCNTs-COOH/Au@CeO_2 nanocomposite for immobilization of antibody (Fig. 5.2). GO/MWCNTs-COOH provided an increased surface area and conductivity for loading CeO_2 nanoparticles, while CeO_2 nanoparticles worked as ECL luminescent materials. After functionalization of anti-CEA on the surface of AuNPs, the immunosensor exhibited an excellent performance toward CEA detection with a wide linear range and a low detection

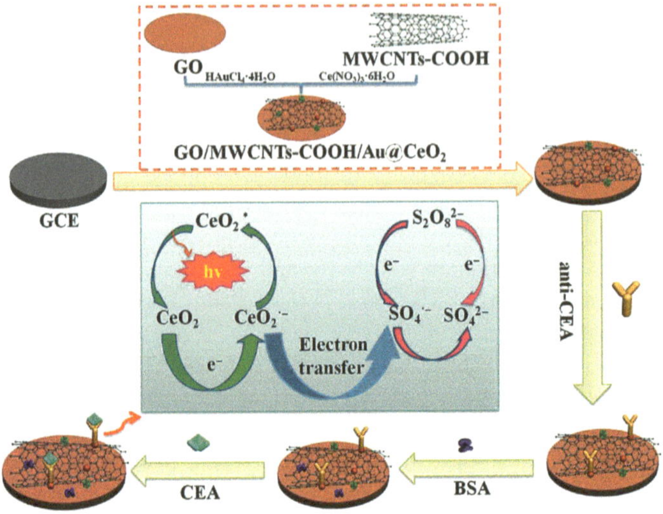

FIGURE 5.2 Schematic diagram for fabrication of the label-free ECL immunosensor. *Recreated from Pang, X., Li, J., Zhao, Y., Wu, D., Zhang, Y., Du, B., et al., 2015. Label-free electrochemiluminescent immunosensor for detection of carcinoembryonic antigen based on nanocomposites of GO/MWCNTs-COOH/Au@CeO2. ACS Appl. Mater. Interfaces 7, 19260–19267, with kind permission of American Chemical Society.*

limit of 0.02 ng mL^{-1}. Li et al. (2015) synthesized MWCNT-NH_2 supported PdPt nanocages to label Ab2. The nanocages showed enhanced electrocatalytic activity toward reduction of H_2O_2 and aroused signal amplification for CEA detection with a low detection limit of 0.2 pg mL^{-1}. Gao et al. (2011) functionalized the electrode with a nanomultilayer film using layer-by-layer (LBL) assembly of poly(diallyldimethylammonium chloride)-wrapped carbon nanotubes and poly(sodium-p-styrene-sulfonate). The membrane offered increased surface area for immobilization of anti-CEA with good biocompatibility, thereby achieving a detection limit of 0.06 ng mL^{-1}.

Carbohydrate antigen (CA) is a big family of glycoproteins relating to various tumors. CA15-3, encoded by the human MUC-1 gene, is known as a biomarker for breast cancer. de Oliveira et al. (2017) reported a disposable microfluidic immunoarray device (DμID) for detection of CA15-3 (Fig. 5.3). The device was fabricated based on a simple and rapid prototyping technique, which was low cost and fast response. A layer-by-layer technique was employed to immobilize primary antibody, while a polyclonal antibody and peroxidase functionalized magnetic particles were used to produce signal. The device realized eight simultaneous detections of CA15-3 with a low detection limit of 6 μU mL^{-1}. CA19-9 is a tumor marker for pancreatic cancer, which is approved by the Food and Drug Administration of the United States. Soares et al. (2017) fabricated an immunosensor by making use of

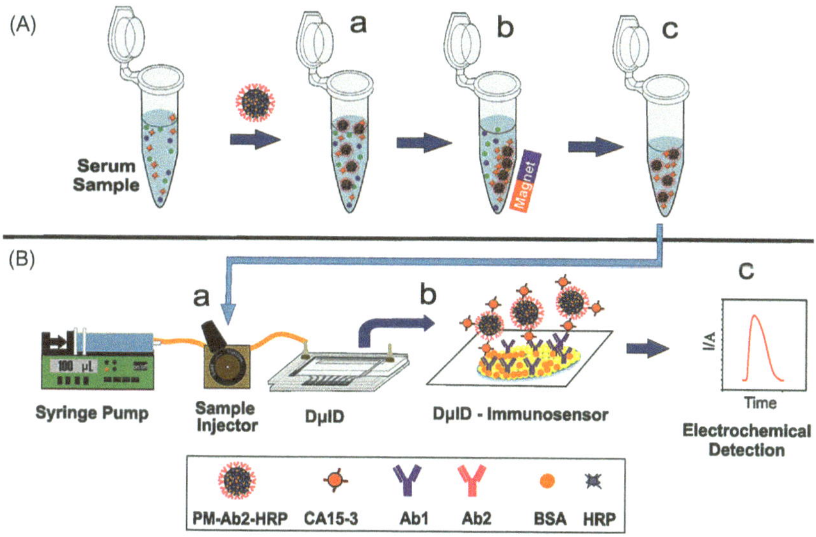

FIGURE 5.3 Schematic representation of the assay method for CA15-3 detection. (A) Off-line CA15-3 capture: addition of MP-Ab_2-HRP to the sample solution followed by (a) biomarker capture via the incubation step, (b) magnetic separation and washing steps, and (c) resuspension of the bioconjugate in PBS. (B) On-line CA15-3 detection: (a) injection of the bioconjugate in the DμID by the injection valve, followed by (b) the incubation and washing steps to form the immunosandwich structure, and (c) amperometric detection by the injection of H_2O_2 and HQ. *Recreated from de Oliveira, R.A.G., Materon, E.M., Melendez, M.E., Carvalho, A.L., Faria, R.C., 2017. Disposable microfluidic immunoarray device for sensitive breast cancer biomarker detection. ACS Appl. Mater. Interfaces 9, 27433–27440, with kind permission of American Chemical Society.*

electrospun nanofibers of polyamide 6 and poly(allylamine hydrochloride). Coated with either multiwalled carbon nanotubes (MWCNTs) or gold nanoparticles (AuNPs), the nanofiber offered a three-dimensional structure for loading anti-CA 19-9, and attained a detection limit of 1.84 or 1.57 U mL^{-1} using impedance spectroscopy. The same group also fabricated a thin film using both polyethyleneimine and carbon nanotubes for functionalization of interdigitated gold electrode (Thapa et al., 2017). After immobilization of anti-CA 19-9, the electrode was used to detect CA19-9 with a detection limit of 0.35 U mL^{-1}.

Platelet-derived growth factor (PDGF) regulates the formation and growth of blood vessels, which are usually overexpressed in several cancers. Feng et al. (2017) reported an immuno-NRCA (net-like rolling circle amplification) strategy for ultrasensitive detection of PDGF (Fig. 5.4). The immunoassay was combined with a novel NRCA procedure, which integrated primer extension, strand displacement, and nicking reaction. The immune-binding was conducted on an interface, while signal amplification process was performed in a solution. The transfer from interface to solution displayed an improved sensitivity with a detection limit of 5.5 fg L^{-1}. Masud et al. (2017) synthesized gold-loaded nanoporous ferric oxide nanocubes (Au-NPFe_2O_3NC) as a mimic of horseradish peroxidase (HRP).

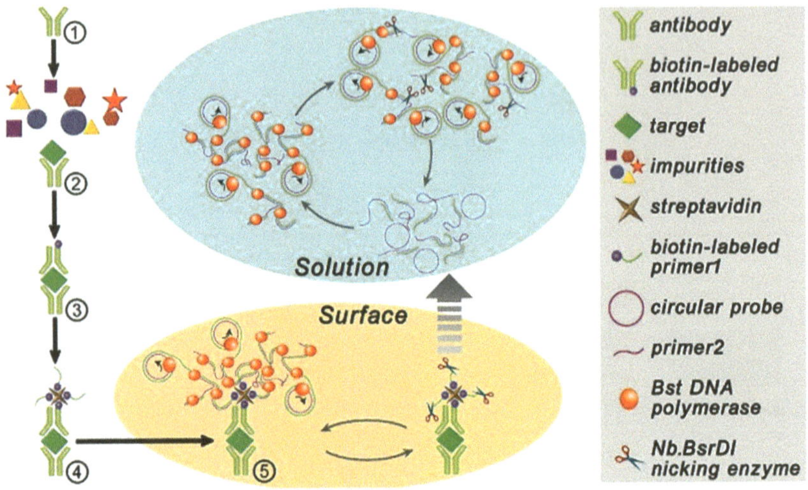

FIGURE 5.4 Schematic presentation of the principle of immuno-NRCA assay. *Recreated from Feng, C., Bo, B., Mao, X., Shi, H., Zhu, X., Li, G., 2017. From interface to solution: integrating immunoassay with netlike rolling circle amplification for ultrasensitive detection of tumor biomarker. Theranostics 7, 31–39, with kind permission of Ivyspring International Publisher.*

The immunosensor achieved satisfactory sensitivity and reproducibility toward the detection of p53-specific autoantibodies, by means of enhanced electrocatalytic and colorimetric activity of Au-NPFe_2O_3NC, which was also applied in plasma samples from patients with epithelial ovarian cancer high-grade serous subtype. Munge et al. (2011) utilized a glutathione-protected gold nanoparticle (GSH-AuNP) to capture primary antibody, and multilabeled HRP–carbon nanotube (CNT) to label Ab_2 for signal amplification. The immunosensor was successfully applied in detection of tumor marker interleukin 8 (IL-8) with a detection limit of 1 fg mL^{-1}, which was also available for determination of IL-8 in head and neck squamous cell carcinoma (HNSCC) cells.

Since one cancer usually may have more than one type of biomarker, the simultaneous determination of different tumor markers contributes to enhanced accuracy of cancer detection. In this sense, several immunosensors are developed for quantitative determination of different tumor markers. Li et al. (2010) reported a sensitive electrochemical immunoassay for determination of breast cancer cells MCF-7 based on simultaneous recognition of two coexpressing tumor markers mucin 1 and CEA (Fig. 5.5). MCF-7 cells were immobilized on the electrode through binding with mucin-1 aptamer, while anti-CEA functionalized CdS nanoparticles were attached on the surface of MCF-7 cells for signal output. Therefore, the immunosensor demonstrated good reproducibility and low detection limit for detection of cancer cells. Similarly, Tang et al. (2016) designed microchip-based microfluidic

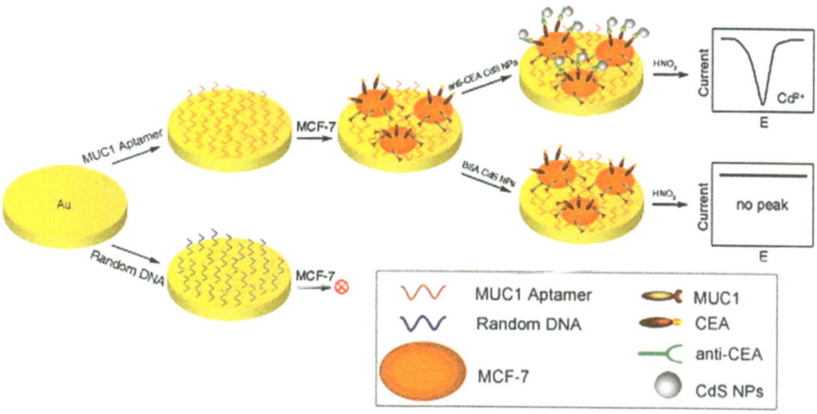

FIGURE 5.5 Schematic illustration of the method to detect breast cancer cells through simultaneous recognition of two different tumor markers. *Recreated from Li, T., Fan, Q., Liu, T., Zhu, X., Zhao, J., Li, G., 2010. Detection of breast cancer cells specially and accurately by an electrochemical method. Biosens. Bioelectron. 25, 2686–2689, with kind permission of Elsevier.*

electrochemical arrays to enable 256 simultaneous measurements in less than 1 h through immunoreaction. The multiplexing electrochemical immunosensor was applied for detection of several tumor biomarkers, including prostate specific antigen (PSA), prostate specific membrane antigen (PSMA), interleukin-6 (IL-6), and platelet factor-4 (PF-4) in clinical samples.

Ren et al. (2017) reported a tri-antibody dual-channel biosensing strategy using graphene oxide/polyaniline nanostructures for functionalization of an electrode and mesoporous NKF-5-3 for adsorption of neutral red or toluidine blue as signals. The fabrication of nanostructure on the electrode transferred the electrode interface to a sensor space. The immunosensor achieved a detection limit of 25 fg mL^{-1} for nuclear matrix protein 22 (NMP22) and 30 fg mL^{-1} for CEA, respectively. Chen et al. (2013) synthesized two different nanoprobes by immobilizing toluidine blue (TB) labeled anti-CEA ($Ab_{2,1}$) and Prussian blue (PB) labeled anti-alpha-fetoprotein (AFP, $Ab_{2,2}$) onto the surface of carboxyl graphene nanosheets(CGS), respectively (Fig. 5.6). Anti-CEA ($Ab_{1,1}$) and anti-AFP ($Ab_{1,2}$) were captured on the surface of chitosan-Au nanoparticles (CHIT-AuNPs)-modified electrode through covalent interaction. After the formation of both sandwich-like structures on the electrode surface, simultaneous detection of CEA and AFP was realized by using two nanoprobes with detection limits of 0.1 ng mL^{-1} and 0.05 ng mL^{-1}, respectively. They also reported simultaneous detection of CEA and AFP using two different metal ions tagged CHIT-AuNPs as signal output. Differential pulse voltammetry (DPV) was applied to determine metal ions for quantitative measurements of tumor markers even in clinical samples (Xu et al., 2014). Fang et al. (2015) also employed metal-containing nanoparticles (Zn and Cd) as signal tags using ordered mesoporous carbon as

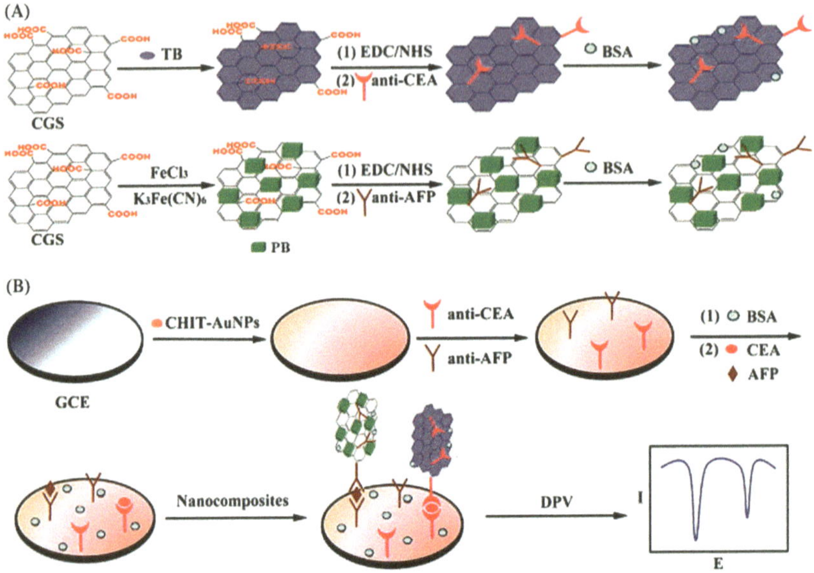

FIGURE 5.6 Schematic illustration of (A) the preparation of biofunctional CGS nanocomposites and (B) the multiplexed electrochemical immunoassay protocol. *Recreated from Chen, X., Jia, X., Han, J., Ma, J., Ma, Z., 2013. Electrochemical immunosensor for simultaneous detection of multiplex cancer biomarkers based on graphene nanocomposites. Biosens. Bioelectron. 50, 356–361, with kind permission of Elsevier.*

matrix. Metallic nanocrystallites incorporated OMC materials were labeled antibodies for recognition of target biomarkers, and led to greatly amplified electrochemical responses for determination of both AFP and human epidermal growth factor receptor type-2 (HER-2) in real human serum samples. Tang et al. (2011) prepared the distinguishable nanoprobes by encapsulating HRP-thionine or HRP-ferrocene conjugations into nanogold hollow microspheres. Two nanoprobes were used to label anti-AFP and anti-CEA, respectively. The electrochemical nanoprobes displayed different electrochemical peak potentials. Therefore, sandwich-like structures were formed on the surface of biofunctionalized magnetic graphene nanosheets (MGO) upon binding with AFP and CEA. The flow-through multiplexed electrochemical immunoassays realized the detection limits of 1.0 pg mL^{-1} for both analytes. Liu et al. (2015) reported the synthesis of POPD/Au and poly(VFc-ATP)/Au nanocomposites using a one-step method. Two nanocomposites were used to label anti-CEA and anti-AFP for fabrication of immunonanoprobes, and realized simultaneous determination of CEA and AFP based on immunoreaction. The immunosensor demonstrated a detection limit of 0.006 ng mL^{-1} for CEA and 0.003 ng mL^{-1} for AFP. Xu et al. (2015) synthesized a carbon–gold nanocomposite (CGN) for adsorption of three different signal molecules, such as thionin (Thi), 2,3-diaminophe-nazine (DAP), and Cd^{2+}. The signal tags were used to label CEA, PSA, and AFP, respectively. The combination of ionic liquid-reduced graphene oxide with poly(sodium-p-styrenesulfonate) was used as a substrate to immobilize three antibodies for immunoreactions. By tracing three separate signals, the simultaneous detection of CEA, PSA, and AFP was realized with detection limits of 2.7, 4.8, and 3.1 pg mL^{-1}, respectively.

5.2 TUBERCULOSIS DIAGNOSTICS AND MONITORING

Tuberculosis (TB) is a deadly infectious disease, which is caused by *Mycobacterium tuberculosis* (MTB). Rapid and early detection of TB is critical to the prevention of airborne contagiousness. The existing TB diagnostic methods suffer from several drawbacks. For example, the microscopic examination has the defects of low sensitivity and accuracy, while culture always needs longer time for growth of MTB. Therefore, it is of great importance to develop more active methods for effective TB screening. Barroso et al. (2018) proposed a magnetoresistive biosensor for direct determination of BCG bacteria (Fig. 5.7). Magnetically-labeled Abs specifically bound to target bacteria, which was magnetically separated from the samples. After being injected into the biochip microfluidics, the bacteria were captured onto the surface of a MR-biochip, inducing changes of MR signals. The magnetoresistive biosensor displayed a low detection limit and high potential for point-of-care TB diagnosis. Prabowo et al. (2018) reported a portable organic light-emitting diode- (OLED)-based surface plasmon resonance (SPR) biosensor by means of a nested PCR technique. IS6110 DNA

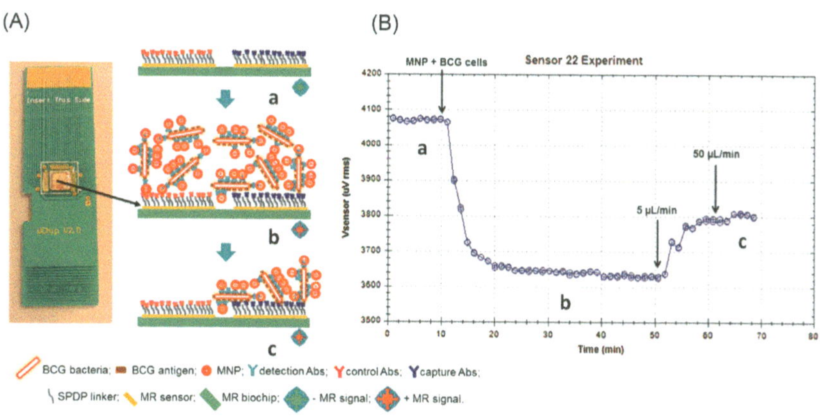

FIGURE 5.7 (A) schematic explanation of the sandwich-immunoassay reactions on MR-biochip and (B) corresponding average voltage variation curves over time. (a) baseline - negative MR signal (no label), (b) voltage drop - positive MR signal (MNP@Abs@BCG in contact with the MR-biochip surface), and (c) washing steps - positive MR signal (MNP@Abs@BCG binded to specic Abs on the biochip surface). *Recreated from Barroso, T.G., Martins, R.C., Fernandes, E., Cardoso, S., Rivas, J., Freitas, P.P., 2018. Detection of BCG bacteria using a magnetoresistive biosensor: a step towards a fully electronic platform for tuberculosis point-of-care detection. Biosens. Bioelectron. 100, 259–265, with kind permission of Elsevier.*

was chosen as target DNA for identification of MTB complex, and DIG-labeled PCR products were immobilized on the sensing surface through interaction with anti-DIG. SPR sensor demonstrated a detection limit of about 63 pg mL^{-1}, 96.6% of sensitivity, and 98.4% of specificity in 600 clinical sputum specimens. Mdluli et al. (2014) designed a rapid dual channel lateral flow assay to detect TB using MTB 38 kDa monoclonal antibody as a target. The target antibody was captured by antigen on the test line, and was then labeled with gold-Protein A conjugate, showing a red colored line. Gold-Protein A conjugate was also captured by anti-Protein A on the control line, proving the validity of the target test. A detection limit of 5 ng mL^{-1} was achieved using MTB lateral flow assay with an ESE Quanti reader. CFP-10 (culture filtrate antigen 10 kDa) is a T cell antigen secreted by MTB, which is also a potent biomarker for TB detection. Kim et al. (2013) reported a magnetophoretic immunoassay for colorimetric detection of CFP-10. The detection is based on formation of a sandwich-type immunocomplex with anti-CFP10 modified Au nanoparticles (NPs) and magnetic microparticles (MMPs), which could be finished in 10 min with a wide linear range and a low detection limit of 10 pg mL^{-1}. Gopinath et al. (2016) utilized 16 kDa heat shock protein (16 kDa HSP) as a target for TB detection. The binding of antibody and antigen was characterized using nanogapped dielectric impedance sensor, which displayed a sensitive detection as low as 100 fM.

Commercial interferon-γ (IFN-γ) release assays have now been introduced to diagnose latent tuberculosis infection (LTBI). Kim et al. (2014) developed a rapid, sensitive, and label-free method for detection of IFN-γ

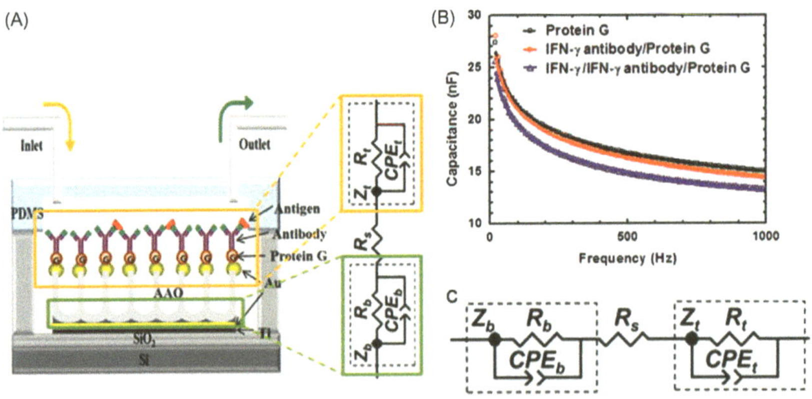

FIGURE 5.8 (A) Schematic diagram of the AAO-based capacitive sensor. Each biomolecule is assembled from bottom to top. (B) Frequency dependence of the capacitance measured in PBS for the sensor with immobilized protein G (black squares), anti-IFN-γantibody/protein G (red circles), and IFN-γ/anti-IFN-γantibody/protein G (blue triangles). (C) Equivalent circuit for the AAO-based capacitive sensor. *Recreated from Kim, J.H., Chang, Y.W., Bok, E., Kim, H.J., Lee, H., Cho, S.N., et al., 2014. Detection of IFN-gamma for latent tuberculosis diagnosis using an anodized aluminum oxide-based capacitive sensor. Biosens. Bioelectron. 51, 366–370, with kind permission of Elsevier.*

based on the capacitance changes upon binding to its antibody (Fig. 5.8). The AAO membrane provided an interface for sensing target antigen after deposition of a gold film. Anti-IFN-γ was immobilized on the substrate through the interaction with Protein G. The binding of antigen and antibody induced signal changes of the capacitive sensor. Therefore, IFN-γ was sensitively detected in a wide linear range with a low detection limit of 0.2 pg mL^{-1}, which was also able to trace the target in the real clinical samples. Huang et al. (2015) described an immunosensor by combining the use of CdS quantum dots (QDs) and magnetic beads (MB). A sandwich-type structure was formed on gold-nanoparticle-functionalized glassy carbon electrode in the presence of antihuman IFN-γ antibodies (Ab1), target protein, and MB-QDs conjugated antihuman IFN-γ antibodies (Ab2). By tracing the electrochemical response of metal cadmium with square wave anodic stripping voltammetry (SWASV), IFN-γ was detected with the detection limit of 0.34 pg mL^{-1}. Wang et al. (2017) fabricated another electrochemical immunosensor using electrochemical impedance spectroscopy (EIS). The formation of immunocomplex on ZnO NPs-modified electrode increased impedance values, and achieved a detection limit of 0.12 pg mL^{-1}. Zhou et al. (2017) prepared an electrochemiluminescence (ECL) immunosensor for simultaneous determination of triple TB biomarkers, including IFN-γ, tumor necrosis factor-alpha (TNF-α), and interleukin (IL)-2. Luminol, carbon quantum dots, and CdS quantum dots were used to functionalize gold nanoparticles as signal labeling, respectively. Accordingly, three sensitive and potential-resolved ECL signals were generated for detection of different biomarkers in the range of 1.6–200 pg mL^{-1}.

Ag85B is a major secretory antigen from active MTB. Jing et al. (2014) fabricated an integrated microfluidic system for capture, enrichment, and rapid detection of MTB based on a specially designed "sieve valve" (Fig. 5.9). Ag85B protein was used as a target for TB identification, while microspheres coated with Protein A were used to capture target antigen. The rabbit polyclonal antibody and FITC labeled antimouse secondary antibody were then attached to the substrate for signal readout. The detection is very fast at less than 50 min, including both enrichment and analysis. Kim et al. (2017) modified the surface of gold nanorods (AuNRs) and silica-coated quantum dots (SiQDs) by the immobilization of the genetically engineered recombinant antibodies GBP-50B14 and SiBP-8B3, respectively. A sandwich immunocomplex was formed upon binding with the target antigen for a quenching effect. The method achieved a limit of detection of 13.0 pg mL^{-1} for Ag85B, and easily distinguished the control proteins (CFP10 and BSA).

The early secreted antigenic target 6-kDa protein (ESAT-6) is the virulent factor of MTB, which is usually used for active TB diagnosis. Diouani et al. (2017) reported a label-free immunosensor for detection of ESAT-6 based on its specific interaction anti-ESAT-6 monoclonal antibody. The immunocomplex on the surface of a gold screen-printed electrode prevented the electron transfer with $[Fe(CN)_6]^{3-/4-}$, which was evaluated by square wave

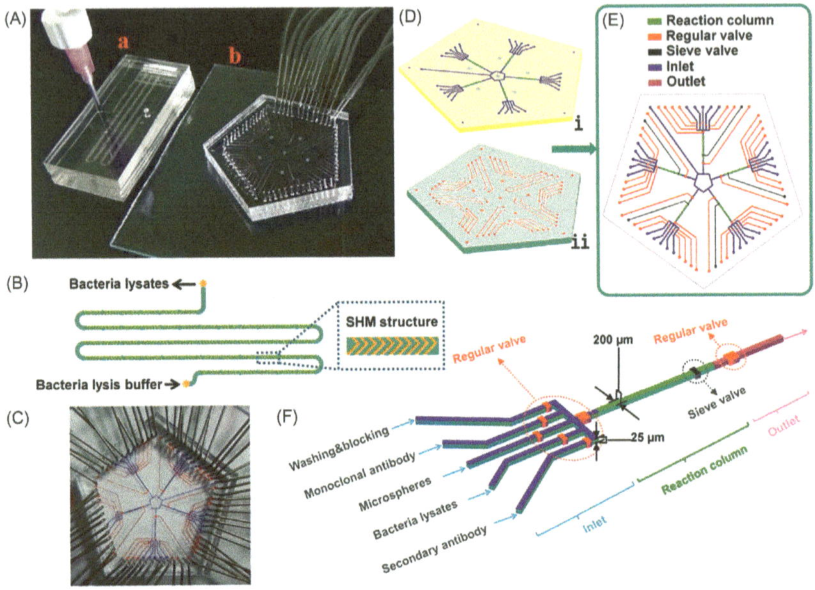

FIGURE 5.9 (A) Image of the system for airborne bacteria rapid enrichment and bacteriological diagnosis: (a) airborne bacteria enrichment microfluidic chip, (b) immunoassay microfluidic chip. (B) Schematic illustration and detailed structure of enrichment microfluidic chip. (C) Image of the microfluidic immunoassay chip. The various channels had been loaded with food dyes to visualize the structure of the microfluidic chip. (D,E) Schematic illustrations of designed immunoassay microfluidic chip. Valves and columns were illustrated by different colors: (i) fluidic layer, (ii) control layer. (F) Enlarged diagram showing detailed structure of microfluidic immunoassay chip. *Recreated from Jing, W., Jiang, X., Zhao, W., Liu, S., Cheng, X., Sui, G., 2014. Microfluidic platform for direct capture and analysis of airborne Mycobacterium tuberculosis. Anal. Chem. 86, 5815–5821, with kind permission of American Chemical Society.*

voltammetry with a detection limit of 7 ng mL^{-1}. Dhakane and Patil (2013) introduced a piezoresistive method for detection of ESAT-6 using microelectromechanical systems (MEMS)/nanoelectromechanical systems (Fig. 5.10). The immunoreaction was conducted on the top of a microcantilever based on antigen–antibody interaction, leading to bending of a microcantilever for signal output. Kim et al. (2016) prepared a liquid crystal (LC)-based biosensor for detection of anti-ESAT-6 antibody in the solution. The anisotropic nanostructures were manufactured to provide polymeric surfaces for immobilization of tuberculous antigens (ESAT-6). The interaction of antigen and antibody transformed the LCs (4-cyano-4′-pentylbiphenyl, 5CB) from a uniform orientation to a random one, realizing a label-free detection of antibody.

5.3 DIABETES DIAGNOSTICS AND MONITORING

An average glucose concentration over the past 2–3 months can be reflected by glycated hemoglobin (HbA1c) in the blood with fewer variations due to

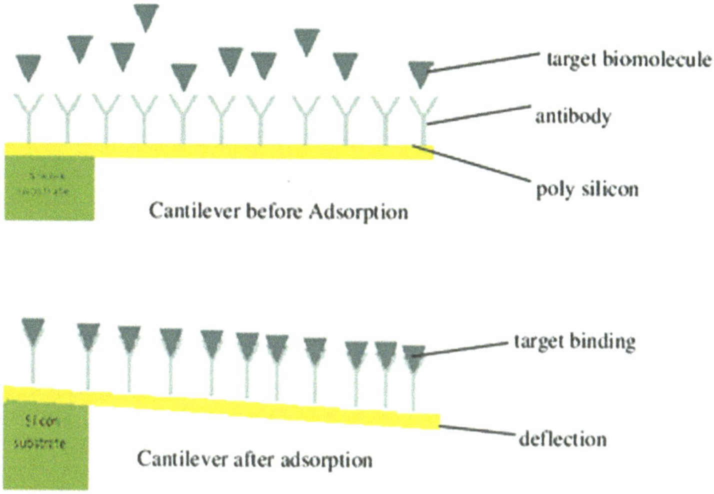

FIGURE 5.10 Working principle of microcantilever beam. *Recreated from Dhakane, S., Patil, W.V., 2014. Piezo-resistive method for tuberculosis detection using microcantilever biosensor. Microsyst. Tech. 20, 457–462, with kind permission of Springer Nature.*

nonglycemic factors. So the accurate measurement of the HbA1c provides a more effective route to diagnose and to monitor DM. An integrated microfluidic system has been developed for the automatic measurement of HbA1c levels (Wu et al., 2014). As shown in Fig. 5.11, the pretreated blood was transported into a mixing chamber by the micropump, resulting in the capture of the target Hb and HbA1c by the anti-Hb-antibody-coated beads. Subsequently, as the target bead complexes themselves are captured by an external magnet, nonreactive molecules are all washed away. After washing, acridinium ester-labeled HbA1c antibodies were added into the chamber and reacted with the HbA1c. After removing unbound secondary antibodies, the indicator reaction was developed by mixing with H_2O_2 solution containing HNO_3 and NaOH solution containing Triton X-100, to give the chemiluminescence signals.

Moreover, a micropotentiometric immunosensor has been established for HbA1c level detection based on mixed SAMs-wrapped nanospheres array (Xue et al., 2011). The mixed SAMs-wrapped nanospheres array shows better biocompatibility, uniform distribution without aggregation, and stable bonding with biomolecules. Moreover, the sensitivity of the immunosensor is enhanced because the amount of antibodies immobilization is increased by using the mixed SAMs-wrapped nanospheres. HbA1c can also be analyzed through a single-use disposable electrochemical label-free immunosensor using differential pulse voltammetry (Molazemhosseini et al., 2016). Moreover, a homogeneous LRET-based immunosensor has also been designed to detect HbA1c levels by quenching the luminescence signals of the UCNPs (donor) (Jo et al., 2016). For LRET-based immunosensor, NaYF4:Yb^{3+}, Er^{3+} UCNPs, and HbA1c are used as the donor and acceptor

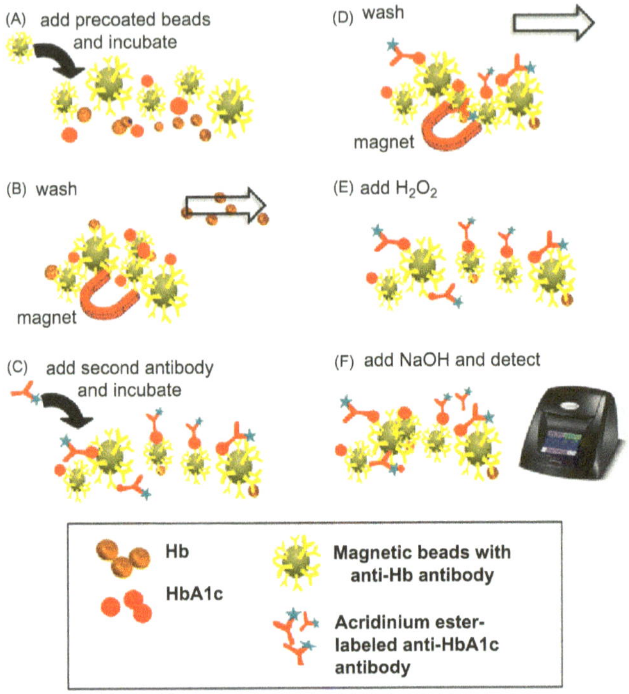

FIGURE 5.11 Illustration of the steps for performing the two-antibody assay on a microfluidic chip. (A) Transport the magnetic beads pre-coated with Hb antibodies to the mixing chamber and incubate with the pre-treated blood. (B) wash with the PBS buffer. (C) transport the acridinium ester-labeled HbA1c antibodies into the mixing chamber and reacted with captured HbA1c. (D) another washing with PBS buffer. (E) transport H_2O_2 into the mixing chamber. (F) inject NaOH to react with the acridinium and detect the resulting chemiluminescence signals with a portable luminometer. *Recreated from Wu, C.C., Lin, H.-I., Chang, K.W., Mai, J.D., Shiesh, S.C., Lee, G.B., 2015. Measurement of glycated hemoglobin levels using an integrated microfluidic system. Microfluid. Nanofluid. 18, 613–621, with kind permission of Springer Nature.*

of the LRET pair, respectively. In the presence of HbA1c, the specific recognition between the anti-HbA1c monoclonal antibody-functionalized UCNPs and HbA1c results in positioning of the energy donor and acceptor in close proximity, ultimately quenching the upconversion luminescence of the UCNPs through LRET (Fig. 5.12). Additionally, a simple diagnostic system that uses MPBA-SAM modified gold screen printed sensor as an affinity layer, has also been developed to capture HbA1c by diol linkages (Chopra et al., 2014). For the system, two receptor molecules including boronic acid and specific electroactive antibodies, are used as a capture and a detector molecule, respectively.

The level of a single glycated albumin can provide a window for monitoring overall diabetic control during the preceding 10–20 days. A highly sensitive one-step homogeneous immunoassay has been developed for

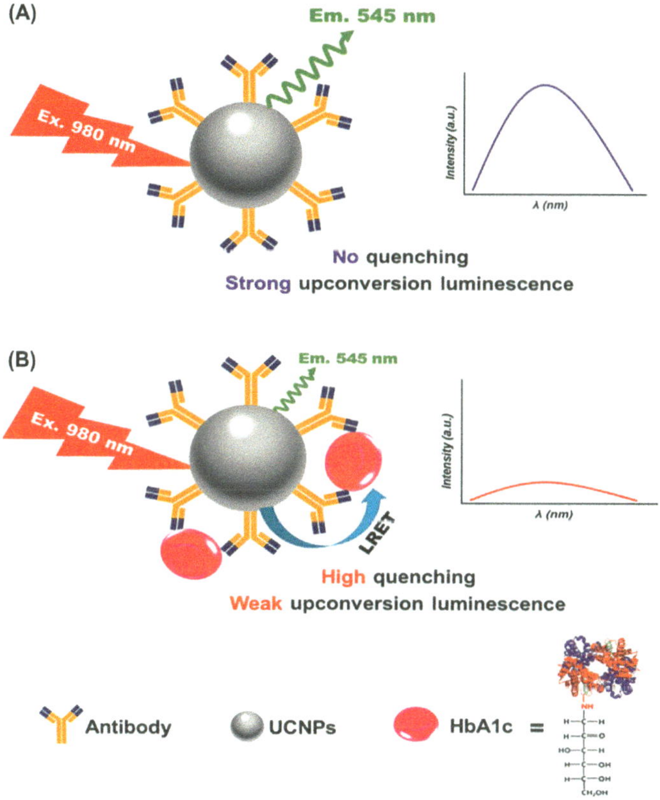

FIGURE 5.12 Schematic illustration of the LRET-based immunosensor: (A) in the absence of HbA1c, no quenching and strong upconversion luminescence; (B) in the presence of HbA1c, high quenching and weak upconversion luminescence. *Recreated from Jo, E.J., Mun, H., Kim, M.G., 2016. Homogeneous immunosensor based on luminescence resonance energy transfer for glycated hemoglobin detection using upconversion nanoparticles. Anal. Chem. 88, 2742–2746, with kind permission of American Chemical Society.*

glycated albumin detection using gold nanoparticle probes coupled with dynamic light scattering analysis for the diagnosis of diabetes (Wangoo et al., 2016). The sensitivity of the assay is excellent with a detection limit of 1 μg mL^{-1} utilizing 40-nm-sized gold nanoparticles.

As a polypeptide hormone, insulin is vital for glucose metabolism. The levels of insulin can reflect the type of diabetic condition or insulinoma. Serum insulin can be tested through a voltammetric immunosensor for diagnosis of type 1 and type 2 diabetic disorders (Singh and Krishnan, 2015). For the immunosensor, the edge-plane pyrolytic graphite electrodes (PGEs) were modified by π-π stacked MWNT with 1-pyrenebutyric acid units (denoted as PGE/MWNT/Py) to covalently immobilize antiinsulin antibody. Serum insulin binding to the surface antibody can be detected at picomolar levels by monitoring the decrease in voltammetric current signals of a redox probe taken in the electrolyte solution. The method has great sensitivity because that serum insulin conjugates with poly(acrylic acid)-functionalized magnetite nanoparticles and MNP-serum insulin conjugate can bind onto the surface insulin–antibody on PGE/MWNT/Py electrodes.

5.4 HUMAN IMMUNODEFICIENCY VIRUS (HIV)

The outer layer of the HIV viral core is composed of proteins called p17, while the inner core is made up of p24 proteins. The p24 Ag antigens can promote an immune response from the infected host. A novel amperometric immunosensor has been constructed to detect the p24 antigen from HIV-1 by using gold nanoparticles (GNP), multiwalled carbon nanotubes (MWCNTs), and an acetone-extracted propolis film (Kheiri et al., 2011). For this immunosensor, amino-functionalized MWCNTs ($MWCNTNH_2$) were prepared and dispersed in a $HAuCl_4$ solution to synthesize GNPs in situ, followed by the preparation of GNP/CNT/AEP nanocomposite through mixing an AEP solution and the GNP/CNT powder. After dripping the nanocomposite onto a gold electrode (GE), p24 antibody was immobilized on the resulting modified gold electrode to construct the immunosensor. Moreover, poly[G]/S-NPs biological labels have been exploited for the innovative PEC immunoassay of HIV-1 p24 antigen (Fig. 5.13) (Zhao et al., 2015). Along with the sandwich immune-binding, the DNA tags could be released. Subsequently,

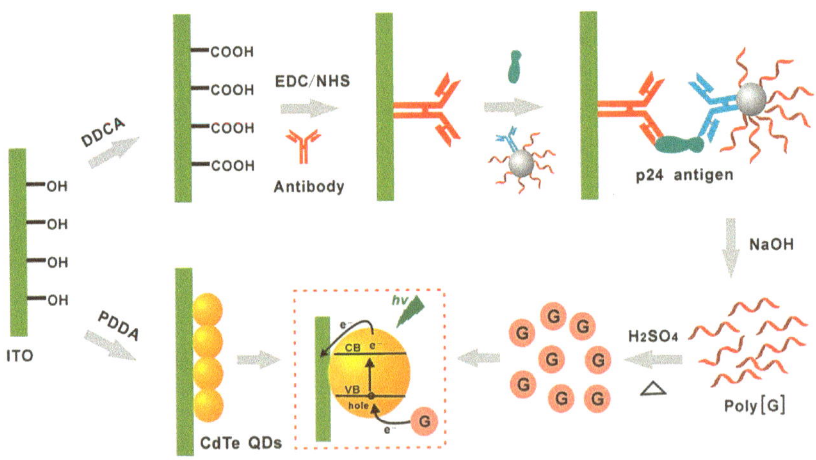

FIGURE 5.13 Schematic illustration of the novel PEC immunoassay strategy using the poly[G]/S-NPs biological label for the sensitive detection of HIV-1 p24 antigen. *Recreated from Zhao, W.W., Han, Y.M., Zhu, Y.C., Zhang, N., Xu, J.J., Chen, H.Y., 2015. DNA labeling generates a unique amplification probe for sensitive photoelectrochemical immunoassay of HIV-1 p24 antigen. Anal. Chem. 87, 5496–5499, with kind permission of American Chemical Society.*

dipurinization of the oligonucleotide strands favored the easy oxidation of free nucleobases at a CdTe quantum dots (QDs)-modified ITO transducer. Such DNA tags induced PEC amplification and readout contributing to the exquisite assay of HIV-1 p24 antigen. In addition, HIV-1 p24 antigen can be analyzed through a system of magnetic beads (MBs) coupled with catalytic fluorescent immunoassay (Zhang et al., 2016). p24 was captured by antibody-immobilized MBs, and the detection antibody was then linked to horseradish peroxidase (HRP) through biotin–streptavidin recognition. The oxidation of o-phenylenediamine (OPD) and hydrogen peroxide was catalyzed so as to produce a fluorescent product.

HIV-1 antibody (anti-HIV-1) is generated after HIV-1 infection, and the window of opportunity to detect anti-HIV-1 is generally 4–8 weeks. A rapid and cost-effective sandwich electrochemiluminescence (ECL) immunosensor has been constructed for the ultrasensitive detection of anti-HIV-1 (Zhou et al., 2014). For the immunosensor, magnetic molecularly imprinted polymers (MMIPs) served as capture probes by combining surface and epitope imprinting techniques and antigen conjugated with horseradish peroxidase (HRP-HIV-1) as labels. As the functional monomer and cross-linking reagent, 3-aminobenzeneboronic acid (APBA) was first polymerized on the surface of silicate-coated magnetic iron oxide nanoparticles in the presence of human immunoglobulin G. Subsequently, a novel sandwich ECL immunosensor was formed through the immunoreaction between MMIPs conjugated with varied concentrations of anti-HIV-1 and HRP-HIV-1.

The HIV virus infects the cells of the immune system, primarily $CD4^+$ T lymphocytes decreasing CD4 levels from the normal values. An electrochemical magneto-actuated biosensor has been established for CD4 count in whole-blood (Carinelli et al., 2015). The $CD4^+$ T lymphocytes were isolated, preconcentrated, and labeled from whole-blood through immunomagnetic separation with anti-CD3 antibodies modified magnetic particles. After being labeled with a biotinylated anti-CD4 antibody, the captured cells reacted with the electrochemical reporter streptavidin–peroxidase conjugate.

5.5 URINARY TRACT INFECTION DIAGNOSIS

UTIs are mainly caused by bacteria. The most common UTI-causing organism is *Escherichia coli*, with 80%–85% of the cases originating from these bacteria. *Staphylococcus saprophyticus* are responsible for 5%–10% of UTI cases, and UTIs can also be caused by viral or fungal infections in some rare cases. The combination of pathogen identification with antimicrobial susceptibility testing (AST) was employed on a concept microfluidic simulator (Fig. 5.14; Dong and Zhao, 2015). The microfluidic device employs a fiberglass membrane with immobilized capture antibodies. In the microfluidic device, the chambers share the same geometric distribution in a standard 384-well microplate, resulting in compatibility with common microplate

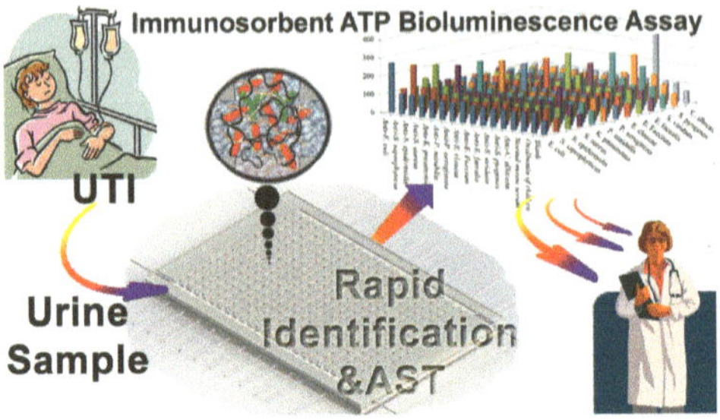

FIGURE 5.14 Rapid identification and susceptibility testing of uropathogenic microbes via immunosorbent ATP-bioluminescence assay on a microfluidic simulator for antibiotic therapy. *Recreated from Dong, T., Zhao, X., 2015. Rapid identification and susceptibility testing of uropathogenic microbes via immunosorbent ATP-bioluminescence assay on a microfluidic simulator for antibiotic therapy. Anal. Chem. 87, 2410–2418, with kind permission of American Chemical Society.*

readers. The microbes can be specifically captured by the antibodies and then quantified through an ATP bioluminescence assay. Owing to the design of the microfluidic device, the ATP-BLA sensitivity, and the antibody specificity, the simulator can identify UTI pathogen species and verify the antiseptic effects of eight antibiotic drugs reliably and simultaneously. The measurement range of the device changes from 1×10^3 to 1×10^5 cells mL^{-1} in urine samples.

Urinary lactoferrin (LTF), a 80 kDa iron-binding protein secreted by polymorphonuclear white blood cells, is a predictive urinary tract infection (UTI) biomarker since the average concentration of LTF was found to be 30.4 ng mL^{-1} in health urine and 3300 ng mL^{-1} in infected urine. An electrochemical immunosensor has been designed for direct detection of the UTI biomarker lactoferrin from infected clinical samples (Pan et al., 2010). Alkanethiolate self-assembled monolayer (SAM) was used for an electrochemical biosensor array. Meanwhile, a sandwich amperometric immunoassay was established for detection of lactoferrin from urine, with a detection limit of 145 pg mL^{-1}.

5.6 PREGNANCY SCREENING

Human chorionic gonadotropin (hCG) is a glycoprotein hormone composed of 244 amino acids with a molecular mass of 36.7 kDa. The hormone is produced by normal trophoblast cells of the placenta during pregnancy. A novel chemically-modified epitaxial graphene diagnostic sensor has been developed for ultrasensitive detection of the biomarker hCG (Teixeira et al., 2014). Multilayer epitaxial graphene (MEG) was grown on silicon carbide substrates and patterned using electron beam lithography to produce channel-based devices. Amine terminated MEG channels were obtained by using

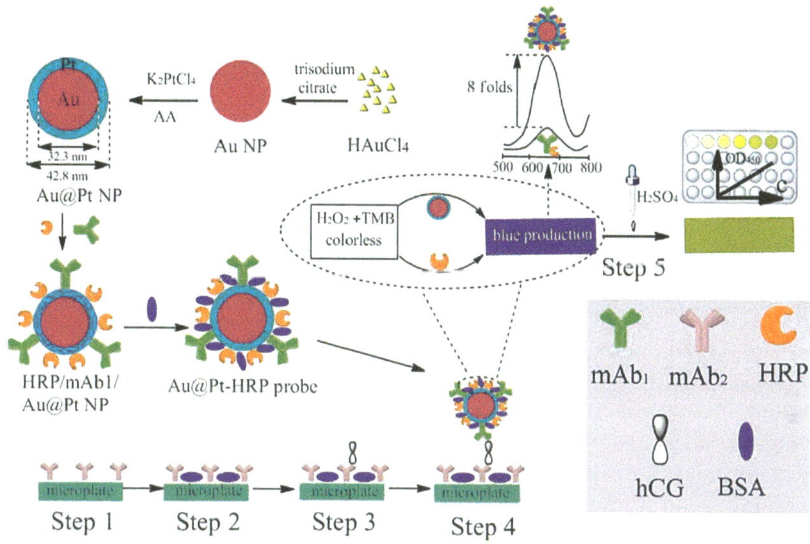

FIGURE 5.15 The colorimetric immunoassay protocol of Au@Pt-HRP probe. *Recreated from Wang, W., Zou, Y., Yan, J., Liu, J., Chen, H., Li, S., et al., 2018. Ultrasensitive colorimetric immunoassay for hCG detection based on dual catalysis of Au@Pt core–shell nanoparticle functionalized by horseradish peroxidase. Spectrochim. Acta A Mol. Biomol. Spectrosc. 193, 102–108, with kind permission of Elsevier.*

3-aminopropyl-triethoxysilane. After that, the anti-hCG antibody was attached to the channel through an amide bond in order to further bind with hCG. Binding of hCG was monitored by measuring the reduction of the channel current of the graphene biosensor. For the investigation of sensitivity of the sensor device, the changes in the channel resistance of the sensor were observed upon exposure to hCG. The detection limit of the sensor was 0.62 ng mL^{-1} with a linear response to hCG in the range 0.62–5.62 ng mL^{-1}.

Moreover, an ultrasensitive colorimetric biosensor for hCG detection has been constructed based on the dual catalysis of the horseradish peroxidase (HRP) and Au@Pt nanoparticles (NPs) related to H_2O_2-TEM system (Fig. 5.15; Wang et al., 2018). HRP and monoclonal mouse anti-hCG antibody were coimmobilized onto the Au@PtNP surface to form a dual functionalized Au@Pt-HRP probe. The Au@Pt-HRP probe showed high sensitivity in the concentration range of 0.4–12.8 IU L^{-1} with a low limit of detection of 0.1 IU L^{-1}.

In addition, hCG can be analyzed by a supramolecular amperometric immunosensor (Ortiz et al., 2014). For the supramolecular strategy, a carboxymethylcellulose carrier was synthesized and trifunctionalized with anti-hCG antibody, horseradish peroxidase (HRP), and ferrocene (Fc) moieties. The anti-hCG antibody was attached to a carboxymethylcellulose (CMC) carrier linked to the electrode via host–guest interactions with the cyclodextrin cavities. The ferrocene was linked with the functionalized CMC within the cavities of electrode surface immobilized cyclodextrin, via host–guest interactions. hCG was detected via a sandwich format, which forms an immunocomplex between the surface immobilized antibody and a glucose oxidase/lactate oxidase labeled secondary antibody. Following formation of the immunocomplex, lactate/glucose was added and the hydrogen peroxide was detected at the electrode surface via the HRP-Fc enzyme-mediator couple.

5.7 CONCLUSION

Various immuno-nanobiosensors have been extensively explored for monitoring and diagnosis of different diseases including cancer, tuberculosis, diabetes, human immunodeficiency virus, urinary tract infection, and pregnancy screening so on. Currently, the sandwich-type biosensor is mainly developed for the detection of protein biomarkers. In the future, there will be great opportunities for the exploration of other types of biosensors.

REFERENCES

Barroso, T.G., Martins, R.C., Fernandes, E., Cardoso, S., Rivas, J., Freitas, P.P., 2018. Detection of BCG bacteria using a magnetoresistive biosensor: a step towards a fully electronic platform for tuberculosis point-of-care detection. Biosens. Bioelectron. 100, 259–265.

Carinelli, S., Xufré Ballesteros, C., Martí, M., Alegret, S., Pividori, M.I., 2015. Electrochemical magneto-actuated biosensor for CD4 count in AIDS diagnosis and monitoring. Biosens. Bioelectron. 74, 974–980.

Chen, X., Jia, X., Han, J., Ma, J., Ma, Z., 2013. Electrochemical immunosensor for simultaneous detection of multiplex cancer biomarkers based on graphene nanocomposites. Biosens. Bioelectron. 50, 356–361.

Chopra, A., Rawat, S., Bhalla, V., Suri, C.R., 2014. Point-of-care amperometric testing of diabetic marker (HbA1c) using specific electroactive antibodies. Electroanalysis 26, 469–472.

de Oliveira, R.A.G., Materon, E.M., Melendez, M.E., Carvalho, A.L., Faria, R.C., 2017. Disposable microfluidic immunoarray device for sensitive breast cancer biomarker detection. ACS Appl. Mater. Interfaces 9, 27433–27440.

Dhakane, S., Patil, W.V., 2014. Piezo-resistive method for tuberculosis detection using microcantilever biosensor. Microsyst. Tech. 20, 457–462.

Diouani, M.F., Ouerghi, O., Refai, A., Belgacem, K., Tlili, C., Laouini, D., et al., 2017. Detection of ESAT-6 by a label free miniature immuno-electrochemical biosensor as a diagnostic tool for tuberculosis. Mat. Sci. Eng. C-Mater. 74, 465–470.

Dong, T., Zhao, X., 2015. Rapid identification and susceptibility testing of uropathogenic microbes via immunosorbent ATP-bioluminescence assay on a microfluidic simulator for antibiotic therapy. Anal. Chem. 87, 2410–2418.

Fang, Y., Huang, X., Zeng, Q., Wang, L., 2015. Metallic nanocrystallites-incorporated ordered mesoporous carbon as labels for a sensitive simultaneous multianalyte electrochemical immunoassay. Biosens. Bioelectron. 73, 71–78.

Feng, C., Bo, B., Mao, X., Shi, H., Zhu, X., Li, G., 2017. From interface to solution: integrating immunoassay with netlike rolling circle amplification for ultrasensitive detection of tumor biomarker. Theranostics 7, 31–39.

Gao, X., Zhang, Y., Chen, H., Chen, Z., Lin, X., 2011. Amperometric immunosensor for carcinoembryonic antigen detection with carbon nanotube-based film decorated with gold nanoclusters. Anal. Biochem. 414, 70–76.

Gopinath, S.C.B., Perumal, V., Kumaresan, R., Lakshmipriya, T., Rajintraprasad, H., Rao, B.S., et al., 2016. Nanogapped impedimetric immunosensor for the detection of 16 kDa heat shock protein against Mycobacterium tuberculosis. Microchim. Acta 183, 2697–2703.

Huang, H.X., Li, J., Shi, S.S., Yan, Y.Q., Zhang, M.Y., Wang, P.C., et al., 2015. Detection of interferon-gamma for latent tuberculosis diagnosis using an immunosensor based on CdS quantum dots coupled to magnetic beads as labels. Int. J. Electrochem. Sci. 10, 2580–2593.

Jing, W., Jiang, X., Zhao, W., Liu, S., Cheng, X., Sui, G., 2014. Microfluidic platform for direct capture and analysis of airborne *Mycobacterium tuberculosis*. Anal. Chem. 86, 5815–5821.

Jo, E.J., Mun, H., Kim, M.G., 2016. Homogeneous immunosensor based on luminescence resonance energy transfer for glycated hemoglobin detection using upconversion nanoparticles. Anal. Chem. 88, 2742–2746.

Kheiri, F., Sabzi, R.E., Jannatdoust, E., Shojaeefar, E., Sedghi, H., 2011. A novel amperometric immunosensor based on acetone-extracted propolis for the detection of the HIV-1 p24 antigen. Biosens. Bioelectron. 26, 4457–4463.

Kim, J., Lee, J., Lee, K.I., Park, T.J., Kim, H.J., Lee, J., 2013. Rapid monitoring of CFP-10 during culture of *Mycobacterium tuberculosis* by using a magnetophoretic immunoassay. Sens. Actuat. B 177, 327–333.

Kim, J.H., Chang, Y.W., Bok, E., Kim, H.J., Lee, H., Cho, S.N., et al., 2014. Detection of IFN-gamma for latent tuberculosis diagnosis using an anodized aluminum oxide-based capacitive sensor. Biosens. Bioelectron. 51, 366–370.

Kim, H.J., Rim, J., Jang, C.H., 2016. Diagnosis of tuberculosis using a liquid crystal-based optical sensor. Macromol. Res. 24, 123–130.

Kim, E.J., Kim, E.B., Lee, S.W., Cheon, S.A., Kim, H.J., Lee, J., et al., 2017. An easy and sensitive sandwich assay for detection of Mycobacterium tuberculosis Ag85B antigen using quantum dots and gold nanorods. Biosens. Bioelectron. 87, 150–156.

Li, C., Ma, J., Fan, Q., Tao, Y., Li, G., 2016. Dynamic light scattering (DLS)-based immunoassay for ultra-sensitive detection of tumor marker protein. Chem. Commun. 52, 7850–7853.

Li, N., Wang, Y., Cao, W., Zhang, Y., Yan, T., Du, B., et al., 2015. An ultrasensitive electrochemical immunosensor for CEA using MWCNT-NH_2 supported PdPt nanocages as labels for signal amplification. J. Mater. Chem. B 3, 2006–2011.

Li, T., Fan, Q., Liu, T., Zhu, X., Zhao, J., Li, G., 2010. Detection of breast cancer cells specially and accurately by an electrochemical method. Biosens. Bioelectron. 25, 2686–2689.

Liu, D., Huang, X., Wang, Z., Jin, A., Sun, X., Zhu, L., et al., 2013. Gold nanoparticle-based activatable probe for sensing ultralow levels of prostate-specific antigen. ACS Nano 7, 5568–5576.

Liu, D., Yang, J., Wang, H.F., Wang, Z., Huang, X., Wang, Z., et al., 2014. Glucose oxidase-catalyzed growth of gold nanoparticles enables quantitative detection of attomolar cancer biomarkers. Anal. Chem. 86, 5800–5806.

Liu, Z., Rong, Q., Ma, Z., Han, H., 2015. One-step synthesis of redox-active polymer/AU nanocomposites for electrochemical immunoassay of multiplexed tumor markers. Biosens. Bioelectron. 65, 307–313.

Masud, M.K., Yadav, S., Islam, M.N., Nguyen, N.T., Salomon, C., Kline, R., et al., 2017. Gold-loaded nanoporous ferric oxide nanocubes with peroxidase-mimicking activity for electrocatalytic and colorimetric detection of autoantibody. Anal. Chem. 89, 11005–11013.

Mdluli, P., Tetyana, P., Sosibo, N., van der Walt, H., Mlambo, M., Skepu, A., et al., 2014. Gold nanoparticle based tuberculosis immunochromatographic assay: the quantitative ESE quanti analysis of the intensity of test and control lines. Biosens. Bioelectron. 54, 1–6.

Molazemhosseini, A., Magagnin, L., Vena, P., Liu, C.C., 2016. Single-use disposable electrochemical label-free immunosensor for detection of glycated hemoglobin (HbA1c) using differential pulse voltammetry (DPV). Sensors 16, 1024.

Munge, B.S., Coffey, A.L., Doucette, J.M., Somba, B.K., Malhotra, R., Patel, V., et al., 2011. Nanostructured immunosensor for attomolar detection of cancer biomarker interleukin-8 using massively labeled superparamagnetic particles. Angew. Chem. Int. Ed. 50, 7915–7918.

Ortiz, M., Wilson, L., Botero, M.L., Baker, P., Iwuoha, E., Fragoso, A., et al., 2014. Supramolecular amperometric immunosensor for detection of human chorionic gonadotropin. Electroanalysis 26, 1481–1487.

Pan, Y., Sonn, G.A., Sin, M.L.Y., Mach, K.E., Shih, M.-C., Gau, V., et al., 2010. Electrochemical immunosensor detection of urinary lactoferrin in clinical samples for urinary tract infection diagnosis. Biosens. Bioelectron. 26, 649–654.

Pang, X., Li, J., Zhao, Y., Wu, D., Zhang, Y., Du, B., et al., 2015. Label-free electrochemiluminescent immunosensor for detection of carcinoembryonic antigen based on nanocomposites of GO/MWCNTs-COOH/Au@CeO_2. ACS Appl. Mater. Interfaces 7, 19260–19267.

Prabowo, B.A., Chang, Y.-F., Lai, H.-C., Alom, A., Pal, P., Lee, Y.-Y., et al., 2018. Rapid screening of *Mycobacterium tuberculosis* complex (MTBC) in clinical samples by a modular portable biosensor. Sens. Actuat. B 254, 742–748.

Ren, X., Ma, H., Zhang, T., Zhang, Y., Yan, T., Du, B., et al., 2017. Sulfur-doped graphene-based immunological biosensing platform for multianalysis of cancer biomarkers. ACS Appl. Mater. Interfaces 9, 37637–37644.

Singh, V., Krishnan, S., 2015. Voltammetric immunosensor assembled on carbon-pyrenyl nanostructures for clinical diagnosis of type of diabetes. Anal. Chem. 87, 2648–2654.

Soares, J.C., Iwaki, L.E.O., Soares, A.C., Rodrigues, V.C., Melendez, M.E., Fregnani, J.H.T.G., et al., 2017. Immunosensor for pancreatic cancer based on electrospun nanofibers coated with carbon nanotubes or gold nanoparticles. ACS Omega 2, 6975–6983.

Tang, C.K., Vaze, A., Shen, M., Rusling, J.F., 2016. High-throughput electrochemical microfluidic immunoarray for multiplexed detection of cancer biomarker proteins. ACS Sens. 1, 1036–1043.

Tang, J., Tang, D., Niessner, R., Chen, G., Knopp, D., 2011. Magneto-controlled graphene immunosensing platform for simultaneous multiplexed electrochemical immunoassay using distinguishable signal tags. Anal. Chem. 83, 5407–5414.

Teixeira, S., Burwell, G., Castaing, A., Gonzalez, D., Conlan, R.S., Guy, O.J., 2014. Epitaxial graphene immunosensor for human chorionic gonadotropin. Sens. Actuat. B 190, 723–729.

Thapa, A., Soares, A.C., Soares, J.C., Awan, I.T., Volpati, D., Melendez, M.E., et al., 2017. Carbon nanotube matrix for highly sensitive biosensors to detect pancreatic cancer biomarker CA19-9. ACS Appl. Mater. Interfaces 9, 25878–25886.

Wang, W., Zou, Y., Yan, J., Liu, J., Chen, H., Li, S., et al., 2018. Ultrasensitive colorimetric immunoassay for hCG detection based on dual catalysis of Au@Pt core–shell nanoparticle functionalized by horseradish peroxidase. Spectrochim. Acta A Mol. Biomol. Spectrosc. 193, 102–108.

Wang, X., Han, X.X., Ma, A.G., Chen, L., Liang, H., Litifu, A., et al., 2017. Fabrication of electrochemical immunosensor for interferon-γ determination and its application of tuberculosis diagnosis. Int. J. Electrochem. Sci. 7262–7271.

Wangoo, N., Swami, A., Kaur, S., Bansal, K., Sharma, R.K., 2016. Development of a colloidal gold-based nanobioprobe for the detection of glycated albumin. BioNanoScience 6, 132–138.

Wu, C.C., Lin, H.-I., Chang, K.W., Mai, J.D., Shiesh, S.C., Lee, G.B., 2015. Measurement of glycated hemoglobin levels using an integrated microfluidic system. Microfluid. Nanofluid. 18, 613–621.

Xu, S., Liu, Y., Wang, T., Li, J., 2011. Positive potential operation of a cathodic electrogenerated chemiluminescence immunosensor based on luminol and graphene for cancer biomarker detection. Anal. Chem. 83, 3817–3823.

Xu, T., Jia, X., Chen, X., Ma, Z., 2014. Simultaneous electrochemical detection of multiple tumor markers using metal ions tagged immunocolloidal gold. Biosens. Bioelectron. 56, 174–179.

Xu, T., Liu, N., Yuan, J., Ma, Z., 2015. Triple tumor markers assay based on carbon-gold nanocomposite. Biosens. Bioelectron. 70, 161–166.

Xue, Q., Bian, C., Tong, J., Sun, J., Zhang, H., Xia, S., 2011. A micro potentiometric immunosensor for hemoglobin-A1c level detection based on mixed SAMs wrapped nano-spheres array. Biosens. Bioelectron. 26, 2689–2693.

Zhang, Y., Yang, H., Yu, J., Wei, H., 2016. Rapid and sensitive detection of HIV-1 p24 antigen by immunomagnetic separation coupled with catalytic fluorescent immunoassay. Anal. Bioanal. Chem. 408, 6115–6121.

Zhao, W.W., Han, Y.M., Zhu, Y.C., Zhang, N., Xu, J.J., Chen, H.Y., 2015. DNA labeling generates a unique amplification probe for sensitive photoelectrochemical immunoassay of HIV-1 p24 antigen. Anal. Chem. 87, 5496–5499.

Zhou, B., Zhu, M., Hao, Y., Yang, P., 2017. Potential-resolved electrochemiluminescence for simultaneous determination of triple latent tuberculosis infection markers. ACS Appl. Mater. Interfaces 9, 30536–30542.

Zhou, J., Gan, N., Li, T., Hu, F., Li, X., Wang, L., et al., 2014. A cost-effective sandwich electrochemiluminescence immunosensor for ultrasensitive detection of HIV-1 antibody using magnetic molecularly imprinted polymers as capture probes. Biosens. Bioelectron. 54, 199–206.

Zhu, Y., Wang, H., Wang, L., Zhu, J., Jiang, W., 2016. Cascade signal amplification based on copper nanoparticle-reported rolling circle amplification for ultrasensitive electrochemical detection of the prostate cancer biomarker. ACS Appl. Mater. Interfaces 8, 2573–2581.

Chapter 6

Aptasensors

Juan Zhang and Ya Cao

Center for Molecular Recognition and Biosensing, School of Life Sciences, Shanghai University, Shanghai, P. R. China

Chapter Outline

Aptamers are oligonucleotides (DNA or RNA) that can be obtained in vitro by Systematic Evolution of Ligands by Exponential enrichment (SELEX) from the libraries of random RNA or DNA sequences. The selection starts with a large population of single-stranded nucleic acid molecules, followed by an iterative in vitro selection by challenging with certain targets. After subsequent amplifications through polymerase chain reactions, the aptamers are enriched and have high affinity and specificity to their targets (Tombelli et al., 2005).

Aptamers can form many three-dimensional structures through intramolecular interactions, such as hydrophobic interactions, hydrogen bonds, and van der Waals interactions (Gijs et al., 2016). They show high and robust affinity to their targets that are comparable or even better than the affinity of monoclonal antibodies. So the aptamers have been recognized as "chemical antibodies" due to many of the following advantages: high specificity, affinity, and solubility, ease of chemical modifications, high reproducibility, and low costs for mass production (Bouchard et al., 2010; Mayer, 2009).

Nano-inspired Biosensors for Protein Assay with Clinical Applications.
DOI: https://doi.org/10.1016/B978-0-12-815053-5.00006-4

As oligonucleotides that can bind to specific target molecules, aptamers have been widely applied as diagnostic and therapeutic tools. As a kind of molecular recognition unit, aptamers have been extensively applied for the construction of biosensors. This kind of biosensors can be also called as aptasensors. The aptasensors have shown great potential for the diagnostics and treatments of different diseases. According to the recognition mechanism, aptasensors can be classified into (A) sandwich aptasensor, (B) aptasensor based on aptamer conformational switch, (C) aptasensor based on competitive binding of target with aptamer, (D) split-type assay aptasensor, and (E) aptasensor based on aptamer–cell interaction.

6.1 SANDWICH APTASENSOR

In a sandwich structured aptasensor, the target should have two or more recognition elements including aptamer; one is utilized as the capturing element and catches target molecules, and the other one serves as a probing element to mark the target with active molecules or nanoparticles. Sandwich structure strategy offers the advantages of high sensitivity and simple operation for biosensor fabrication when compared to the strategy by using only one recognition element to capture and label the target molecules.

6.1.1 Cancers

Cancer is the second largest disease throughout the world with an increasing mortality rate over the past few years. The patient's survival rate is uncertain due to the limitations of cancer diagnosis and therapy. Early diagnosis of cancer is decisive for its successful treatment. A biomarker-based cancer diagnosis may significantly improve the early diagnosis and subsequent treatment. Nanobiosensors play a crucial role in the detection of biomarkers as they are easy to use, portable, and can carry out analysis in real time.

Carcinoembryonic antigen (CEA), a glycoprotein, is a wide-spectrum tumor marker for cancer diagnosis. A label-free and lectin-based sandwich electrochemical aptasensor for CEA detection has been developed based on concanavalin A and a DNA aptamer against CEA (Wang et al., 2018). A general electrochemical signal transduction strategy for aptasensor has also been proposed based on an aptamer-induced electrochemical current due to the reaction of phosphate groups contained in the aptamer with molybdate to form redox molybdophosphate precipitate (Si et al., 2017).

Human epidermal growth factor receptor-2 (HER2) plays a crucial role in promoting breast cancer cell proliferation and malignant growth. A novel and sensitive electrochemical aptasensing platform has been established for the detection of the breast cancer biomarker (Yang et al., 2018). For this electrochemical HER2 aptasensing platform, RecJf exonuclease-catalyzed target recycling strategy was applied. Furthermore, a capacitive aptasensor has also been explored based on an interdigitated electrode (Arya et al., 2018). For the biosensor development, gold IDμE chips were functionalized with DNA aptamers via self-assembly.

As one of the critical growth factors found in human platelets, platelet-derived growth factor (PDGF), has increased importance due to its role in the regulation of cell growth and division. PDGF is composed of two disulfide-linked polypeptide chains (A and B), and occurs in three isoforms (PDGF-BB, PDGF-AB, and PDGF-AA). Among these isoforms, PDGF-BB is known to be directly implicated in the cell transformation process and in tumor growth and progression. Moreover, PDGF-BB is considered as a good predictor for early deterioration of renal function. An electrochemiluminescence aptasensor has been fabricated based on in situ growth of gold nanoparticles on layered molybdenum disulfide for sensitive detection of platelet-derived growth factor-BB by using MoS_2-AuNP composites as the matrix and quantum dots as the signal probe (Liu et al., 2014). Meanwhile, a luminol electrochemiluminescence aptasensor has also been reported based on glucose oxidase modified gold nanoparticles for measurement of platelet-derived growth factor BB (Zhang et al., 2015).

It is well known that highly sensitive and selective detection of cancer cells is still a challenge for the diagnosis and treatment of cancer. For cellular study, aptamers have been used to distinguish different cell lines and identify the native protein targets on the cell surface. Sandwich aptasensors have been extensively explored for detection of cancer cells. A flow injection amperometric sandwich-type electrochemical aptasensor has been explored for the determination of adenocarcinoma gastric cancer cell using aptamer-Au@Ag nanoparticles as labeled aptamer (Amouzadeh Tabrizi et al., 2017). For the aptasensor, MWCNT-Au nano was used as a nanoplatform to immobilize primary thiolated aptamer and the secondary aptamer-Au@Ag nanoparticles as the labeled aptamers to increase electrocatalysis for reduction of H_2O_2 in a flow injection microchannel device. An aptasensor has also been constructed for sensitive detection of human breast cancer cells by using porous GO/Au composites and porous PtFe alloy as effective sensing platform and signal amplification labels (Yan et al., 2013). Moreover, human liver hepatocellular carcinoma cells were detected using a label-free aptasensor (Kashefi-Kheyrabadi et al., 2014). The target cells were captured in a sandwich architecture using TLS11a aptamer covalently attached to a gold surface and a secondary TLS11a aptamer. The application of TLS11a aptamer as a recognition layer resulted in a sensor with high affinity to HepG2 cancer cells (Fig. 6.1).

Furthermore, a repeatable assembling and disassembling electrochemical aptamer cytosensor was proposed for the sensitive detection of human liver hepatocellular carcinoma cells (HepG2) based on a dual recognition and signal amplification strategy (Fig. 6.2; Sun et al., 2015). With the assistance of TLS11a aptamer and G-quadruplex/hemin/aptamer-AuNPs-HRP nanoprobes,

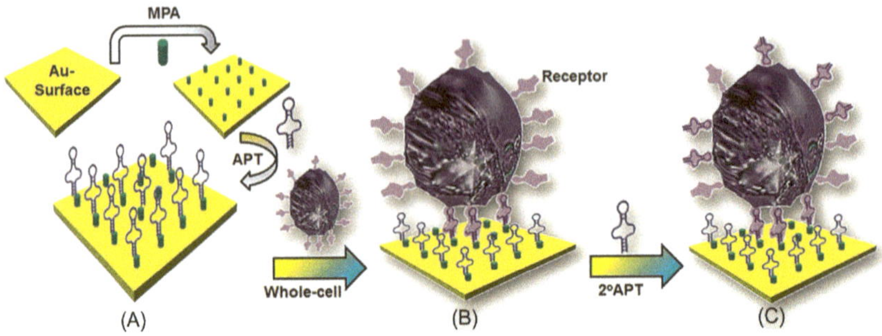

FIGURE 6.1 (A) TLS11a aptamer conjugated Au surface via coupling of amino-labeled aptamer to the MPA–Au surface. (B) Interaction of HepG2 cells on Au surface-immobilized aptamer. (C) Binding of secondary aptamer to complete the HepG2 cells sandwich format. *Recreated from Kashefi-Kheyrabadi, L., Mehrgardi, M.A., Wiechec, E., Turner, A.P.F., Tiwari, A., 2014. Ultrasensitive detection of human liver hepatocellular carcinoma cells using a label-free aptasensor. Anal. Chem. 86, 4956–4960, with kind permission of American Chemical Society.*

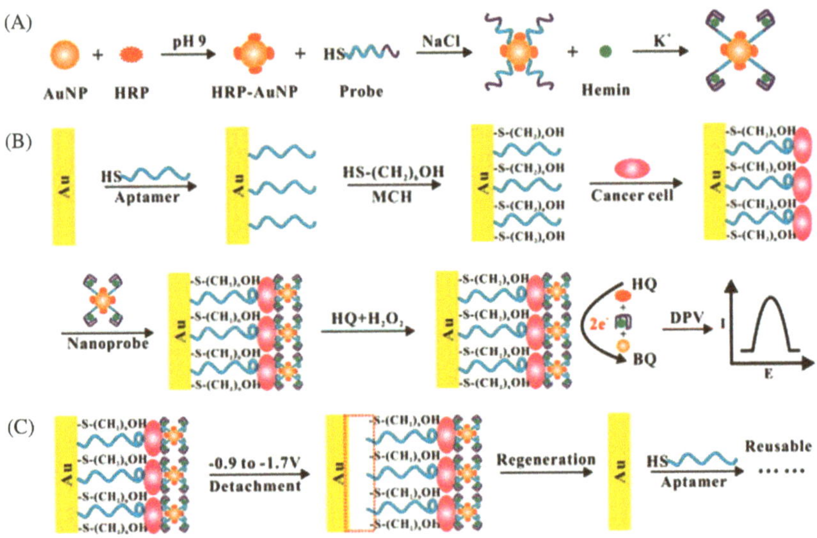

FIGURE 6.2 (A) Procedures for the fabrication of G-quadruplex/hemin/aptamer–AuNPs–HRP nanoprobe. (B) Schematic illustration of the novel electrochemical aptamer cytosensor for cancer cell detection. (C) Procedures of Au–thiol bond detached from a gold electrode surface with a negative voltage pulse, and the regeneration of the gold electrode surfaces for the reused repeatedly. *Recreated from Sun, D., Lu, J., Chen, Z., Yu, Y., Mo, M., 2015. A repeatable assembling and disassembling electrochemical aptamer cytosensor for ultrasensitive and highly selective detection of human liver cancer cells. Anal. Chim. Acta 885, 166–173, with kind permission of Elsevier.*

the HepG2 cancer cells were captured to fabricate an aptamer–cell–nanoprobes sandwich-like superstructure on a gold electrode surface. Both HRP and G-quadruplex/hemin HRP-mimicking DNAzyme could be used for the enzyme catalysis and electrochemical signal amplification. The signal of the electrochemical aptasensor can be obtained by measuring the increasing reduction current generated by hydroquinone and benzoquinone, which was catalyzed by HRP and hemin/G-quadruplex DNAzyme in the presence of H_2O_2.

Additionally, an electrochemical aptasensor was developed by fabricating an aptamer–cell–aptamer sandwich architecture on an SBA-15-3-aminopropyltriethoxysilane (SBA-15-pr-NH2) and Au nanoparticles (AuNPs)-modified graphite screen-printed electrode (GSPE) surface for the label-free detection of CT26 cancer cells (Fig. 6.3; Hashkavayi et al., 2017). The thiol-terminated aptamer, with special binding tendency to colorectal cancer cells, was self-assembled onto AuNPs/SBA-15-pr-NH2/SPE surface and applied in sandwich architecture. Apt/AuNPs/SBA-15-pr-NH_2/SPE was used as a nanoscale anchorage substrate to efficiently grab cells on SPE surface. Finally, the cells trapped by the aptamer were treated again with aptamer to form a sandwich structure.

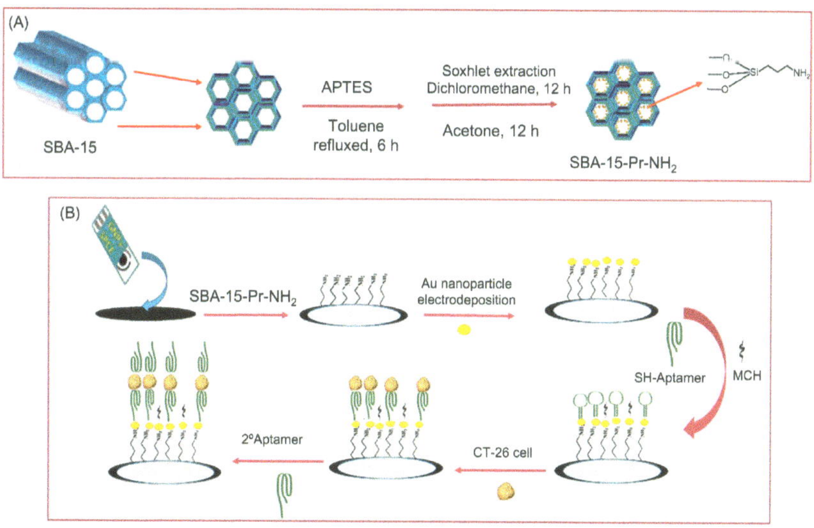

FIGURE 6.3 (A) Synthesis steps of SBA-15-pr-NH_2. (B) Schematic presentation of different steps of aptasensor construction. *Recreated from Hashkavayi, A.B., Raoof, J.B., Ojani, R., Kavoosian, S., 2017. Ultrasensitive electrochemical aptasensor based on sandwich architecture for selective label-free detection of colorectal cancer (CT26) cells. Biosens. Bioelectron. 92, 630–637, with kind permission of Elsevier.*

6.1.2 Tuberculosis

Tuberculosis (TB) is an infectious disease mainly caused by *Mycobacterium tuberculosis*. The emergence of multidrug resistance and human immunodeficiency virus infection are increasing tuberculosis prevalence rapidly. The specific IS6110 DNA fragment of MTB has been validated to be an important molecular marker for the diagnosis of TB. A polyaniline-reduced graphene oxide nanocomposite has been constructed as a redox nanoprobe in a voltammetric DNA biosensor for the detection of MTB (Chen et al., 2017). The composite was decorated with gold nanoparticles onto which the signal probe was immobilized to form the tracer label. The polyaniline-reduced graphene oxide redox probe can give the voltammetric signal after hybridization between target DNA and tracer label.

6.1.3 Diabetes

As one of the worldwide health problems, diabetes is classed as a metabolism disorder. According to International Diabetes Federation, one person dies from diabetes related causes every 6 s. Blood glycated hemoglobin (HbA1c) reflects the average blood glucose level in the proceeding 2–3 months, and has been recommended for screening/diagnosis and patient management of diabetes. An automated microfluidic system capable of rapid measurements of HbA1c concentrations in blood has been explored (Li et al., 2016). The system features a dual-aptamer assay through the usage of high affinity and high specificity to HbA1c. The dual-aptamer assay was implemented on magnetic beads, leading to the realization for the capturing of target molecules and fluorescent simultaneous detection of the hemoglobins and glycated hemoglobins (HbA1c) in a complete diagnosis procedure.

6.2 ASSAY BASED ON APTAMER CONFORMATIONAL SWITCH

A classic biosensor directly transduces ligand-target binding events into a measurable physical readout. The biosensing strategies have been explored based on the coupled ligand-induced structural switching of biomolecules with advanced transducers. Through the usage of target-responsive DNA structures, a series of nucleic acid sensors can employ surface-confined DNA structures with appropriate labels. These sensors can monitor target-induced structural switching of DNA or aptamer-specific small molecule probes.

6.2.1 Cancers

For the assay of cancer protein marker, an ultrasensitive electrochemical aptasensor dependent on shell-encoded AuNPs, has been proposed for the specific dual disease biomarkers detection (Zhao et al., 2018). Through the deposition of different amounts of Cu and Ag precursors, shell-encoded AuNPs involves Au@Cu_2O core–shell NPs and Au@Ag core–shell NPs. Cu_2O shell and Ag shell-encoded AuNPs exhibited noninterfering DPV responses at −0.08 V and 0.26 V, respectively. Meanwhile, the increasing thickness of shells can largely amplify the currents. The adoption of shell-encoded Au@Cu_2O core–shell NPs and Au@Ag core–shell NPs could dramatically improve the performance of electrochemical aptasensors. Shell-encoded AuNPs engineered electrochemical aptasensors offered multiplexing capability and demonstrated exquisite specificity owing to high affinity and specificity of aptamers to the targets. Moreover, a turn-on and label-free aptasensor for protein detection has also been developed based on target-induced formation of Ag nanoclusters for the detection of PDGF-BB (Fig. 6.4; Liu et al., 2012b). The region I consisting of the aptamer sequence of PDGF-BB, was blocked by hybridization with region II. P2 also was composed of two domains named as III and IV. The region IV was AgNCs nucleation sequence and could be blocked by hybridization with region III. Meanwhile, the region III of P2 was partially complementary to the region II of P1. Both probes were in the closed form without PDGF-BB so as to prevent the formation of the fluorescent AgNCs. In contrast, in the presence of PDGF-BB, the binding of the target molecule induced the aptamer region of the P1 to fold into an expected three-way helix junction, leading to the release of the region II at the 3-end. Subsequently, the region II of P1 can hybridize to the region III of P2 and release the AgNCs nucleation sequence. In this situation, AgNCs can be formed via the reduction of Ag^+ by $NaBH_4$.

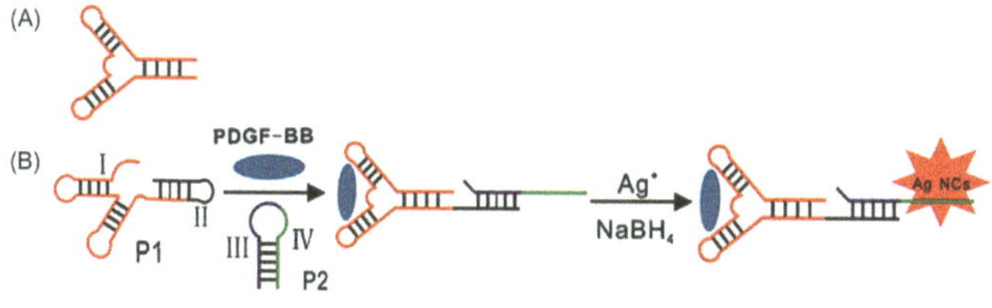

(C) Aptamer: 5'-CAGGCTACGGCACGTAGAGCATCACCATG ATCCTG-3'
P1: 5'-CAGGCTACGGCACGTAGAGCATCACCATG ATCCTGATTAGTTCCTCAGCATCAGG-3'
P2: 5'-CCCACCCTCCCACCTGATGCTGAGGGT-3'

FIGURE 6.4 (A) Secondary structure of anti-PDGF aptamer. (B) Schematic illustration of the sensing procedure for analysis of PDGF-BB based on the target induced formation of Ag NCs. (C) The sequences of the anti-PDGF-BB aptamer and the used nucleic acid strands. *Recreated from Liu, J.J., Song, X.R., Wang, Y.W., Zheng, A.X., Chen, G.N., Yang, H.H., 2012b. Label-free and fluorescence turn-on aptasensor for protein detection via target-induced silver nanoclusters formation. Anal. Chim. Acta 749, 70–74, with kind permission of Elsevier.*

Vascular endothelial growth factor (VEGF) is a key regulator of physiological angiogenesis during embryogenesis, skeletal growth, and reproductive functions. VEGF has also been implicated in pathological angiogenesis. For the detection of cancer protein marker, VEGF, a variety of nanobiosensors have been developed. Cho et al. (2012) presented a single-step aptamer-based surface-enhanced fluorescent optical sensor, through the combination of an aptamer-target interaction for target recognition and a nanoplasmonic fluorophore interaction for signal enhancement. As shown in Fig. 6.5, unfolded VEGF165 aptamer was electrostatically bound to a positively charged PLL-coated gold nanoparticle (GNP) surface. The surface-enhanced fluorescence (SEF) of Cy3B conjugated with the VEGF165 aptamer was created. The interaction of the VEGF165 aptamer with its target induced the reversible conformation change of the aptamer so as to decrease the electrostatic binding force. As a result, the target-binding interaction of the aptamer caused the irreversible detachment of the aptamer from the GNP surface and avoided the SEF effect of Cy3B. Additionally, Fu et al. (2016) synthesized the DNA-templated Ag/Pt bimetallic nanoclusters and detected VEGF by using the electrochemical biosensor based on the nanocluster.

Human prostatic epithelial cells constitutively secrete prostate-specific antigen (PSA). As a kallikrein-like serine protease, PSA is a normal component of the seminal plasma. PSA is currently used as a specific diagnostic marker for the early detection of prostate cancer. Liu et al. (2012a) detected the PSA using an aptasensor with gold nanoparticles encapsulated by graphitized mesoporous carbon. Argoubi et al. (2018) reported the development of a novel label-free electrochemical aptasensing platform to detect PSA based on mesoporous silica thin film-coated gold electrodes as sensing interface. The aptasensing approach relied on the inhibition of the diffusion of $[Fe(CN)_6]^{3/4-}$ redox probe through the nanochannels of the mesoporous film gated by the recognition of the target biomarker via its anti-PSA specific

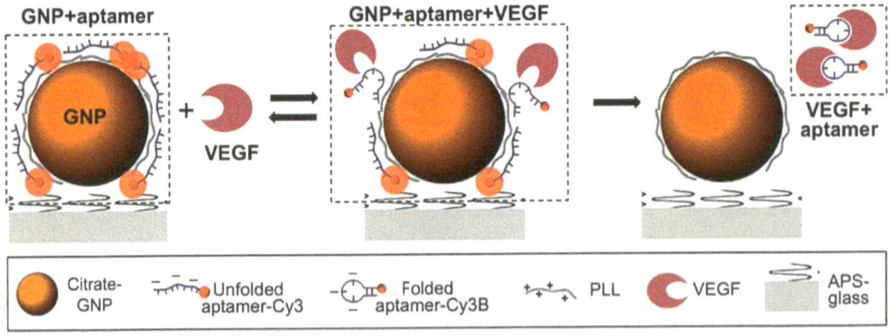

FIGURE 6.5 Single-step nanoplasmonic VEGF165 aptasensor for early cancer diagnosis. *Recreated from Cho, H., Yeh, E.C., Sinha, R., Laurence, T.A., Bearinger, J.P., Lee, L.P., 2012. Single-step nanoplasmonic VEGF165 aptasensor for early cancer diagnosis. ACS Nano 6, 7607–7614, with kind permission of American Chemical Society.*

DNA aptamer attached onto the outer surface of the silica nanopores. Benvidi et al. (2018) developed an impedimetric PSA aptasensor based on the usage of a glassy carbon electrode modified with titanium oxide nanoparticles and silk fibroin nanofibers.

As a protein marker of cancer, carcinoembryonic antigen (CEA) has been extensively investigated by using the nanobiosensor based on aptamer conformational switch. Wu et al. (2015) constructed an aptasensor for CEA on the basis of fluorescence resonance energy transfer between upconversion phosphors (UCPs) and carbon nanoparticles (CNPs). Polyacrylic acid (PAA)-capped UCPs (NaYF4:Yb, Er) covalently tagged with an amino group modified CEA aptamer, were employed as the energy donor. The fluorescence of UCPs was dramatically quenched by CNPs due to the stacking interaction between CEA aptamer and CNPs. The interaction brought UCPs and CNPs in close proximity and triggered the energy transfer from UCPs to CNPs. With the introduction of CEA protein into the UCPs−aptamer−CNPs complex, the CEA aptamer preferentially interacted with the protein, leading to a structural change of the aptamer which was unfavorable for the stacking. As a result, the energy transfer process was blocked and the fluorescence of UCPs was recovered. Huang et al. (2018) developed an electrochemical aptasensor for high-sensitivity determination of CEA based on lead ion (Pb^{2+})-dependent DNAzyme-assisted signal amplification and graphene quantum dot-ionic liquid-Nafion (GQDs-IL-NF) composite film (Fig. 6.6). In the presence of CEA, hairpin DNA recognized the target and performed a

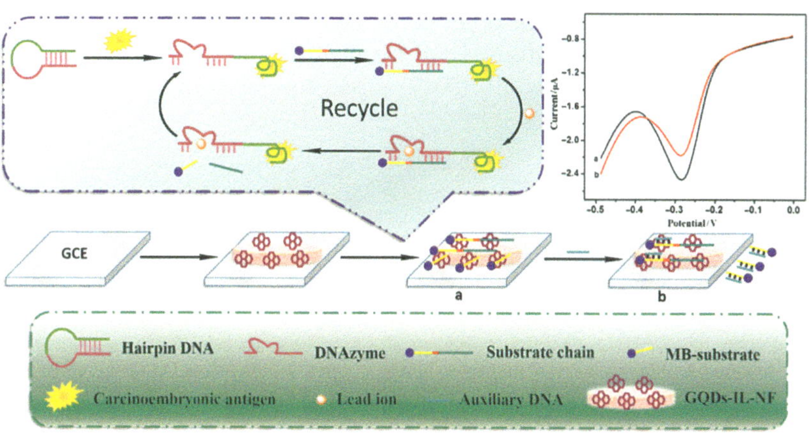

FIGURE 6.6 Schematic diagram of the GQDs-IL-NF nanomatrix and DNAzyme-assisted recycling of target-aptamer complex for sensitive electrochemical detection of CEA. *Recreated from Huang, J.Y., Zhao, L., Lei, W., Wen, W., Wang, Y.J., Bao, T., et al., 2018. A high-sensitivity electrochemical aptasensor of carcinoembryonic antigen based on graphene quantum dots-ionic liquid-nafion nanomatrix and DNAzyme-assisted signal amplification strategy. Biosens. Bioelectron. 99, 28–33, with kind permission of Elsevier.*

DNAzyme-assisted signal amplification reaction to give a large number of single-stranded DNA. The GQDs-IL-NF composite film was immobilized on the glassy carbon electrode through noncovalent $\pi-\pi$ stacking interaction with single-stranded DNA. Moreover, the methylene blue-labeled substrate DNA was fixed on the electrode and exhibited an initial electrochemical signal.

Aptamer conformational switch has also been utilized for the construction of colorimetric cytosensor. Yu et al. (2016) developed an easy and intuitive dispersion dominated colorimetric strategy for cancer cell detection based on combining multi-DNA released from an aptamer scaffold with cyclic enzymatic amplification, which was triggered by aptamer DNA conformational switch and demonstrated by noncross-linking gold nanoparticles (AuNPs) aggregation. Zhang et al. (2014) detected cancer cells by a colorimetric aptasensor based on cell-triggered cyclic enzymatic signal amplification (Fig. 6.7). In the absence of target cells, hairpin aptamer probes (HAPs) and linker DNAs stably coexisted in solution, and the linker DNA assembled DNA-AuNPs, giving a purple solution. In the presence of target cells, the specific binding of HAPs to the target cells triggered a conformational switch, leading to the linker DNA hybridization and cleavage by nicking endonuclease-strand scission cycles. Consequently, the cleaved fragments of linker DNA can no longer assemble into DNA AuNPs, causing the appearance of a red color.

Except for colorimetric cytosensor, fluorescent cytosensor has been numerously established based on fluorescence quenching, Förster resonance energy transfer (FRET), and fluorescence enhancement effect with the aid of nanoparticles. Xie et al. (2014) developed a GO-based aptasensor for cancer cell detection by taking advantage of the high fluorescence quenching efficiency of GO. Xiao et al. (2017) also developed a label-free and

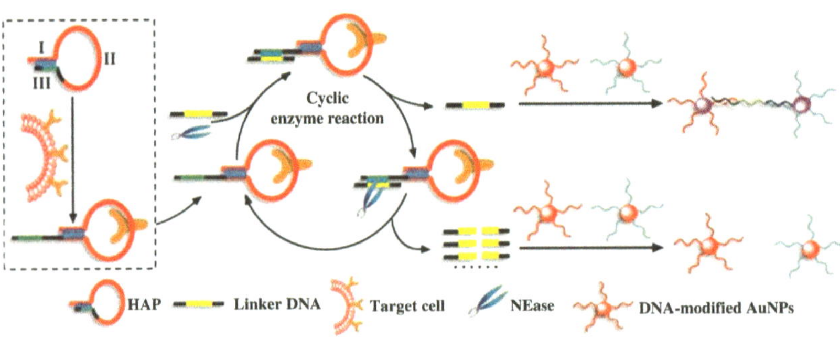

FIGURE 6.7 Schematic illustration of a highly sensitive colorimetric method for the detection of rare cancer cells based on CTCESA. *Recreated from Zhang, X., Xiao, K., Cheng, L., Chen, H., Liu, B., Zhang, S., et al., 2014. Visual and highly sensitive detection of cancer cells by a colorimetric aptasensor based on cell-triggered cyclic enzymatic signal amplification. Anal. Chem. 86, 5567–5572, with kind permission of American Chemical Society.*

high-efficient GO-based aptasensor for the detection of low quantity cancer cells based on cell-triggered cyclic enzymatic signal amplification (Fig. 6.8). Without target cells, hairpin aptamer probes (HAPs) and dye-labeled linker DNAs stably coexisted in solution, and the fluorescence was quenched by the GO-based FRET process. On contrary, the specific binding of HAPs with the target cells triggered a conformational alternation, resulting in linker DNA complementary pairing and cleavage by nicking endonuclease-strand scission cycles. Thus, the enhanced fluorescence can be obtained as a result of the formation of more cleaved fragments of linker DNAs with more of the

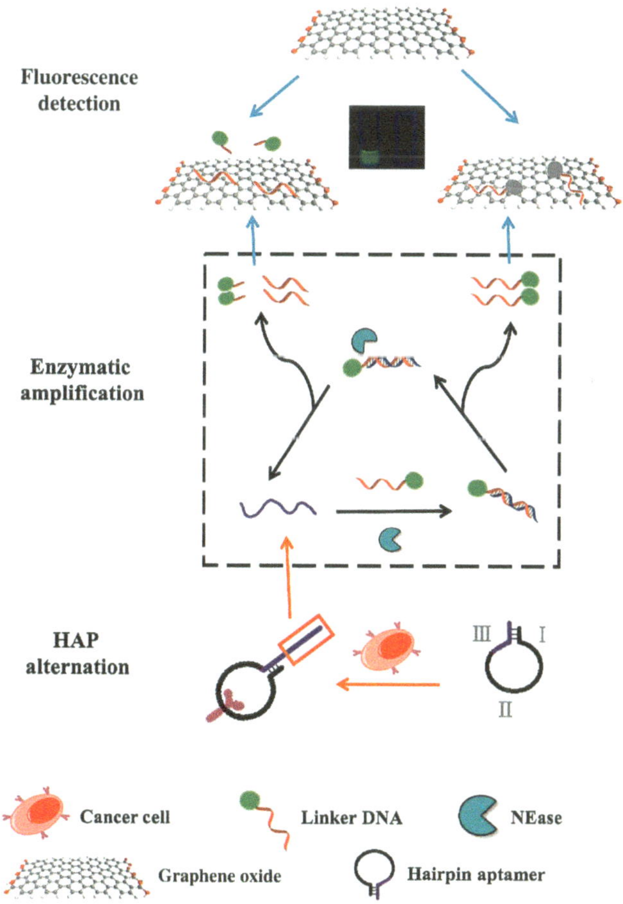

FIGURE 6.8 Schematic illustration of a label-free and high-efficient GO-based aptasensor for the detection of low quantity cancer cells based on CTCESA. *Recreated from Xiao, K., Liu, J., Chen, H., Zhang, S., Kong, J., 2017. A label-free and high-efficient GO-based aptasensor for cancer cells based on cyclic enzymatic signal amplification. Biosens. Bioelectron. 91, 76–81, with kind permission of Elsevier.*

terminal labeled dyes, because these cleaved DNA fragments hardly combined with GOs and prevented the FRET process.

Yin et al. (2013) presented a label-free and turn-on aptamer strategy for cancer cell detection based on the recognition-induced conformation alteration of aptamer and hybridization-induced fluorescence enhancement effect of DNA-silver nanoclusters (DNA-AgNCs) in the proximity of guanine-rich DNA sequences (Fig. 6.9). This strategy involved two tailored DNA probes. One was designed as a hairpin shaped structure composed of a target specific aptamer sequence at the 3′-end, a guanine-rich DNA sequence, and an arm segment at the 5′-end. The other was served as a signal probe, and contained a sequence for AgNCs templated synthesis and a link sequence complementary to the arm segment of the recognition probe. Recognizing and binding of the aptamer to cancer cells enforced the recognition probe to undergo a conformational alteration, so as to initiate hybridization between the arm segment of the recognition probe and the link sequence of the signal probe. Finally, an enhanced fluorescence readout was obtained through the approaching of AgNCs to the guanine-rich DNA.

NIR-driven PEC biosensor has also been further explored. NaYF4:Yb, Er particles were combined with $CdTe/TiO_2$ heterostructure to form a NIR-light-motivated PEC interface (Fig. 6.10; Wang et al., 2016). The upconversion nanoparticles and $TiO_2/CdTe$ heterostructure were combined to prepare the film electrode, and the high-affinity aptamer AS1411 was conjugated to the electrode to recognize MCF-7 cells. The upconversion nanoparticles

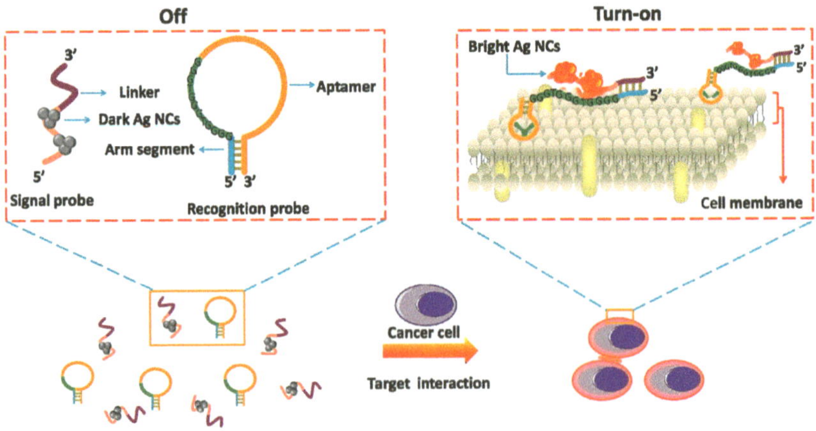

FIGURE 6.9 Schematic representation of the label-free and turn-on aptamer strategy for cancer cell detection based on DNA–silver nanocluster fluorescence upon recognition-induced hybridization. *Recreated from Yin, J., He, X., Wang, K., Xu, F., Shangguan, J., He, D., et al., 2013. Label-free and turn-on aptamer strategy for cancer cells detection based on a DNA-silver nanocluster fluorescence upon recognition-induced hybridization. Anal. Chem. 85, 12011–12019, with kind permission of American Chemical Society.*

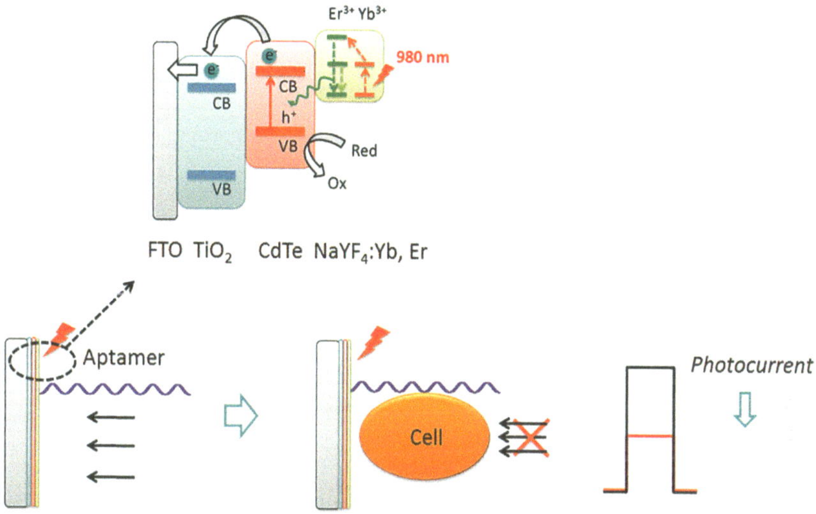

FIGURE 6.10 Illustration of the NIR-light-excited PEC interface and the MCF-7 cell detection principle. *Recreated from Wang, K., Zhang, R., Sun, N., Li, X., Wang, J., Cao, Y., et al., 2016. Near-infrared light-driven photoelectrochemical aptasensor based on the upconversion nanoparticles and TiO_2/CdTe heterostructure for detection of cancer cells. ACS Appl. Mater. Interfaces 8, 25834–25839, with kind permission of American Chemical Society.*

transformed the near-infrared light to visible light, so as to excite the semiconductor to enhance the current response. The affinity interaction between the immobilized aptamers and the target cells was explored for the cell detection, and the resultant photocurrent intensity was related directly with the number of cells.

As well as colorimetric cytosensor and fluorescent cytosensor, electrochemical cytosensor has also been proposed. A novel sensing strategy for sensitive detection of mucin 1 protein (MUC1) and MCF-7 cells has been established based on electrochemiluminescence (ECL) resonance energy transfer (ERET) from bis(2,20-bipyridine)-(5-aminophenanthroline)ruthenium(II) (Ru1) to GO (Wei et al., 2012). Moreover, an electrochemical impedance aptasensor by MUC-1 aptamer conjugated nanomaterials of CNSs has also been explored as a sensing platform for tumor targeting (Cao et al., 2014).

6.2.2 Tuberculosis

Tuberculosis is a global pandemic. Rapid identification of *Mycobacterium tuberculosis* (MTB) culture isolate is very important for drug susceptibility testing. Thakur et al. established a specific electrochemical aptaelectrode for the detection of MTB surface antigen MPT64 using a hybrid nanocomposite of chitosan, iron oxide nanoparticles, and graphene (Thakur et al., 2017). Bai

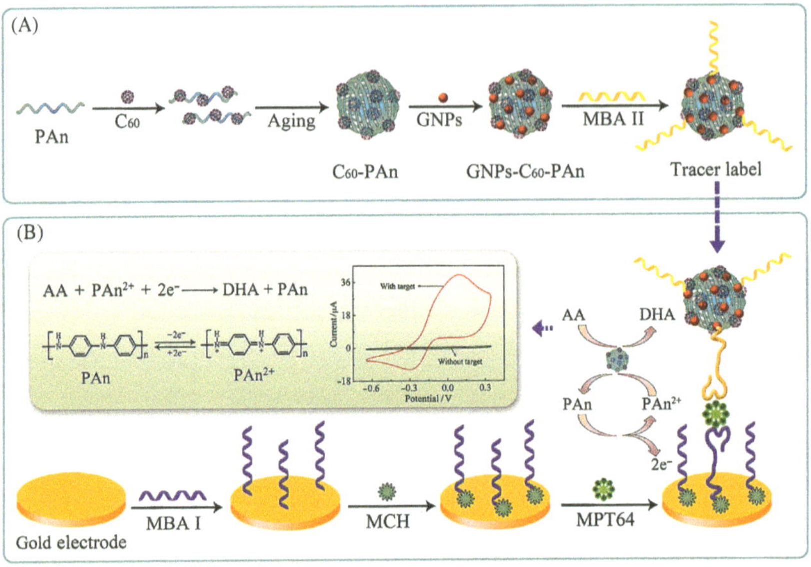

FIGURE 6.11 (A) Illustration of the preparation procedure of the tracer label. (B) Schematic diagram of the preparation of the electrochemical aptasensor. *Recreated from Bai, L., Chen, Y., Bai, Y., Chen, Y., Zhou, J., Huang, A., 2017. Fullerene-doped polyaniline as new redox nanoprobe and catalyst in electrochemical aptasensor for ultrasensitive detection of Mycobacterium tuberculosis MPT64 antigen in human serum. Biomaterials. 133, 11–19, with kind permission of Elsevier.*

et al. (2017) used fullerene-doped polyaniline as new redox nanoprobe and catalyst in an electrochemical aptasensor for ultrasensitive detection of MTB MPT64 antigen in human serum (Fig. 6.11). The electrochemical response signal was produced from PAn in C60-PAn after the sandwich reaction between MPT64 antigen and the tracer label. The large conjugated structure in polymer chain and reversible electron transfer between PAn and PAn2þ contributed to the outstanding peak current response.

Interferon-gamma (IFN-γ) is a biological marker for diagnosis of latent tuberculosis in serum sample. Taghdisi et al. (2017) established an amplified fluorescent aptasensor based on single-stranded DNA binding protein, copper, and silica nanoparticles for sensitive detection of IFN-γ.

6.2.3 Human Immunodeficiency Virus (HIV)

HIV has caused great horror all over the world with the emergence of one of the most deadly diseases, acquired immune deficiency syndrome (AIDS). The genes of the HIV group are two same RNA strands, which can transcribed into DNA for further gene expression in host cells through reverse transcription. It attacks the host's immune system and causes the destruction

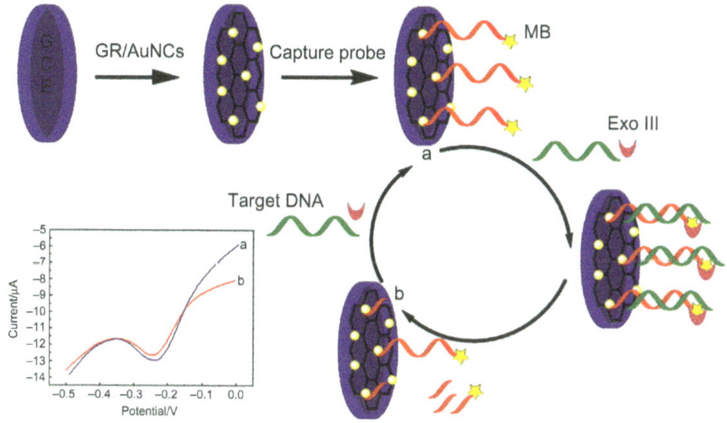

FIGURE 6.12 Schematic illustration of the proposed biosensor fabrication process. *Recreated from Wang, Y., Bai, X., Wen, W., Zhang, X., Wang, S., 2015. Ultrasensitive electrochemical biosensor for HIV gene detection based on graphene stabilized gold nanoclusters with exonuclease amplification. ACS Appl. Mater. Interfaces 7, 18872–18879, with kind permission of American Chemical Society.*

of T4 lymphocytes, resulting in the breakdown of the human immune system. In this case, the human body is not able to defend itself against many diseases, leading to the occurrence of death. An electrochemical biosensor based on GR/AuNCs platform has been developed with Exo III-assisted DNA recycling amplification. With super high specific area, GR/AuNCs provide a good conductive platform with more fixed sites for C-rich capture probes, giving the good electrochemical results (Fig. 6.12; Wang et al., 2015). Meanwhile, it was found that the employment of ERGO as an immobilization platform efficiently accelerated the electron transfer and enhanced the EIS response of the DNA biosensor. The ssDNA/ERGO-based DNA biosensor was sensitive and selective for the target DNA, with a low detection limit down to 0.3 pM. Also, the biosensor could well discriminate the mismatched or noncomplementary sequences (Gong et al., 2015).

6.2.4 Diabetes

Diabetes mellitus (DM) has become one of the most common metabolic diseases all over the world. Diabetes causes most cases of blindness and renal failure. The higher RBP4 levels in serum may contribute to the onset of obesity or type 2 diabetes, and these levels can be used as a potential indicator of health risks. An SPR detection system has been reported by using the binding of an aptamer to RBP4 with the aid of magnetic nanoparticles.

Blood glycated hemoglobin (HbA1c) can reflect the average blood glucose level in the proceeding 2–3 months and has been recommended for screening/diagnosis and patient management of diabetes. Chang et al. (2015)

have developed an integrated microfluidic system for measurement of glycated hemoglobin levels by using an aptamer–antibody assay on magnetic beads.

Glycated albumin (GA) is a much more accurate marker for diabetes mellitus, especially for patients who are undergoing hemodialysis or having hematologic disorders. An optical sensor composed of DNA aptamer, semiconductor quantum dot, and gold (Au) nanoparticle has been designed for the detection of GA (Ghosh et al., 2017). The system functioned as a "turn-on" sensor because an increase in photoluminescence intensity was observed with the addition of GA. The DNA aptamer can fold to form a large hairpin loop before the addition of the analyte and open up after the addition of target so as to bind to GA, resulting in the leaving of the quantum dot and the Au nanoparticle and a responding increase in photoluminescence.

6.3 ASSAY BASED ON COMPETITIVE BINDING OF TARGET WITH APTAMER

Aptamers require the formation of a three-dimensional structure for target binding. They can have a much higher affinity for binding its target rather than other substances. This phenomenon can be exploited for the development of an assay based on competitive binding of the target with aptamer.

6.3.1 Cancer

Anterior gradient homolog 2 (AGR2) is a secreted protein and shows key association with proliferation and migration in the development of tumor cells. AGR2 is overexpressed in many adenocarcinomas. Recent clinical studies have identified AGR2 protein as a potential biomarker of multiple cancers, including breast, prostate, ovarian, intestinal, and colon cancers. The aptasensor has been constructed based on gold nanoparticles/DNA/magnetic beads for analysis of AGR2 (Fig. 6.13; Hu et al., 2015). AuNPs and magnetic beads were joined together by unoccupied anti-AGR2 aptamer and its partially complementary DNA probe. As catchers, the aptamers on the MBs can well capture the target proteins and the DNA probes on the AuNPs competed with proteins to hybridize with catchers. Consequently, the increasing number of target proteins would reduce the possibility of the sandwich-structure formation. Meanwhile, the formed AuNPs/DNA/MBs can be easily removed by the magnet, and the unpaired AuNPs in the supernatant were served for a spectrometer quantitation.

Based on competitive binding of target with aptamer, a simple and sensitive electrochemical impedimetric aptasensor has been developed through gold nanoparticles (AuNPs) signal amplification for the ultrasensitive detection of tumor markers (Liu et al., 2015). Gold electrode was modified with the designed cDNA, which is partly complementary to the aptamer of

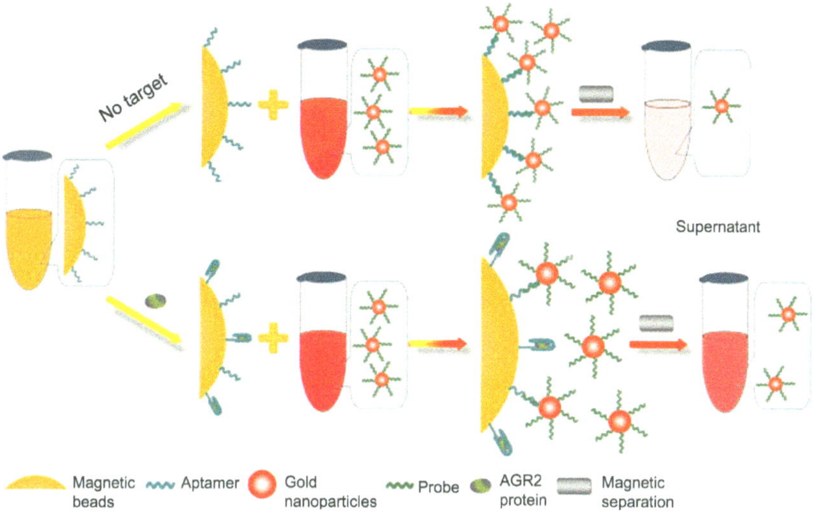

FIGURE 6.13 Schematic representation of AGR2 protein detection procedure. *Recreated from Hu, Y., Li, L., Guo, L., 2015. The sandwich-type aptasensor based on gold nanoparticles/DNA/ magnetic beads for detection of cancer biomarker protein AGR2. Sens. Actuat. B 209, 846–852, with kind permission of Elsevier.*

MUC1. MUC1 could be analyzed through switching structures of aptamers from DNA/DNA duplex to DNA/target complex. As the signal enhancer, the aptamer-modified AuNPs conjugate was modified onto the surface of the electrode through the hybridization of cDNA with aptamer. Moreover, an mesoporous carbon nanospheres featured fluorescent aptasensor has been constructed for multiple diagnosis of cancer in vitro and in vivo (Fig. 6.14; Li et al., 2015). The Cy3 fluorescence was quickly and totally quenched due to π-π stacked interaction of a dye (Cy3)-labeled ssDNA probe (P0-Cy3) on the surface of OMCN. With the addition of malignant tumors with the over-expressed MUC1, P0-Cy3 was gradually released from OMCN owing to the much stronger interaction between P0-Cy3 and MUC1 than that between P0-Cy3 and MUC1, giving an obvious fluorescence recovery. This OMCN featured fluorescent aptasensor not only can quantify the MUC1 molecules and MCF-7 tumor cells, but also can clearly image the cancer cells, ex vivo tissues, and solid tumors.

The aptamer–cancer cell interaction has been applied to DNA-based assay to detect cancer cells indirectly by detecting released DNA through the aptamer-based competition strategy with the aid of MNPs. As shown in Fig. 6.15, the large amounts of G-quadruplex DNAzyme and electron mediator MB were loaded onto the supersandwich structure (Lu et al., 2015). Meanwhile, the G-quadruplex DNAzyme catalyzes the reduction of H_2O_2 with the aid of MB, leading to a dramatically amplified electronic signal.

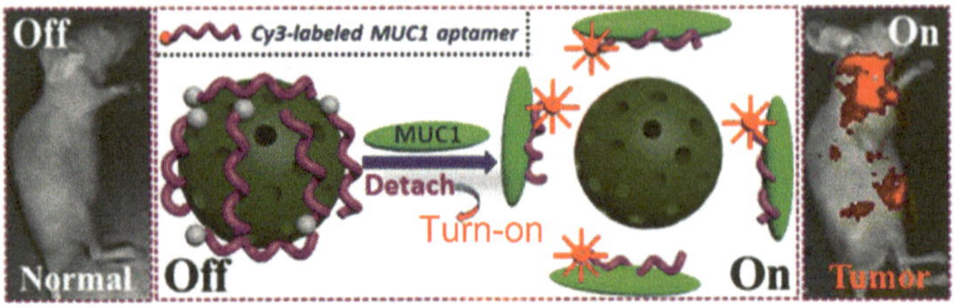

FIGURE 6.14 Mesoporous carbon nanospheres featured fluorescent aptasensor for multiple diagnosis of cancer in vitro and in vivo. *Recreated from Li, C., Meng, Y., Wang, S., Qian, M., Wang, J., Lu, W., et al., 2015. Mesoporous carbon nanospheres featured fluorescent aptasensor for multiple diagnosis of cancer in vitro and in vivo. ACS Nano 9, 12096–12103, with kind permission of American Chemical Society.*

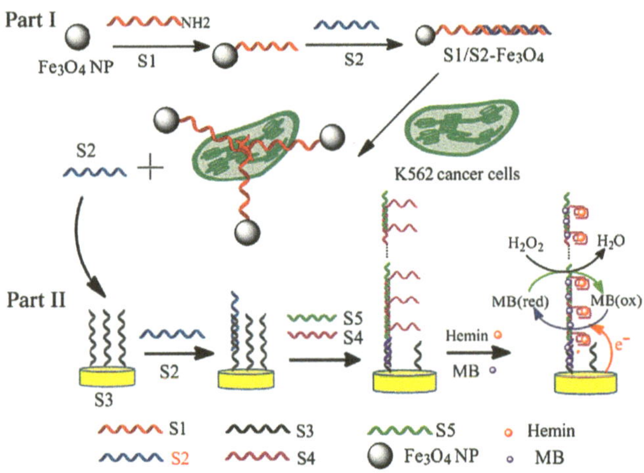

FIGURE 6.15 Schematic representation of electrochemical detection of K562 cells. *Recreated from Lu, C.Y., Xu, J.J., Wang, Z.H., Chen, H.Y., 2015. A novel signal-amplified electrochemical aptasensor based on supersandwich G-quadruplex DNAzyme for highly sensitive cancer cell detection. Electrochem. Commun. 52, 49–52, with kind permission of Elsevier.*

The free-running DNA walker has also served for the establishment of a signal amplification electrochemical aptasensor for the detection of breast cancer cell (Fig. 6.16; Cai et al., 2016). Released by the addition of target MCF-7 cell into double chain modified magnetic bead, DNA walker (DNAzyme) repeatedly binded and cleaved D-RNA in the presence of Mg^{2+}. In this strategy, one DNA walker can scissor all D-RNA anchored on gold electrode into short products so as to bring the considerable signal suppression.

A novel, sensitive and versatile SPR sensor for miRNA and cancer cell has been explored based on multiple signal amplification strategy (Fig. 6.17; Liu et al., 2017). In the strategy, Au film was functionalized with

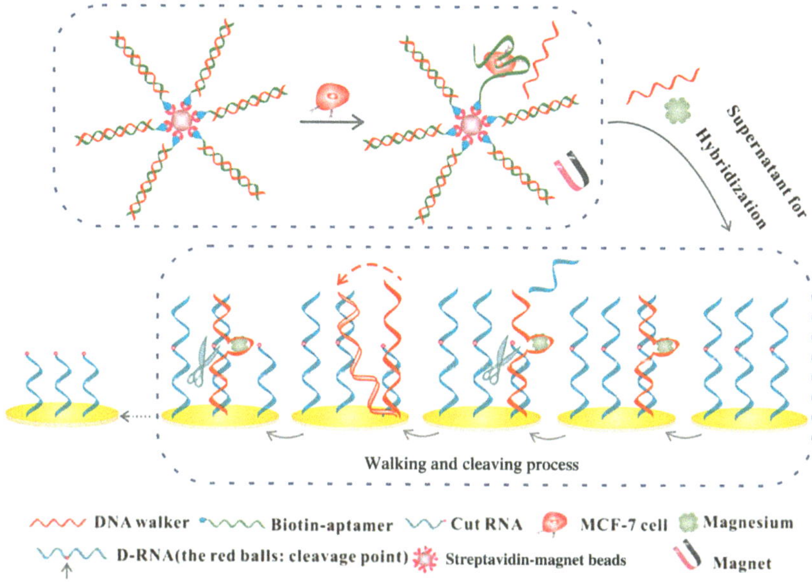

FIGURE 6.16 Schematic illustration of the signal amplification electrochemical aptasensor via free-running DNA walker. The upper one shows the specific recognition and binding between MCF-7 cell and its corresponding aptamer. The lower one depicts the walking and cleaving process. *Recreated from Cai, S., Chen, M., Liu, M., He, W., Liu, Z., Wu, D., et al., 2016. A signal amplification electrochemical aptasensor for the detection of breast cancer cell via free-running DNA walker. Biosens. Bioelectron. 85, 184–189, with kind permission of Elsevier.*

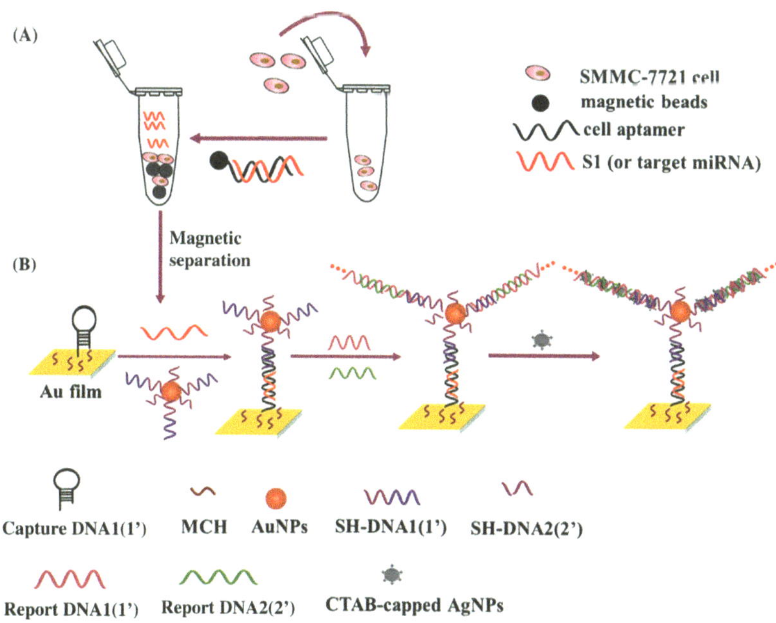

FIGURE 6.17 Schematic representation of SPR biosensor based on multiple amplification strategy for miRNA and cancer cells detection. *Recreated from Liu, R., Wang, Q., Li, Q., Yang, X., Wang, K., Nie, W., 2017. Surface plasmon resonance biosensor for sensitive detection of microRNA and cancer cell using multiple signal amplification strategy. Biosens. Bioelectron. 87, 433–438, with kind permission of Elsevier.*

thiol-modified hairpin probe including a sequence complementary to the target miRNA, which was used as capture DNA. In the presence of target miRNA, DNA-linked gold nanoparticles (AuNPs) were bound to Au film by hybridization with the terminus of capture DNA with unfolded stem-loop structure. After the introduction of report DNA1 and report DNA2, DNA-linked AuNPs initiated the formation of DNA supersandwich structure. As a result of the electronic coupling between localized plasmon of AuNPs and the surface plasmon wave associated with Au film, as well as the enhancement of the refractive index of the medium next to the Au film induced by DNA supersandwich structure, the shift of resonance angle was enhanced numerously. After that, the positively charged silver nanoparticles (AgNPs) were attached to the long-range negatively charged DNA surpersandwich, causing a further increase of resonance angle shift. The concentration of target miRNA could be quantified according to the resonance angle shift.

Furthermore, a cascade signal amplification strategy has been developed for sensitive detection of cancer cell by coupling target triggered self-assembly of polycatenated DNA scaffold and rolling circle amplified synthesis of DNAzymes (Li et al., 2012).

6.3.2 Tuberculosis

A SWCNT/aptamer/Au-IDE MSPQC H37Rv sensor has been constructed by using a new *Mycobacterium tuberculosis* H37Rv aptamer (Zhang et al. 2017). In SWCNT/aptamer/Au-IDE probe, the Au-IDE electrode surface was modified with the selected aptamer against H37Rv by Au-S bonding, and SWCNT was further binded to aptamer through $\pi-\pi$ stacking interaction between the carbon nanotubes walls and the puric and pyrimidic bases. Moreover, Au-IDE/CFP10-ESAT6 aptamer/DNA-AuNPs MSPQC was constructed for rapid detection of *Mycobacterium tuberculosis* (Fig. 6.18; He et al., 2016). The Au-IDE/CFP10-ESAT6 aptamer/DNA-AuNPs electrode probe was obtained through the modification of the complementary DNA-AuNPs onto interdigital array microelectrode with CFP10-ESAT6 aptamer. The aptamer could catch CFP10-ESAT6 protein to form a tight complex on the electrode surface, resulting in the leaving of DNA-AuNPs fragments from the electrode surface. In addition, FbpA aptamers were used for the construction of a highly sensitive and selective GO-based aptasensor for FbpA analysis (Ansari et al., 2018). In this biosensor, the generated aptamer (Apt22) was labeled with the fluorophore ATTO647N as the energy donor and GO acted as the energy acceptor.

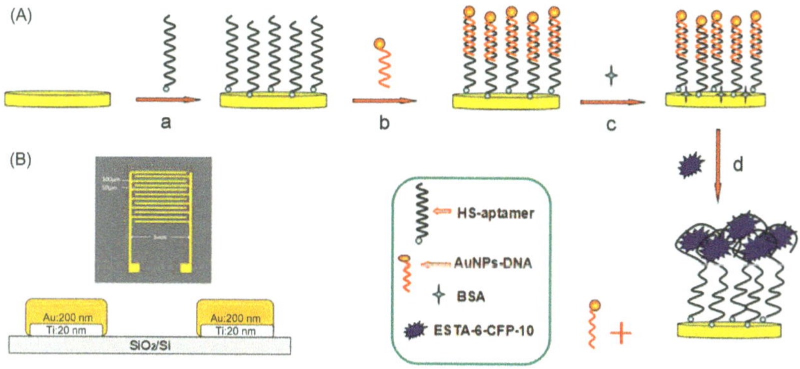

FIGURE 6.18 (A) Modification of the interdigital gold electrodes and response process to CFP10-ESAT6. (a) modification of thiol-aptamer on Au-IDE; (b) hybridization of aptmer with DNA-AuNPs; (c) blocking of active sites; (d) Binding of CFP10-ESAT6 to aptmer. (B) The diagram of IDE electrode. *Recreated from He, F., Xiong, Y., Liu, J., Tong, F., Yan, D., 2016. Construction of Au-IDE/CFP10-ESAT6 aptamer/DNA-AuNPs MSPQC for rapid detection of Mycobacterium tuberculosis. Biosens. Bioelectron. 77, 799–804, with kind permission of Elsevier.*

6.3.3 Human Immunodeficiency Virus (HIV)

An impedimetric HIV-1 gene biosensor has been developed based on graphene–Nafion composite film (Fig. 6.19; Gong et al., 2017). The single-stranded DNA (ssDNA) adsorbed on graphene–Nafion modified on the surface of glassy carbon electrode via the $\pi-\pi$ stacking interactions. The electron transfer of the electrodes toward the $[Fe(CN)_6]^{3/4}$ redox couple was difficult due to the negative ssDNA and the steric hindrance, so the electron transfer resistance value increased. In the presence of HIV gene, ssDNA probe with the target DNA to form double-stranded DNA (dsDNA) so as to induce the release of dsDNA from the surface of the biosensor.

6.3.4 Diabetes

Glycated human serum albumin (GHSA) is formed through nonenzymatic glycation of HSA by excess circulating sugar and it influences the normal HAS functions. Diabetic patients own 2–5 times higher levels of a typical diabetic GHSA in glucose metabolism of both adipocyte cells and skeletal muscle cells, so the pathogenic implication of GHSA formation has been confirmed. GHSA can be analyzed through the sensitive graphene-based aptasensors (Apiwat et al., 2016). The hairpin-loop structure with 23 nucleotides length containing triple G-C hairpins and 15-nucleotide loop, plays a key role in GHSA binding. The aptamers with high binding affinity were used in combination with GO.

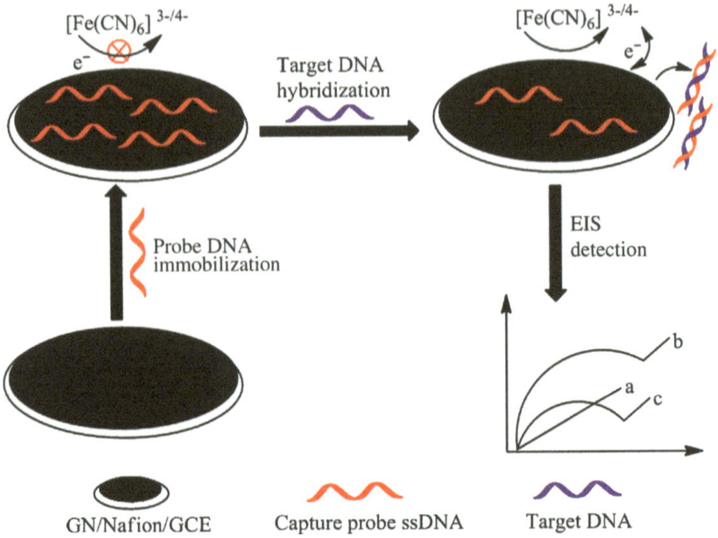

FIGURE 6.19 Schematic diagram of the fabrication of the impedimetric DNA biosensor and the detection of target. *Recreated from Gong, Q., Wang, Y., Yang, H., 2017. A sensitive impedimetric DNA biosensor for the determination of the HIV gene based on graphene-Nafion composite film. Biosens. Bioelectron. 89 (Pt 1), 565–569, with kind permission of Elsevier.*

6.4 SPLIT-TYPE ASSAY

As a dominant strategy for protein determination, the sandwich assay is highly specific and sensitive resulting from the dual recognition mechanism. However, the approach is poorly suited to detection of targets with low molecular weight, which are unlikely to bind to two aptamers simultaneously due to steric hindrance. To circumvent this drawback, a split aptamer strategy has been proposed and developed. For split-type assay, nucleic acid aptamers were split into two fragments that could specifically form a ternary assembly in the presence of ligand. This elegant approach has been extensively adopted for detection of various targets, with different transduction methods including colorimetric, fluorescence, and electrochemical techniques.

For the detection of CEA, a near-infrared-to-ultraviolet light-mediated photoelectrochemical aptasensing platform has been established based on core–shell NaYF4: Yb, Tm@TiO_2 upconversion microrods (Fig. 6.20; Qiu et al., 2018). In this system, Apt1-MB was synthesized through covalent conjugation of capture aptamer-1 with the carboxylated magnetic bead through a classical carbodiimide coupling. In the presence of CEA, the sandwiched complex was formed between Apt1-MB and Apt2-pDNA. Then the carried primer DNA with the aptamer-2 underwent an unbiased RCA reaction

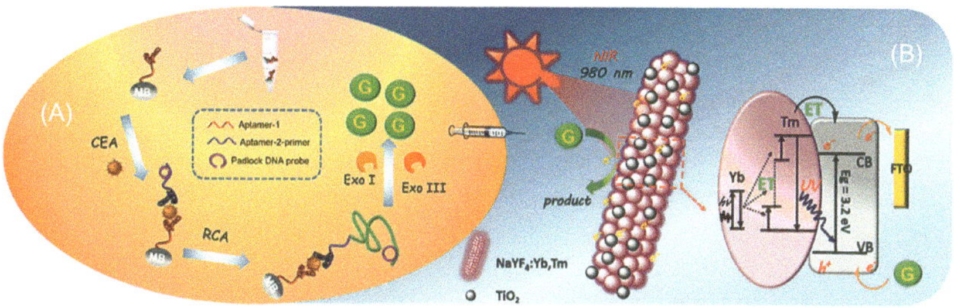

FIGURE 6.20 (A) Aptamer-CEA-aptamer reaction between capture aptamer-1-functionalized magnetic bead (Apt1-MB) and the aptamer-2-primer DNA conjugate (Apt2-pDNA) with a sandwich-type assay format, and the released process of guanine bases followed by the RCA reaction. (B) Diagram of energy transfer among lanthanide ion and TiO_2 under near-IR irradiation leading to the formation of hole-electron pairs. *Recreated from Qiu, Z., Shu, J., Tang, D., 2018. Near-infrared-to-ultraviolet light-mediated photoelectrochemical aptasensing platform for cancer biomarker based on core-shell NaYF4: Yb, Tm@TiO2 upconversion microrods. Anal. Chem. 90, 1021–1028, with kind permission of American Chemical Society.*

through the linear padlock-based circular DNA template with the assistance of ligase and polymerase. With the addition of Exo I and Exo III, the RCA product was digested to generate the numerous free guanine bases, resulting in the amplification of the photocurrent on NaYF4: Yb, Tm@TiO_2 microrods-modified electrode. By monitoring the change in the photocurrent, the concentration of target CEA in the sample can be quantitatively evaluated. The establishment of the split-type assay mode for the aptamer–target reaction and photocurrent measurement can efficiently avoid the damage of biomolecules.

6.5 ASSAY BASED ON APTAMER–CELL INTERACTION

Aptamer–cell interaction has been used to visually test cancer cells (Fig. 6.21; Borghei et al., 2016). Through affinity interaction between AS 1411 and nucleolin receptors, cancer cells were able to capture nucleolin aptamers (AS 1411). The specific binding of AS 1411 to the target cells triggered the removal of aptamers from the solution. In this case, no aptamer remained in the solution to hybridize with complementary ssDNA-AuNP probes, so the solution color is red. Without target cells or with normal cells, ssDNA-AuNP probes and aptamers coexisted in solution and the aptamers assembled to form DNA-AuNPs, producing a purple solution. The hybridization-based method exhibited selective colorimetric responses to the presence or absence of target cells, with detectable results by naked eye.

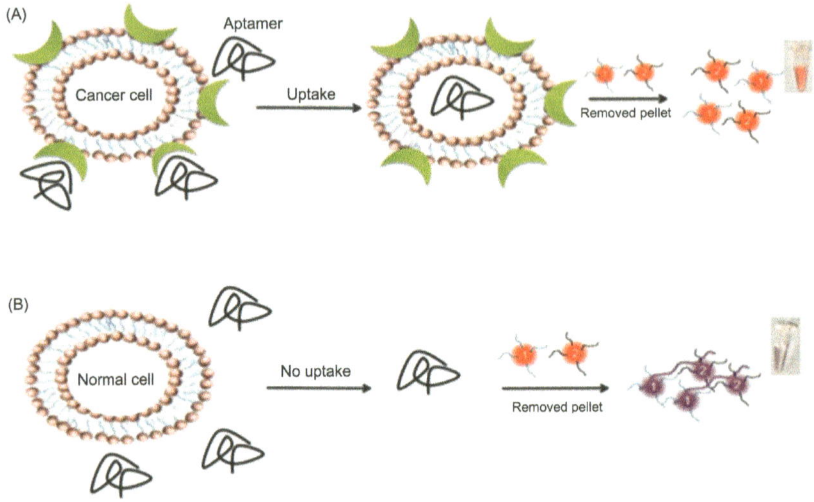

FIGURE 6.21 Schematic representation of selective colorimetric method for detection of cancer cells by employing DNA probe 1,2-functionalized gold nanoparticles and AS1411 aptamer. *Recreated from Borghei, Y.-S., Hosseini, M., Dadmehr, M., Hosseinkhani, S., Ganjali, M.R., Sheikhnejad, R., 2016. Visual detection of cancer cells by colorimetric aptasensor based on aggregation of gold nanoparticles induced by DNA hybridization. Anal. Chim. Acta 904, 92–97, with kind permission of Elsevier.*

6.6 CONCLUSION

In view of its advantages including high specificity, affinity, and solubility, ease of chemical modifications, high reproducibility, and low costs for mass production, aptamer can be well utilized as a molecular recognition element for construction of biosensors. For protein assay, it has been explored for a variety of aptasensors including (A) sandwich aptasensor, (B) aptasensor based on aptamer conformational switch, (C) aptasensor based on competitive binding of target with aptamer, (D) split-type assay aptasensor, and (E) aptasensor based on aptamer–cell interaction. Aptasensors have a great potential for development and application in the future.

REFERENCES

Amouzadeh Tabrizi, M., Shamsipur, M., Saber, R., Sarkar, S., Sherkatkhameneh, N., 2017. Flow injection amperometric sandwich-type electrochemical aptasensor for the determination of adenocarcinoma gastric cancer cell using aptamer-Au@Ag nanoparticles as labeled aptamer. Electrochim. Acta 246, 1147–1154.

Ansari, N., Ghazvini, K., Ramezani, M., Shahdordizadeh, M., Yazdian-Robati, R., Abnous, K., et al., 2018. Selection of DNA aptamers against Mycobacterium tuberculosis Ag85A, and its application in a graphene oxide-based fluorometric assay. Microchim. Acta 185, 21.

Apiwat, C., Luksirikul, P., Kankla, P., Pongprayoon, P., Treerattrakoon, K., Paiboonsukwong, K., et al., 2016. Graphene based aptasensor for glycated albumin in diabetes mellitus diagnosis and monitoring. Biosens. Bioelectron. 82, 140–145.

Argoubi, W., Sánchez, A., Parrado, C., Raouafi, N., Villalonga, R., 2018. Label-free electrochemical aptasensing platform based on mesoporous silica thin film for the detection of prostate specific antigen. Sens. Actuat. B 255, 309–315.

Arya, S.K., Zhurauski, P., Jolly, P., Batistuti, M.R., Mulato, M., Estrela, P., 2018. Capacitive aptasensor based on interdigitated electrode for breast cancer detection in undiluted human serum. Biosens. Bioelectron. 102, 106–112.

Bai, L., Chen, Y., Bai, Y., Chen, Y., Zhou, J., Huang, A., 2017. Fullerene-doped polyaniline as new redox nanoprobe and catalyst in electrochemical aptasensor for ultrasensitive detection of *Mycobacterium tuberculosis* MPT64 antigen in human serum. Biomaterials. 133, 11–19.

Benvidi, A., Banaei, M., Tezerjani, M.D., Molahosseini, H., Jahanbani, S., 2018. Impedimetric PSA aptasensor based on the use of a glassy carbon electrode modified with titanium oxide nanoparticles and silk fibroin nanofibers. Microchim. Acta 185, 50.

Borghei, Y.S., Hosseini, M., Dadmehr, M., Hosseinkhani, S., Ganjali, M.R., Sheikhnejad, R., 2016. Visual detection of cancer cells by colorimetric aptasensor based on aggregation of gold nanoparticles induced by DNA hybridization. Anal. Chim. Acta 904, 92–97.

Bouchard, P.R., Hutabarat, R.M., Thompson, K.M., 2010. Discovery and development of therapeutic aptamers. Annu. Rev. Pharmacol. 50, 237–257.

Cai, S., Chen, M., Liu, M., He, W., Liu, Z., Wu, D., et al., 2016. A signal amplification electrochemical aptasensor for the detection of breast cancer cell via free-running DNA walker. Biosens. Bioelectron. 85, 184–189.

Cao, H., Ye, D., Zhao, Q., Luo, J., Zhang, S., Kong, J., 2014. A novel aptasensor based on MUC-1 conjugated CNSs for ultrasensitive detection of tumor cells. Analyst 139, 4917–4923.

Chang, K.W., Li, J., Yang, C.H., Shiesh, S.C., Lee, G.B., 2015. An integrated microfluidic system for measurement of glycated hemoglobin levels by using an aptamer-antibody assay on magnetic beads. Biosens. Bioelectron. 68, 397–403.

Chen, Y., Li, Y., Yang, Y., Wu, F., Cao, J., Bai, L., 2017. A polyaniline-reduced graphene oxide nanocomposite as a redox nanoprobe in a voltammetric DNA biosensor for *Mycobacterium tuberculosis*. Microchim. Acta 184, 1801–1808.

Cho, H., Yeh, E.C., Sinha, R., Laurence, T.A., Bearinger, J.P., Lee, L.P., 2012. Single-step nanoplasmonic VEGF165 aptasensor for early cancer diagnosis. ACS Nano 6, 7607–7614.

Fu, X.M., Liu, Z.J., Cai, S.X., Zhao, Y.P., Wu, D.Z., Li, C.Y., et al., 2016. Electrochemical aptasensor for the detection of vascular endothelial growth factor (VEGF) based on DNA-templated Ag/Pt bimetallic nanoclusters. Chin. Chem. Lett. 27, 920–926.

Ghosh, S., Datta, D., Cheema, M., Dutta, M., Stroscio, M.A., 2017. Aptasensor based optical detection of glycated albumin for diabetes mellitus diagnosis. Nanotechnology 28, 435505.

Gijs, M., Aerts, A., Impens, N., Baatout, S., Luxen, A., 2016. Aptamers as radiopharmaceuticals for nuclear imaging and therapy. Nucl. Med. Bio. 43, 253–271.

Gong, Q., Yang, H., Dong, Y., Zhang, W., 2015. A sensitive impedimetric DNA biosensor for the determination of the HIV gene based on electrochemically reduced graphene oxide. Anal. Methods 7, 2554–2562.

Gong, Q., Wang, Y., Yang, H., 2017. A sensitive impedimetric DNA biosensor for the determination of the HIV gene based on graphene-Nafion composite film. Biosens. Bioelectron. 89 (Pt 1), 565–569.

Hashkavayi, A.B., Raoof, J.B., Ojani, R., Kavoosian, S., 2017. Ultrasensitive electrochemical aptasensor based on sandwich architecture for selective label-free detection of colorectal cancer (CT26) cells. Biosens. Bioelectron. 92, 630–637.

He, F., Xiong, Y., Liu, J., Tong, F., Yan, D., 2016. Construction of Au-IDE/CFP10-ESAT6 aptamer/DNA-AuNPs MSPQC for rapid detection of *Mycobacterium tuberculosis*. Biosens. Bioelectron. 77, 799–804.

Hu, Y., Li, L., Guo, L., 2015. The sandwich-type aptasensor based on gold nanoparticles/DNA/ magnetic beads for detection of cancer biomarker protein AGR2. Sens. Actuat. B 209, 846–852.

Huang, J.Y., Zhao, L., Lei, W., Wen, W., Wang, Y.J., Bao, T., et al., 2018. A high-sensitivity electrochemical aptasensor of carcinoembryonic antigen based on graphene quantum dots-ionic liquid-nafion nanomatrix and DNAzyme-assisted signal amplification strategy. Biosens. Bioelectron. 99, 28–33.

Kashefi-Kheyrabadi, L., Mehrgardi, M.A., Wiechec, E., Turner, A.P.F., Tiwari, A., 2014. Ultrasensitive detection of human liver hepatocellular carcinoma cells using a label-free aptasensor. Anal. Chem. 86, 4956–4960.

Li, C., Meng, Y., Wang, S., Qian, M., Wang, J., Lu, W., et al., 2015. Mesoporous carbon nanospheres featured fluorescent aptasensor for multiple diagnosis of cancer in vitro and in vivo. ACS Nano 9, 12096–12103.

Li, J., Chang, K.W., Wang, C.H., Yang, C.H., Shiesh, S.C., Lee, G.B., 2016. On-chip, aptamer-based sandwich assay for detection of glycated hemoglobins via magnetic beads. Biosens. Bioelectron. 79, 887–893.

Li, Y., Zeng, Y., Ji, X., Li, X., Ren, R., 2012. Cascade signal amplification for sensitive detection of cancer cell based on self-assembly of DNA scaffold and rolling circle amplification. Sens. Actuat. B 171-172, 361–366.

Liu, B., Lu, L., Hua, E., Jiang, S., Xie, G., 2012a. Detection of the human prostate-specific antigen using an aptasensor with gold nanoparticles encapsulated by graphitized mesoporous carbon. Microchim. Acta 178, 163–170.

Liu, J.J., Song, X.R., Wang, Y.W., Zheng, A.X., Chen, G.N., Yang, H.H., 2012b. Label-free and fluorescence turn-on aptasensor for protein detection via target-induced silver nanoclusters formation. Anal. Chim. Acta 749, 70–74.

Liu, R., Wang, Q., Li, Q., Yang, X., Wang, K., Nie, W., 2017. Surface plasmon resonance biosensor for sensitive detection of microRNA and cancer cell using multiple signal amplification strategy. Biosens. Bioelectron. 87, 433–438.

Liu, X., Qin, Y., Deng, C., Xiang, J., Li, Y., 2015. A simple and sensitive impedimetric aptasensor for the detection of tumor markers based on gold nanoparticles signal amplification. Talanta 132, 150–154.

Liu, Y.M., Zhou, M., Liu, Y.Y., Shi, G.F., Zhang, J.J., Cao, J.T., et al., 2014. Fabrication of electrochemiluminescence aptasensor based on in situ growth of gold nanoparticles on layered molybdenum disulfide for sensitive detection of platelet-derived growth factor-BB. RSC Adv. 4, 22888.

Lu, C.Y., Xu, J.J., Wang, Z.H., Chen, H.Y., 2015. A novel signal-amplified electrochemical aptasensor based on supersandwich G-quadruplex DNAzyme for highly sensitive cancer cell detection. Electrochem. Commun. 52, 49–52.

Mayer, G., 2009. The chemical biology of aptamers. Angew. Chem. Int. Ed. 48, 2672–2689.

Qiu, Z., Shu, J., Tang, D., 2018. Near-infrared-to-ultraviolet light-mediated photoelectrochemical aptasensing platform for cancer biomarker based on core-shell NaYF4: Yb, Tm@TiO_2 upconversion microrods. Anal. Chem. 90, 1021–1028.

Si, Z., Xie, B., Chen, Z., Tang, C., Li, T., Yang, M., 2017. Electrochemical aptasensor for the cancer biomarker CEA based on aptamer induced current due to formation of molybdophosphate. Microchim. Acta 184, 3215–3221.

Sun, D., Lu, J., Chen, Z., Yu, Y., Mo, M., 2015. A repeatable assembling and disassembling electrochemical aptamer cytosensor for ultrasensitive and highly selective detection of human liver cancer cells. Anal. Chim. Acta 885, 166–173.

Taghdisi, S.M., Danesh, N.M., Ramezani, M., Yazdian-Robati, R., Abnous, K., 2017. An amplified fluorescent aptasensor based on single-stranded DNA binding protein, copper and silica nanoparticles for sensitive detection of interferon-gamma. Anal. Chim. Acta 984, 162–167.

Thakur, H., Kaur, N., Sabherwal, P., Sareen, D., Prabhakar, N., 2017. Aptamer based voltammetric biosensor for the detection of *Mycobacterium tuberculosis* antigen MPT64. Microchim. Acta 184, 1915–1922.

Tombelli, S., Minunni, M., Mascini, M., 2005. Analytical applications of aptamers. Biosens. Bioelectron. 20, 2424–2434.

Wang, K., Zhang, R., Sun, N., Li, X., Wang, J., Cao, Y., et al., 2016. Near-infrared light-driven photoelectrochemical aptasensor based on the upconversion nanoparticles and TiO_2/CdTe heterostructure for detection of cancer cells. ACS Appl. Mater. Interfaces 8, 25834–25839.

Wang, Q.L., Cui, H.F., Song, X., Fan, S.F., Chen, L.L., Li, M.M., et al., 2018. A label-free and lectin-based sandwich aptasensor for detection of carcinoembryonic antigen. Sens. Actuat. B 260, 48–54.

Wang, Y., Bai, X., Wen, W., Zhang, X., Wang, S., 2015. Ultrasensitive electrochemical biosensor for HIV gene detection based on graphene stabilized gold nanoclusters with exonuclease amplification. ACS Appl. Mater. Interfaces 7, 18872–18879.

Wei, W., Li, D.F., Pan, X.H., Liu, S.Q., 2012. Electrochemiluminescent detection of mucin 1 protein and MCF-7 cancer cells based on the resonance energy transfer. Analyst 137, 2101–2106.

Wu, Z., Li, H., Liu, Z., 2015. An aptasensor for carcinoembryonic antigen based on upconversion fluorescence resonance energy transfer. Sens. Actuat. B 206, 531–537.

Xiao, K., Liu, J., Chen, H., Zhang, S., Kong, J., 2017. A label-free and high-efficient GO-based aptasensor for cancer cells based on cyclic enzymatic signal amplification. Biosens. Bioelectron. 91, 76–81.

Xie, Q., Tan, Y., Guo, Q., Wang, K., Yuan, B., Wan, J., et al., 2014. A fluorescent aptasensor for sensitive detection of human hepatocellular carcinoma SMMC-7721 cells based on graphene oxide. Anal. Methods 6, 6809–6814.

Yan, M., Sun, G., Liu, F., Lu, J., Yu, J., Song, X., 2013. An aptasensor for sensitive detection of human breast cancer cells by using porous GO/Au composites and porous PtFe alloy as effective sensing platform and signal amplification labels. Anal. Chim. Acta 798, 33–39.

Yang, S., You, M., Zhang, F., Wang, Q., He, P., 2018. A sensitive electrochemical aptasensing platform based on exonuclease recycling amplification and host-guest recognition for detection of breast cancer biomarker HER2. Sens. Actuat. B 258, 796–802.

Yin, J., He, X., Wang, K., Xu, F., Shangguan, J., He, D., et al., 2013. Label-free and turn-on aptamer strategy for cancer cells detection based on a DNA-silver nanocluster fluorescence upon recognition-induced hybridization. Anal. Chem. 85, 12011–12019.

Yu, T., Dai, P.P., Xu, J.J., Chen, H.Y., 2016. Highly sensitive colorimetric cancer cell detection based on dual signal amplification. ACS Appl. Mater. Interfaces 8, 4434–4441.

Zhang, J.J., Cao, J.T., Shi, G.F., Huang, K.J., Liu, Y.M., Ren, S.W., 2015. A luminol electrochemiluminescence aptasensor based on glucose oxidase modified gold nanoparticles for measurement of platelet-derived growth factor BB. Talanta 132, 65–71.

Zhang, X., Xiao, K., Cheng, L., Chen, H., Liu, B., Zhang, S., et al., 2014. Visual and highly sensitive detection of cancer cells by a colorimetric aptasensor based on cell-triggered cyclic enzymatic signal amplification. Anal. Chem. 86, 5567–5572.

Zhang, X., Feng, Y., Yao, Q., He, F., 2017. Selection of a new *Mycobacterium tuberculosis* H37Rv aptamer and its application in the construction of a SWCNT/aptamer/Au-IDE MSPQC H37Rv sensor. Biosens. Bioelectron. 98, 261–266.

Zhao, Y., Yang, Y., Sun, Y., Cui, L., Zheng, F., Zhang, J., et al., 2018. Shell-encoded Au nanoparticles with tunable electroactivity for specific dual disease biomarkers detection. Biosens. Bioelectron. 99, 193–200.

Chapter 7

Peptide-Based Biosensors

Juan Zhang and Ya Cao

Center for Molecular Recognition and Biosensing, School of Life Sciences, Shanghai University, Shanghai, P. R. China

Chapter Outline

In view of the advantages including mature synthesis protocols, diverse structures, and as highly selective substrates for enzymes, peptides have been used as components in biological analysis and fabrication of novel biosensors.

7.1 BIOSENSORS BASED ON THE BINDING OF PEPTIDE TO TARGET

Due to its same chemical structure with proteins, peptide is an ideal candidate to substitute for protein as the receptor in biosensors. Peptides have been utilized as biorecognition elements to bind to various analytes including proteins, nucleic acid, bacteria, metal ions, enzymes, and antibodies in biosensors. Artificial peptides can be synthesized through the standard solid-phase synthesis protocols. These peptide-based molecular biosensors have been developed for convenient, fast detection of various proteins such as the biomarkers.

Nano-inspired Biosensors for Protein Assay with Clinical Applications.
DOI: https://doi.org/10.1016/B978-0-12-815053-5.00007-6

7.1.1 Cancer

Smo is known as an effector protein of the Hedgehog signaling pathway, a pathway that is frequently upregulated in tumor cells for initiating and sustaining the metastatic transformation. Due to its intimate connection with metastasis, Smo has become a therapeutic target for antimetastatic drugs. A probe has been successfully constructed by using a Smo-specific drug in combination with peptides and other synthetic agents (Fig. 7.1; Li et al., 2015). The peptide-based probe was designed to be able to not only conjugate with cyclopamine but also had a binding site of the nanolabel. With a thiol group, the probe immobilized on the surface of the working electrode can capture target cells via multiple specific interactions with the cell surface Smo. The captured cells were lyzed, followed by the cleavage of thermolysin. Smo-free probes were removed from the electrode surface, while Smo-bound probes can be shielded from cleavage by the captured Smo molecules. Subsequently, the nanolabels loaded with a large number of nanoparticles consisting of electroactive organic molecules, are attached to the Smo-bound probes. The great number of nanoparticles loaded in the nanolabel can then give rise to prominent signal readout.

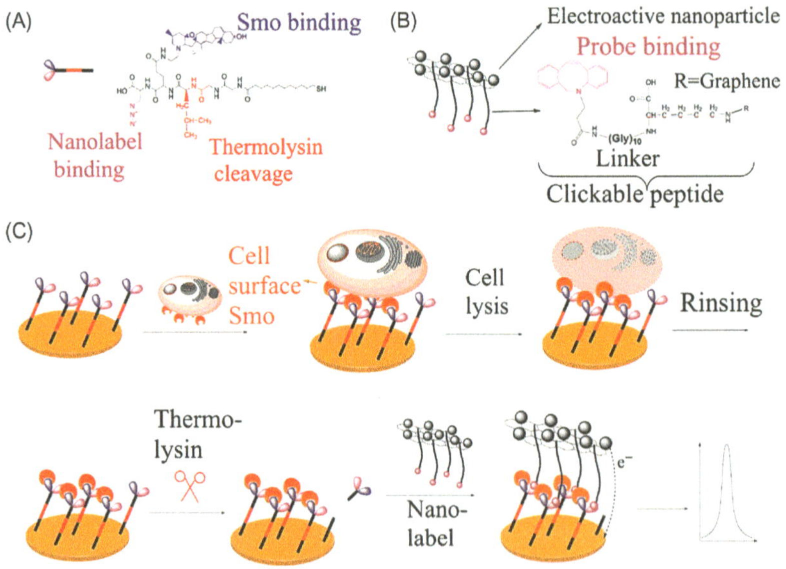

FIGURE 7.1 Peptide-based method for detection of metastatic transformation in primary tumors of breast cancer. (A) Probe; (B) nanolabel; and (C) detection procedure. *Recreated from Li, H., Huang, Y., Yu, Y., Li, W., Yin, Y., Li, G., 2015. Peptide-based method for detection of metastatic transformation in primary tumors of breast cancer. Anal. Chem. 87, 9251–9256, with kind permission of American Chemical Society.*

As a member of the cyclin family, cyclin A2 is critical for the initiation of DNA replication, transcription, and cell cycle regulation through the association of cyclin-dependent kinases (CDK). As a prognostic indicator in early-stage cancers, cyclin A2 has been reported to overexpress in many types of cancers including breast cancer, liver cancer, lung cancer, soft tissue sarcoma, leukemia, and lymphoma. Cyclin A2 can be detected in early-stage cancer using functionalized graphene-based peptide sensors (Fig. 7.2; Feng et al., 2012). For this biosensor, the graphene with the intrinsic properties was functionalized noncovalently with porphyrin through strong $\pi-\pi$ stacking and hydrophobic interactions. The specific hexapeptide P0 was further immobilized on the surface as the detection probe of cyclin A2. The following attachment of cyclin A2 hampered the redox probe $[Fe(CN)_6]^{3-/4-}$ close to the electrode surface so as to change electrochemical impedance signal.

Cysteine-rich intestinal protein 1 (CRIP1) is a transcription factor present in abnormally high amounts in 90% of breast cancer cases. CRIP1 has been identified as an ideal biomarker for detecting breast cancer at earlier and more treatable stages, and it has significant prognostic impact on breast cancer. A cyclic peptide has been selected as the capture probe to bind to the N-terminal one of the two zinc finger domains of CRIP1 (Xie et al., 2014).

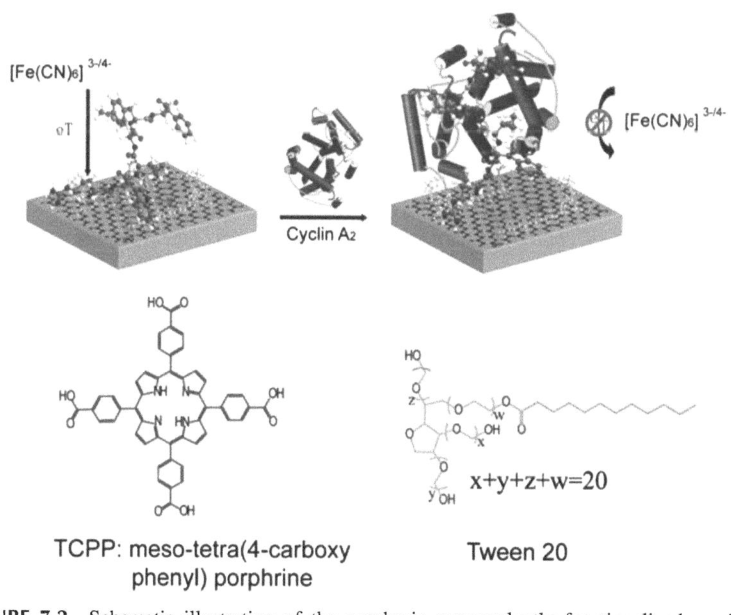

FIGURE 7.2 Schematic illustration of the porphyrin noncovalently functionalized graphene-based peptide sensor for cyclin A2 detection. *Recreated from Feng, L., Wu, L., Wang, J., Ren, J., Miyoshi, D., Sugimoto, N., et al., 2012. Detection of a prognostic indicator in early-stage cancer using functionalized graphene-based peptide sensors. Adv. Mater. 24, 125–131, with kind permission of John Wiley.*

As a primordial regulation protein in apoptosis, X-linked inhibitor of apoptosis protein (XIAP) was found to be overexpressed in many human tumor cells, such as breast cancer, bladder cancer, and prostate cancer cells. Moreover, XIAP plays an important role in inducing resistance for cancer therapies. Among XIAP, the third one, BIR3, has been widely recognized as a valuable anticancer target. The BIR3 domain of X-linked inhibitor of apoptosis protein was electrochemically detected by using a peptide-based conducting polymer sensor (Fig. 7.3; Wang et al., 2014). XIAP-BIR3 can be tested directly and specifically through a peptide-based electrochemical biosensor. The octapeptide, AVPFAQKG, conjugated to a redox transducer, can bind to XIAP-BIR3 without loss of the affinity. This so-called preintegrated conjugate and redox transducer was electro-copolymerized onto a GC electrode to provide a biosensor. The specific recognition between the immobilized peptide probe (AVPFAQKG) and the target protein, XIAP-BIR3 (13.05 kDa), can result in a considerable increase in steric hindrance at the electrode/electrolyte interface and provoked a current decrease.

Epidermal growth factor receptor peptide magnetic nanovesicles with lipid bilayers have been constructed to enhanced capture of liver cancer circulating tumor cells (Ding et al., 2016). Using a peptide amphiphile (PA) as a skeleton material, a peptide magnetic nanovector platform has been acquired with a lipid bilayer. Epidermal growth factor receptor positive liver tumor cells can be recognized by fluorescein isothiocyanate-labeled EGFR peptide nanoparticles (NPs). Meanwhile, EGFR peptide magnetic vesicles could efficiently recognize and separate hepatoma carcinoma cells from cell solutions and treated blood samples. Moreover, fluorescently labeled, EGFR-targeted PVX filaments were prepared using a two-step bioconjugation reaction (Chariou et al., 2015). Briefly, a bifunctional PEG linker with a 24.6 Å

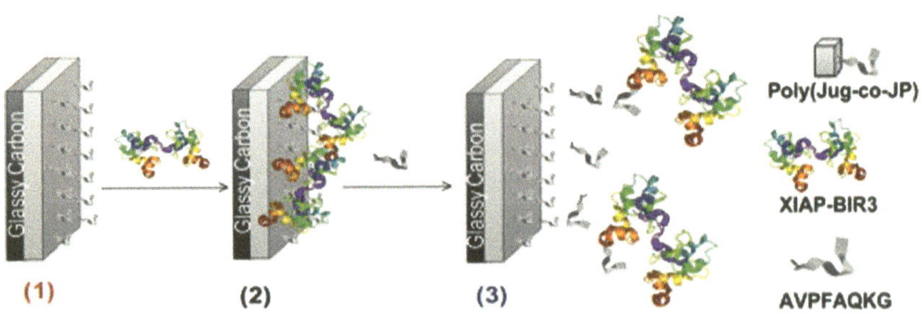

FIGURE 7.3 Strategy for label-free electrochemical detection of XIAP-BIR3 based on surface hindrance of the poly(JUG-co-JP)-modified electrode. (1) poly(JUG-co-JP)-modified electrode; (2) after complexation with XIAP-BIR3 (signal-off); (3) after addition of the AVPFAQKG octapeptide into the assay solution (signal-on). *Recreated from Wang, X., Piro, B., Reisberg, S., Anquetin, G., de Rocquigny, H., Jiang, P., et al., 2014. Direct, reagentless electrochemical detection of the BIR3 domain of X-linked inhibitor of apoptosis protein using a peptide-based conducting polymer sensor. Biosens. Bioelectron. 61, 57–62, with kind permission of Elsevier.*

spacer arm (SM(PEG)4) and Alexa Fluor 647 succinimidyl ester (NHS-A647) have been linked to solvent-exposed lysines on PVX followed by addition of the cysteine-terminated GE11 peptide targeting the maleimide side groups of SM(PEG)4. The proteinaceous particles remained stable after two rounds of bioconjugation, and several hundred copies of fluorophores with the incorporation of EGFR-specific GE11 peptide ligands.

Syk is a 72 kDa nonreceptor tyrosine kinase from bovine thymus and porcine spleen and plays a key role in B lymphocyte development and activation. Dysregulation of the expression or the activity of Syk is related to various disease states, making it a potential therapeutic target. A peptide containing a sequence derived from known Syk substrate preference motifs was designed. Subsequently, a Syk-specific artificial peptide biosensor has been constructed in a cell-based assay for direct detection of intracellular Syk activity and inhibition in both laboratory cell lines and primary splenic B cells (Lipchik et al., 2012).

7.1.2 Microbial Infection

Microbial infection of the bloodstream is a serious infection with a high rate of mortality (20%–70%). Detection of microbes is of great importance in clinical diagnosis by guiding the identification and antibiotic susceptibility of microbes. A rapid microbial detection method has been established by using a pleurocidin/single-walled carbon nanotubes/interdigital electrode-multichannel series piezoelectric quartz crystal sensor (Shi et al., 2017). The pleurocidin antimicrobial peptide was selected and served as a recognition probe that exhibits broad-spectrum antimicrobial activity. Meanwhile, the SWCNT was used as the electronic transducer and cross-linker for the immobilization of the peptide. Pleurocidin is detached from the SWCNT-modified IDE due to its specific interaction with the microbe, resulting in a signal output.

7.1.3 Tuberculosis

Tuberculosis (TB) is the second leading cause for death resulting from infectious diseases worldwide. *Mycobacterium tuberculosis* was detected based on h37rv binding peptides using surface functionalized magnetic microspheres coupled with quantum dots—a nano detection method (Yang et al., 2015).

7.1.4 Pregnancy Screening

Human chorionic gonadotropin (hCG) is a hormone produced by the placenta, and the hormone can be considered as the major indicator of embryo implantation in pregnancy. Additionally, hCG is an important clinical parameter for the early diagnosis of ectopic pregnancy and in prenatal screening for Down's syndrome. Based on a peptide-regulated catalytic AuNP-based mechanism, a simple, sensitive, label-free colorimetric assay has been developed by combining catalytic gold nanoparticles (AuNPs) with an hCG-specific peptide aptamer (Fig. 7.4; Chang et al., 2014). Owing to the presence of arginine, the total charge of the hCG-binding oligopeptide (PPLRINRHILTR) is positive at pH 7.4. AuNPs can bind to a peptide

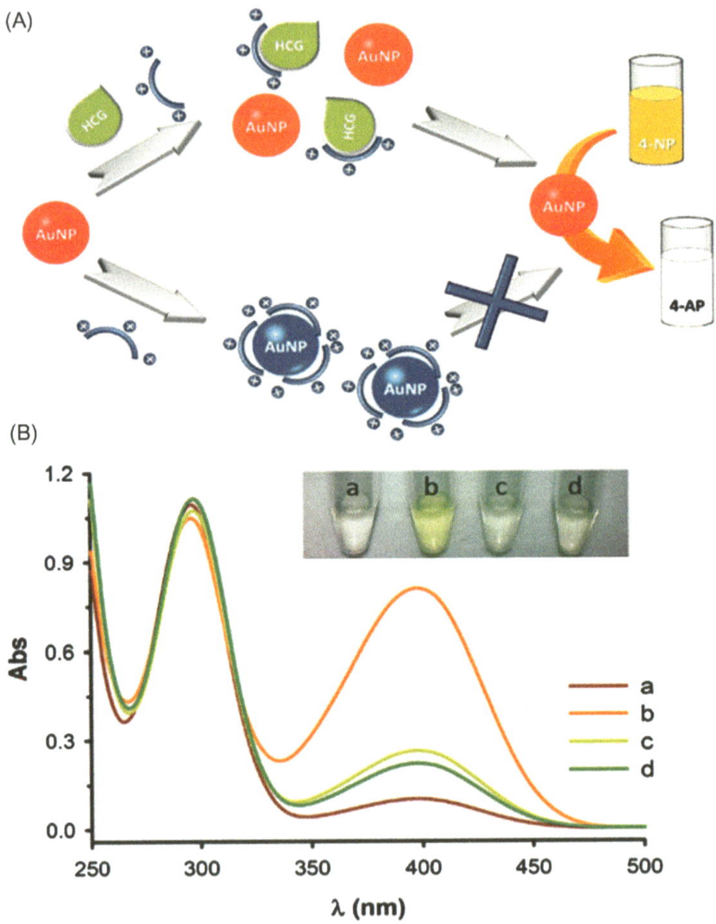

FIGURE 7.4 (A) Schematic illustration of the proposed colorimetric detection of human chorionic gonadotropin (hCG) based on peptide-regulated gold nanoparticles (AuNPs) catalysis. (B) Ultraviolet-visible spectrophotometry absorption spectra of (a) AuNPs, (b) AuNP + peptide aptamer, (c) AuNP + peptide aptamer + hCG, and (d) AuNP + hCG. Inset: the corresponding photographs. *Recreated from Chang, C.C., Chen, C.P., Lee, C.H., Chen, C.Y., Lin, C.W., 2014. Colorimetric detection of human chorionic gonadotropin using catalytic gold nanoparticles and a peptide aptamer. Chem. Commun. 50, 14443–14446, with kind permission of Royal Society of Chemistry.*

aptamer to form the AuNP–aptamer complex through electrostatic interaction. The bound peptide decreases the catalytic ability of AuNPs due to its coating. On the contrary, hCG molecules can competitively and specifically recognize the peptide, render its detachment with citrate-capped AuNPs and recover the catalytic activity.

Moreover, a short oligopeptide sequence, (N-)PPLRINRHILTR(-C), was identified after five rounds of screening by use of a phage library, and an antibody-free and label-free mechanism were reported to detect hCG (Fig. 7.5; Ding and Yang, 2013). After binding with the oligopeptide sequence, liquid crystal (LC) can transduce the binding event into optical signals. However, the captured hCG can disrupt a thin layer ($\sim 6\ \mu m$) of LC covered on the surface. Depending on the initial concentration of hCG, LC gave distinct optical signals visible to the naked eye.

hCG has also been tested through a peptide aptamer-based electrochemical biosensor (Fig. 7.6; Xia et al., 2017). Peptide probes on the electrode surface can trigger the *in situ* formation of the AgNPs-based network architecture. The network architecture of AgNPs-peptide formed on the electrode surface produced an amplified electrochemical signal through the solid-state Ag/AgCl reaction from AgNPs. In the presence of hCG, the peptide probe immobilized on the electrode surface would lose its ability to trigger the in situ formation of AgNPs-based network architecture. Thus, the electrochemical signal decreased with the increase of hCG concentration. Furthermore, the competitive assay for hCG detection was achieved by immersing the sensing electrode with hCG sample before incubation with AgNPs/peptide.

The graphene oxide–peptide-based surface plasmon resonance biosensors have been utilized to detect hCG (Fig. 7.7; Chiu et al., 2017). A GO–peptide-based SPR chip has been investigated to enhance the sensitivity

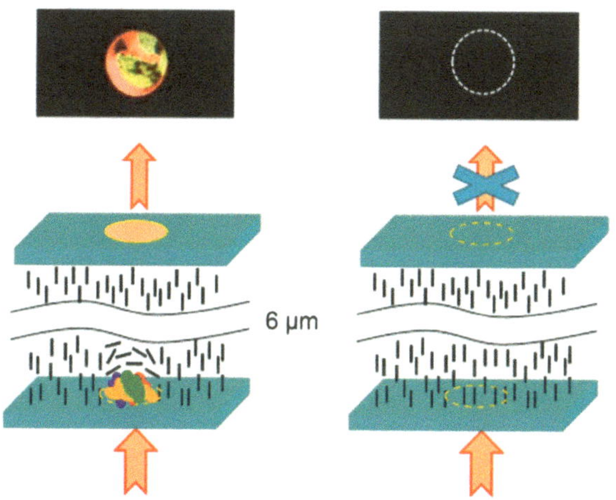

FIGURE 7.5 Antibody-free detection of human chorionic gonadotropin by use of liquid crystals. *Recreated from Ding, X., Yang, K.L., 2013. Antibody-free detection of human chorionic gonadotropin by use of liquid crystals. Anal. Chem. 85, 10710–10716, with kind permission of American Chemical Society.*

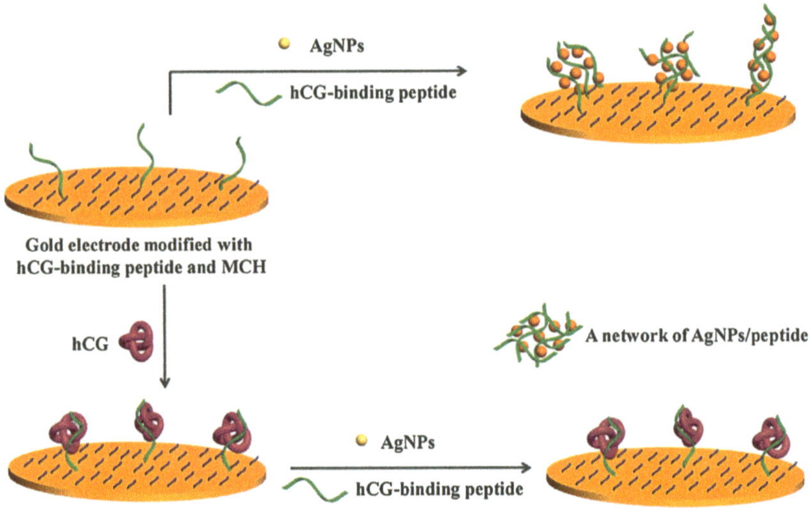

FIGURE 7.6 Schematic illustration of the electrochemical method for hCG detection using AgNPs as the redox reporters and hCG-binding peptide as the receptor of hCG and the inducer of AgNPs aggregation. *Recreated from Xia, N., Chen, Z., Liu, Y., Ren, H., Liu, L., 2017. Peptide aptamer-based biosensor for the detection of human chorionic gonadotropin by converting silver nanoparticles-based colorimetric assay into sensitive electrochemical analysis. Sens. Actuat. B 243, 784–791, with kind permission of Elsevier.*

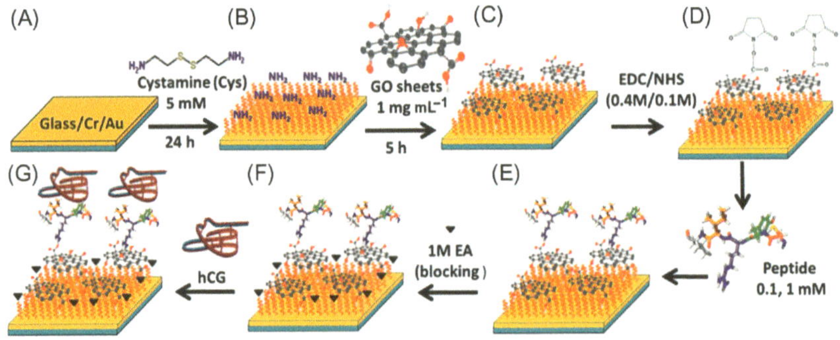

FIGURE 7.7 The GO-peptide-based SPR biochip experimental conditions. *Recreated from Chiu, N.-F., Kuo, C.T., Lin, T.L., Chang, C.C., Chen, C.Y., 2017. Ultra-high sensitivity of the non-immunological affinity of graphene oxide-peptide-based surface plasmon resonance biosensors to detect human chorionic gonadotropin. Biosens. Bioelectron. 94, 351–357, with kind permission of Elsevier.*

of a pregnancy disease assay. A modified peptide aptamer can be covalently attached to the surface of GO sheets, so as to enhance the hCG protein assay with GO–peptide-specific interactions. In addition, the interaction functions and different response surfaces of the peptide aptamer with GO sheets have been explored.

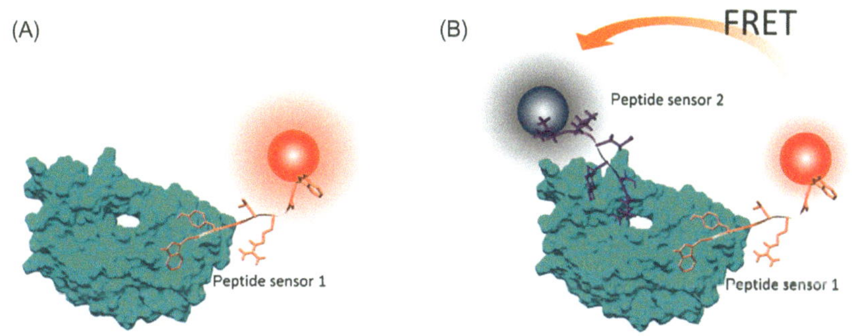

FIGURE 7.8 Schematic of the detection mechanism of HIV-1 protease. (A) A fluorescent sensor binds specifically to the HIV-1 PR molecule (peptide sensor 1 WSRVGYW-AF647). (B) A second sensor (peptide sensor 2 LLEYSL-BHQ-3), tagged with a fluorescent quencher, interacts with the active site of the protease. The biorecognition events trigger an energy transfer process between peptide sensors 1 and 2, and the signal from the fluorescent dye decreases. *Recreated from Herpoldt, K.L., Artzy-Schnirman, A., Christofferson, A.J., Makarucha, A.J., de la Rica, R., Yarovsky, I., et al., 2015. Designing fluorescent peptide sensors with dual specificity for the detection of HIV-1 protease. Chem. Mater. 27, 7187–7195, with kind permission of American Chemical Society.*

7.1.5 Human Immunodeficiency Virus

HIV-1 protease (HIV-1 PR) is a dimeric enzyme from the family of aspartic proteases. The enzyme has been widely exploited as a drug target and exhibits broad substrate recognition. HIV-1 protease can be detected through designed interactions with multiple peptide sensors (Fig. 7.8; Herpoldt et al., 2015). An energy transfer can happen between two peptide sensors with simultaneous mixing with the target protein. The multivalent nature of this assay increases the specificity of the detection by requiring all molecules to be interacting in order to give a FRET signal.

7.2 BIOSENSORS BASED ON CLEAVAGE OF THE PEPTIDE

As the important substrates for proteases and kinases, peptides have been widely used in monitoring enzyme activity, screening for enzyme inhibitors, and controlling release of drugs. The various biosensors for enzyme activity assay have been usually constructed based on the enzymatic catalysis of peptide substrates.

Legumain or asparaginyl endopeptidase (AEP) is a lysosomal cysteine protease with high cleavage specificity after an asparagine residue of substrate proteins. Legumain expression and activity are linked to a number of pathological conditions including cancer, atherosclerosis, and inflammation, yet its biological role in these pathologies is not well-understood. It is overexpressed in a majority of human solid tumors such as carcinomas of the breast, colon, and prostate. Knock-down of legumain in mouse cancer models results in marked decrease in tumor growth and metastasis. A new MRI contrast agent ([Gd-NBCB-TTDA-Leg(L)]) and a NIR fluorescent probe (CyTE777-Leg(L)-CyTE807) are available for in vitro and in vivo detection of legumain activity (Fig. 7.9; Chen et al., 2014). A cyclobutyl and a benzyl group residues were introduced on the carbon backbone of TTDA (3,6,10-tri(carboxymethyl)-3,6,10-triaza-dodecanedioic acid) to increase lipophilicity to the Gd(III) complex upon cleavage of a legumain-specific peptide substrate (Leg(L)) linked to the benzyl group. Legumain-mediated cleavage of the peptide substrate facilitates better wrapping of the Gd(III) ion and increases the water exchange rate (kex).

Matrix metalloproteinase-9 (MMP-9), a subgroup of the MMP family, is a 92 kDa-enzyme that is secreted by most human cancer cells. MMP-9 degrades collagen type IV, a major constituent of cellular basement membrane, and promotes cancer invasion. Increased amounts of MMP-9 have been detected in serum and plasma samples taken from rats and humans with malignant tumors. The peptide has been designed to have the specific site for MMP-9 cleavage (Gly/Leu) with a Cys residue at one end to facilitate interaction with gold ions during the cosynthesis process. A novel self-assembled peptide-MUA/AuNC/GO nanocomplex (pMAG) has been further constructed as a resonance energy transfer-based nanosensor for the development of simple and highly sensitive bioassay for MMP-9 detection (Fig. 7.10; Nguyen et al., 2017). The designed pMAG-based bioassay could selectively and sensitively detect MMP-9 by the dequenched AuNC fluorescence resulting from the peptide cleavage.

By using the cleavage of MMP-2 on peptide, a novel electrochemical biosensor was fabricated on the basis of optimal interenzyme distance regulated by PtNPs for highly efficient enzyme cascade amplification (Kou et al.,

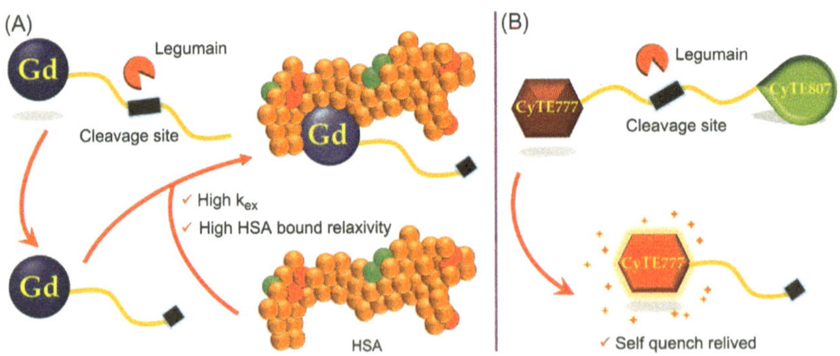

FIGURE 7.9 Systematic representation of (A) MRI contrast agent ([Gd-NBCB-TTDA-Leg (L)]) and (B) NIR fluorescent probe (CyTE777-Leg(L)-CyTE807) for legumain detection. *Recreated from Chen, Y.J., Wu, S.C., Chen, C.Y., Tzou, S.C., Cheng, T.L., Huang, Y.F., et al., 2014. Peptide-based MRI contrast agent and near-infrared fluorescent probe for intratumoral legumain detection. Biomaterials. 35, 304–315, with kind permission of Elsevier.*

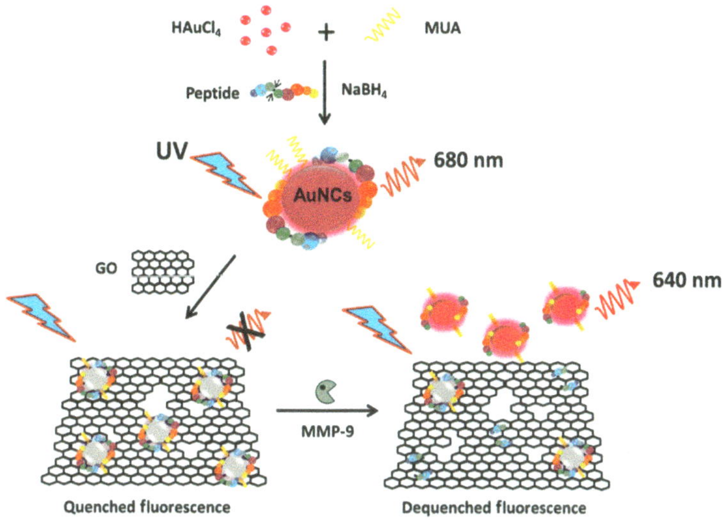

FIGURE 7.10 Schematic diagram of the synthesis and the application of fluorescent peptide-mercaptoundecanoic acid co-stabilized gold nanoclusters (peptide-MUA/AuNCs) for the detection of metalloproteinase-9. *Recreated from Nguyen, P.D., Cong, V.T., Baek, C., Min, J., 2017. Fabrication of peptide stabilized fluorescent gold nanocluster/graphene oxide nanocomplex and its application in turn-on detection of metalloproteinase-9. Biosens. Bioelectron. 89 (Pt 1), 666–672, with kind permission of Elsevier.*

2017). PtNPs can not only be convenient for regulating interenzyme distance for efficient enzyme cascade amplification, but can also as catalysts to accelerate enzyme cascade reactions, resulting in the enhancement of catalytic efficiency.

Meanwhile, a ultrasensitive electrochemiluminescence biosensing platform has been constructed to detect multiple types of biomarkers toward identical cancer on a single interface (Fig. 7.11; Nie et al., 2017). 5'-Thiolated and 3'-phosphate group-blocked probe DNA (pDNA)-modified CdS QDs surface served as the ECL emitters. The pDNA hybridized with the target, which could further hybridize with trigger DNA (tDNA). A long ssDNA nanotail was generated at 3'OH terminal of tDNA with the aid of TdT, and further captured numerous Fc-peptidess DNA conjugates. A signal switch "off" state was obtained owing to the efficient quenching effect of Fc on CdS QDs27. MMP-2 can specifically hydrolyze Fc-peptide-ssDNA conjugates (HOOC-PLGVR-Fc) at a particular site between G and V, resulting in the release of Fc from the sensing surface.

Moreover, MMP-2 can be detected ultrasensitively by upconversion fluorescence resonance energy transfer-based biosensor (Wang et al., 2012). A polypeptide probe was designed containing a specific MMP-2 substrate domain and a π-electron-rich region, tagged to the energy donor. MMP-2 cleaved the probe peptide into two fragments at the G-V bond, leading to the

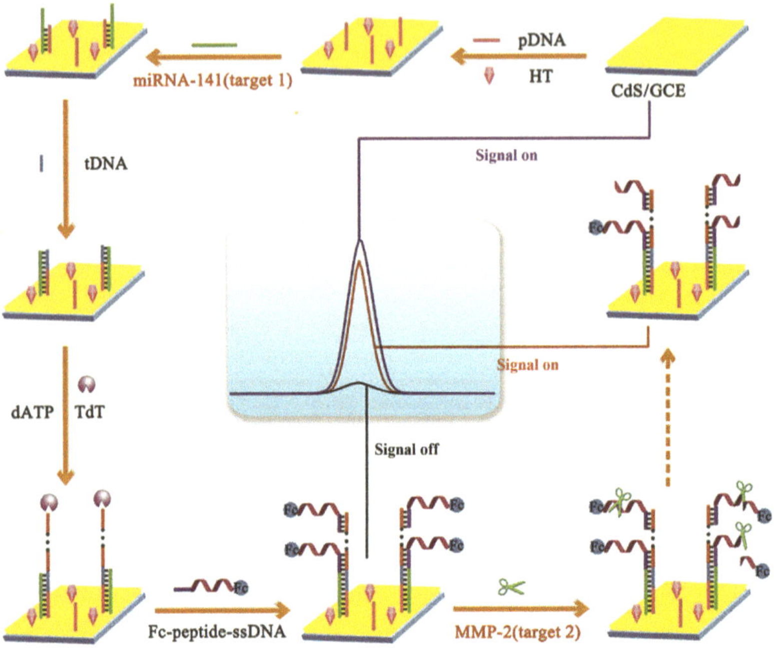

FIGURE 7.11 Fabrication of the ECL biosensing platform for the detection of multiple types of biomarkers on a single interface. *Recreated from Nie, Y., Zhang, P., Wang, H., Zhuo, Y., Chai, Y., Yuan, R., 2017. Ultrasensitive electrochemiluminescence biosensing platform for detection of multiple types of biomarkers toward identical cancer on a single interface. Anal. Chem. 89, 12821–12827, with kind permission of American Chemical Society.*

separation of the donor and acceptor. The fluorescence of UCPs is restored owing to the blocked FRET process. The substrate pentapeptide is preferred as -PLGVR- rather than the reverse sequence, i.e., -RVGLP-, in the probe. Through electrostatic attraction, the positively charged Arg residue was reported to adsorb graphene oxide. Inversely, the adsorption is unlikely to occur in the presence of CNP, because CNPs do not carry negatively charged carboxyl groups as graphene oxide does, and the particles are dispersed by nonionic surfactant in water solutions. So keeping the Arg residue on the UCP surface after cleavage will not prevent the separation of the donor from the acceptor.

Caspase-3 takes part in both intrinsic and extrinsic apoptosis pathways, and it is the most frequently used target to detect apoptosis. An *N*-terminal blocked peptide Ac-Gly-Gly-His-Asp-Glu-Val-Asp-His-Gly-Gly-Gly-Cys was employed as the enzyme substrate. The cleavage of peptide by caspase-3 can result in the exposure of a new free N-terminal amine group (Chen et al., 2015). Moreover, an octapeptide Ac-Gly-Asp-Glu-Val-Asp-Cys-Cys-Arg-NH_2 (GDEVDCCR, GR-8) was designed as the substrate of caspase-3.

The total charge of GR-8 was negative at pH 7.4 because Asp and Glu in the sequence were negatively charged while only Arg was positively charged. The -SH group on each cysteine made it convenient for the peptide to attach to citrate-capped AuNPs through Au-S bonds, and the binding of GR-8 did not induce AuNPs aggregation because of the negatively charged property of the peptide. Inversely, a shorter and positively charged peptide Cys-Cys-Arg-NH_2 (CCR, CR-3) was released due to the enzymatic cleavage of GR-8. The binding of CR-3 decreases the negative charge density on the AuNPs surface and broke the electrostatic stability of AuNPs, resulting in aggregation of AuNPs and corresponding color change from red to violet or blue (Pan et al., 2012; Fig. 7.12).

High-temperature requirement factor A1 (HtrA1) is essential for modulating preeclampsia (PE). The activity of HtrA1 in the pathological progress of PE can be monitored using a designed probe after sequential interactions. HtrA1 can cleave the substrate probe to expose a short sequence and subsequently the sequence can be recognized by TGase 2, resulting in the next conjugation to a receiving sequence of the signal probe. The receiving sequence was at one end of the signal probe and the short sequence on the substrate probe was at the opposite end, which was exposed by HtrA1 cleavage. These two motifs were spaced by the seed sequence. Under the proper conditions, the seed sequence can initiate a cascade of biomineralization to reduce Ag^+ to Ag nanoparticles, followed by the formation of the silver nanoparticle-decorated peptide nanowire (Fig. 7.13; Zhang et al., 2016).

Prostate specific antigen (PSA) is secreted by prostatic epithelial cells and has proved to be the best available tumor marker for early diagnosis of

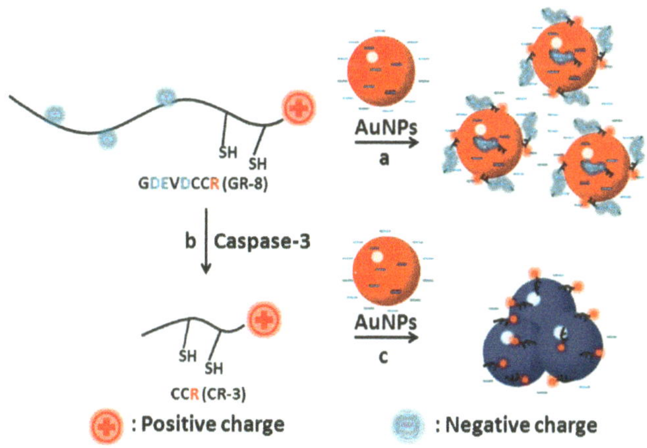

FIGURE 7.12 Schematic representation of the colorimetric assay of apoptosis. *Recreated from Pan, Y., Guo, M., Nie, Z., Huang, Y., Peng, Y., Liu, A., et al., 2012. Colorimetric detection of apoptosis based on caspase-3 activity assay using unmodified gold nanoparticles. Chem. Commun. 48, 997–999, with kind permission of Royal Society of Chemistry.*

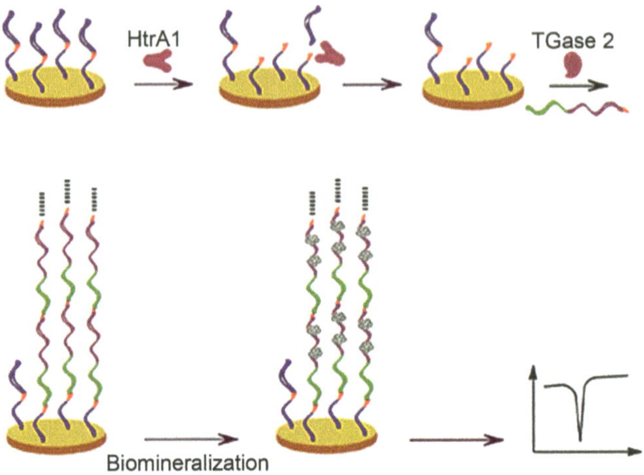

FIGURE 7.13 Peptide-induced biomineralization as a biomimetic means of detecting proteins in a mineralizing biocontext. *Recreated from Zhang, Y., Li, H., Huang, Y., Sun, L., Li, G., 2016. Peptide-induced bio-mineralization as a bio-mimetic means of detecting proteins in a mineralizing bio-context. Nano Res. 9, 1489–1496, with kind permission of Nanotechnology.*

prostate cancer. An electrochemical peptide cleavage-based biosensor has been explored for the determination of PSA by coupling with Au@SiO_2 to amplify the anodic stripping signal of deposition Ag (Fig. 7.14; He et al., 2015). PSA can specifically recognize and cleave the peptide, resulting in release of the tracing tag from electrode surface. Subsequently, silver deposition was performed on the remaining dithiobis (succinimidylpropionate) (DSP)@Au@SiO_2 composites, and its electrochemical stripping signal was used to monitor the PSA activity. Moreover, PSA can also be detected through the induced cleavage of a specific peptide with positively charged gold nanoparticles as signal enhancer (Wang et al., 2015). AuNPs were linked on the peptide as a molecular recognition element, and integrated with the positively charged surfactant cetyltrimethylammonium bromide (CTAB). PSA can specifically hydrolyze and cleave the peptide modified with positively charged AuNPs, resulting in its release from the electrode surface. Furthermore, PSA can be analyzed ultrasensitively by using bovine serum albumin as an effective sensitivity enhancer (Tang et al., 2017). In addition, PSA can also be detected by using an electrochemical peptide cleavage-based biosensor via host–guest interaction between ferrocene and β-cyclodextrin (β-CD) (Xie et al., 2015). For this biosensor, the peptide cleavage events were directly converted into electrochemical signals via the host–guest interaction between Fc and β-CD. PSA selectively cut Fc labeling peptide immobilized onto Fe_3O_4@Au magnetic beads, to release much Fc containing peptide fragments, followed by magnetic separation. The released Fc molecules can be concentrated on the electrode surface through

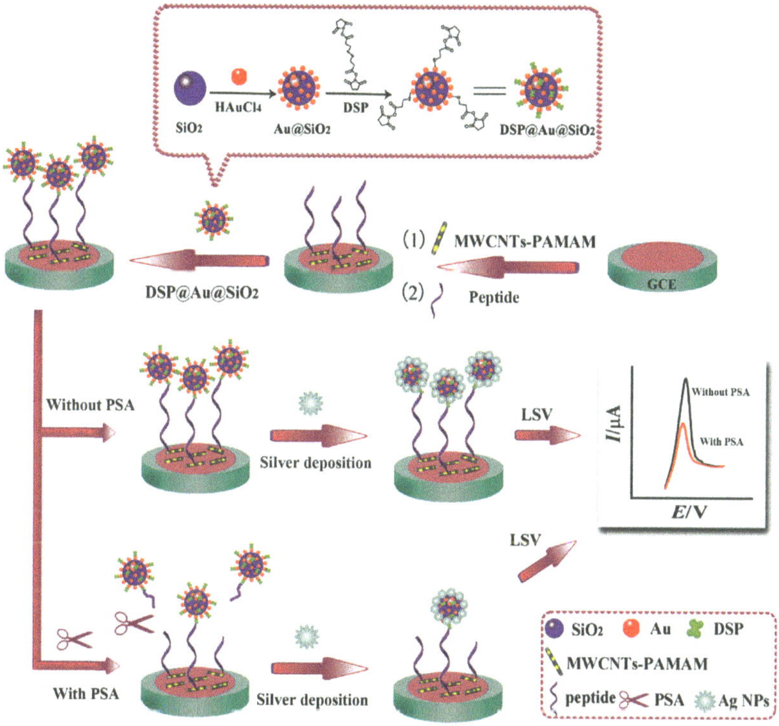

FIGURE 7.14 Schematic diagram of the electrochemical-peptide biosensor for detection of PSA. *Recreated from He, Y., Xie, S., Yang, X., Yuan, R., Chai, Y., 2015. Electrochemical peptide biosensor based on in situ silver deposition for detection of prostate specific antigen. ACS Appl. Mater. Interfaces 7, 13360–13366, with kind permission of American Chemical Society.*

the formation of a host–guest complex with β-CD. Moreover, the usage of MWCNT-polyamidoamine dendrimers (CNT-PAMAM) nanohybrids, can not only enhance the immobilization of β-CD for capturing the released Fc, but also facilitated electron transfer to enhance the sensitivity of the electrochemical biosensor.

Electrogenerated chemiluminescence biosensor has also been used to analyze PSA through target-induced cleavage of peptide within Nafion films incorporated with gold nanoparticles and ECL emitting species (Fig. 7.15; Qi et al., 2014). PSA can recognize and cleave a specific peptide (CHSSKLQK) as a molecular recognition element. Meanwhile, AuNPs were used as the amplification platform, with tris(2,2′-bipyridine) dichlororuthenium(II) ($Ru(bpy)^{32+}$) as an ECL emitting species and ferrocene carboxylic acid (Fc) as the ECL quencher.

Glypican-3 (GPC3) is a valuable biomarker for hepatocellular carcinoma (HCC) patients and its circulating level is very promising in the distinction between HCC and benign hepatic disorders. GPC3 can be successfully

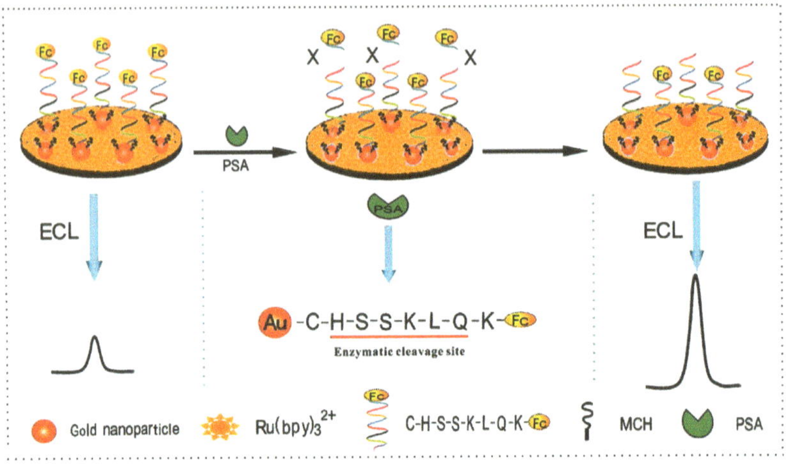

FIGURE 7.15 Schematic diagram of the fabrication process of ECL-PB biosensor for the determination of PSA. *Recreated from Qi, H., Li, M., Dong, M., Ruan, S., Gao, Q., Zhang, C., 2014. Electrogenerated chemiluminescence peptide-based biosensor for the determination of prostate-specific antigen based on target-induced cleavage of peptide. Anal. Chem. 86, 1372–1379, with kind permission of American Chemical Society.*

analyzed by using peptide-based nanolabels (Huang et al., 2014). The capture probe contained a specific GPC3-binding motif, a distal biotin, and a GGG linker between the two, in order to avoid interference of biotin in the GPC3–peptide interaction. The probe was further ferrocenylated and immobilized on the streptavidin-coated gold nanoparticle (GNP) to form a nanolabel. Through incubating with a solution containing GPC3, a fraction of the probes on the electrode surface became GPC3-bound. The access of thermolysin to the cleavage site on the GPC3-bound probes was blocked by the bulky GPC3 molecules, resulting in the inaccessible nanolabel onto the surface of electrode.

7.3 BIOSENSORS BASED ON THE CONFORMATIONAL SHIFT OF PEPTIDE

Unregulated changes in protease activity are linked to many diseases including cancer. A recombinant polypeptide comprised of a stimuli-responsive elastin-like polypeptide, a protease substrate sequence, and a highly charged terminal sequence, has been used for the assay for matrix metalloproteinase-1 (MMP-1) (Fig. 7.16; Ghoorchian et al., 2014). The first segment is a thermally responsive ELP, which was chosen because it has several important properties relevant to the design of the assay. The second component of the modular polypeptide was a protease substrate that was fused to the ELP. We chose the DGQVPMSMRGG sequence because it is a substrate for MMP-1.

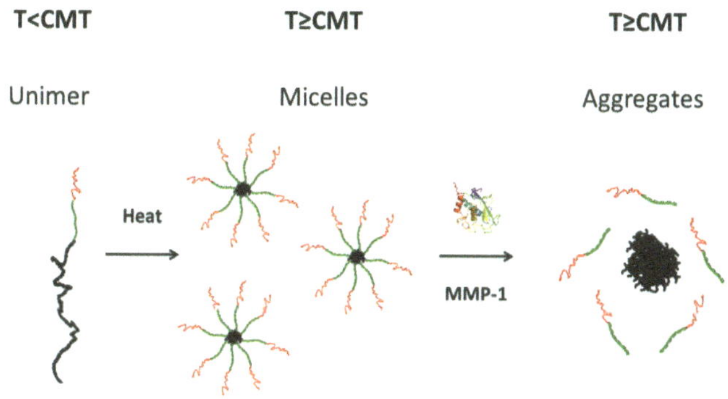

FIGURE 7.16 Sequence of the engineered, stimuli-responsive polypeptide (unimer, herein referred to as ELP – substrate fusion), schematic depiction of its micellization above its critical micellization temperature (CMT), and subsequent aggregation of the ELP after incubation with the enzyme, MMP-1, at a temperature > CMT. *Recreated from Ghoorchian, A., Chilkoti, A., López, G.P., 2014. Simple assay for proteases based on aggregation of stimulus-responsive polypeptides. Anal. Chem. 86, 6103–6110, with kind permission of American Chemical Society.*

The third module was DDEGQQDDEEGY, an anionic peptide that was appended to the hydrophilic protease substrate to provide amphiphilicity, so as to drive the entire ELP–substrate fusion peptide to self-assemble into micelles at the physiological temperature of the assay. Addition of MMP-1 to the solution containing these micelles results in cleavage of the micellar peptide sequences, which abrogates their amphiphilicity and, hence, self-assembly of the micelles.

7.4 CONCLUSION

For protein assay, a variety of biosensors have been explored including (1) biosensors based on the binding of peptide to target, (2) biosensors based on cleavage of the peptide, and (3) biosensors based on the conformational shift of peptide. Considering the fact that peptide possesses great advantages including mature synthesis protocols, diverse structures, and as highly selective substrates for enzymes, peptide-based biosensors have a great potential for further exploration and application in the future.

REFERENCES

Chang, C.C., Chen, C.P., Lee, C.H., Chen, C.Y., Lin, C.W., 2014. Colorimetric detection of human chorionic gonadotropin using catalytic gold nanoparticles and a peptide aptamer. Chem. Commun. 50, 14443–14446.

Chariou, P.L., Lee, K.L., Wen, A.M., Gulati, N.M., Stewart, P.L., Steinmetz, N.F., 2015. Detection and imaging of aggressive cancer cells using an epidermal growth factor receptor (EGFR)-targeted filamentous plant virus-based nanoparticle. Bioconjug. Chem. 26, 262–269.

Chen, Y.J., Wu, S.C., Chen, C.Y., Tzou, S.C., Cheng, T.L., Huang, Y.F., et al., 2014. Peptide-based MRI contrast agent and near-infrared fluorescent probe for intratumoral legumain detection. Biomaterials. 35, 304–315.

Chen, H., Zhang, J., Gao, Y., Liu, S., Koh, K., Zhu, X., et al., 2015. Sensitive cell apoptosis assay based on caspase-3 activity detection with graphene oxide-assisted electrochemical signal amplification. Biosens. Bioelectron. 68, 777–782.

Chiu, N.F., Kuo, C.T., Lin, T.L., Chang, C.C., Chen, C.Y., 2017. Ultra-high sensitivity of the non-immunological affinity of graphene oxide-peptide-based surface plasmon resonance biosensors to detect human chorionic gonadotropin. Biosens. Bioelectron. 94, 351–357.

Ding, X., Yang, K.L., 2013. Antibody-free detection of human chorionic gonadotropin by use of liquid crystals. Anal. Chem. 85, 10710–10716.

Ding, J., Wang, K., Tang, W.J., Li, D., Wei, Y.Z., Lu, Y., et al., 2016. Construction of epidermal growth factor receptor peptide magnetic nanovesicles with lipid bilayers for enhanced capture of liver cancer circulating tumor cells. Anal. Chem. 88, 8997–9003.

Feng, L., Wu, L., Wang, J., Ren, J., Miyoshi, D., Sugimoto, N., et al., 2012. Detection of a prognostic indicator in early-stage cancer using functionalized graphene-based peptide sensors. Adv. Mater. 24, 125–131.

Ghoorchian, A., Chilkoti, A., López, G.P., 2014. Simple assay for proteases based on aggregation of stimulus-responsive polypeptides. Anal. Chem. 86, 6103–6110.

He, Y., Xie, S., Yang, X., Yuan, R., Chai, Y., 2015. Electrochemical peptide biosensor based on in situ silver deposition for detection of prostate specific antigen. ACS Appl. Mater. Interfaces 7, 13360–13366.

Herpoldt, K.L., Artzy-Schnirman, A., Christofferson, A.J., Makarucha, A.J., de la Rica, R., Yarovsky, I., et al., 2015. Designing fluorescent peptide sensors with dual specificity for the detection of HIV-1 protease. Chem. Mater. 27, 7187–7195.

Huang, Y., Li, H., Gao, T., Liu, X., Li, G., 2014. A sensitive method for protein assays using a peptide-based nano-label: human glypican-3 detection for hepatocellular carcinomas diagnosis. Analyst 139, 3744–3747.

Kou, B.B., Chai, Y.Q., Yuan, Y.L., Yuan, R., 2017. PtNPs as scaffolds to regulate interenzyme distance for construction of efficient enzyme cascade amplification for ultrasensitive electrochemical detection of MMP-2. Anal. Chem. 89, 9383–9387.

Li, H., Huang, Y., Yu, Y., Li, W., Yin, Y., Li, G., 2015. Peptide-based method for detection of metastatic transformation in primary tumors of breast cancer. Anal. Chem. 87, 9251–9256.

Lipchik, A.M., Killins, R.L., Geahlen, R.L., Parker, L.L., 2012. A peptide-based biosensor assay to detect intracellular Syk kinase activation and inhibition. Biochemistry 51, 7515–7524.

Nguyen, P.D., Cong, V.T., Baek, C., Min, J., 2017. Fabrication of peptide stabilized fluorescent gold nanocluster/graphene oxide nanocomplex and its application in turn-on detection of metalloproteinase-9. Biosens. Bioelectron. 89 (Pt 1), 666–672.

Nie, Y., Zhang, P., Wang, H., Zhuo, Y., Chai, Y., Yuan, R., 2017. Ultrasensitive electrochemiluminescence biosensing platform for detection of multiple types of biomarkers toward identical cancer on a single interface. Anal. Chem. 89, 12821–12827.

Pan, Y., Guo, M., Nie, Z., Huang, Y., Peng, Y., Liu, A., et al., 2012. Colorimetric detection of apoptosis based on caspase-3 activity assay using unmodified gold nanoparticles. Chem. Commun. 48, 997–999.

Qi, H., Li, M., Dong, M., Ruan, S., Gao, Q., Zhang, C., 2014. Electrogenerated chemiluminescence peptide-based biosensor for the determination of prostate-specific antigen based on target-induced cleavage of peptide. Anal. Chem. 86, 1372–1379.

Shi, X., Zhang, X., Yao, Q., He, F., 2017. A novel method for the rapid detection of microbes in blood using pleurocidin antimicrobial peptide functionalized piezoelectric sensor. J. Microbiol. Methods 133, 69–75.

Tang, Z., Fu, Y., Ma, Z., 2017. Bovine serum albumin as an effective sensitivity enhancer for peptide-based amperometric biosensor for ultrasensitive detection of prostate specific antigen. Biosens. Bioelectron. 94, 394–399.

Wang, Y., Shen, P., Li, C., Wang, Y., Liu, Z., 2012. Upconversion fluorescence resonance energy transfer based biosensor for ultrasensitive detection of matrix netalloproteinase-2 in blood. Anal. Chem. 84, 1466–1473.

Wang, X., Piro, B., Reisberg, S., Anquetin, G., de Rocquigny, H., Jiang, P., et al., 2014. Direct, reagentless electrochemical detection of the BIR3 domain of X-linked inhibitor of apoptosis protein using a peptide-based conducting polymer sensor. Biosens. Bioelectron. 61, 57–62.

Wang, D., Zheng, Y., Chai, Y., Yuan, Y., Yuan, R., 2015. Target protein induced cleavage of a specific peptide for prostate-specific antigen detection with positively charged gold nanoparticles as signal enhancer. Chem. Commun. 51, 10521–10523.

Xia, N., Chen, Z., Liu, Y., Ren, H., Liu, L., 2017. Peptide aptamer-based biosensor for the detection of human chorionic gonadotropin by converting silver nanoparticles-based colorimetric assay into sensitive electrochemical analysis. Sens. Actuat. B 243, 784–791.

Xie, H., Li, H., Huang, Y., Wang, X., Yin, Y., Li, G., 2014. Combining peptide and DNA for protein assay: CRIP1 detection for breast cancer staging. ACS Appl. Mater. Interfaces 6, 459–463.

Xie, S., Zhang, J., Yuan, Y., Chai, Y., Yuan, R., 2015. An electrochemical peptide cleavage-based biosensor for prostate specific antigen detection via host-guest interaction between ferrocene and beta-cyclodextrin. Chem. Commun. 51, 3387–3390.

Yang, H., Qin, L., Wang, Y., Zhang, B., Liu, Z., Ma, H., et al., 2015. Detection of *Mycobacterium tuberculosis* based on H37R(v) binding peptides using surface functionalized magnetic microspheres coupled with quantum dots - a nano detection method for *Mycobacterium tuberculosis*. Int. J. Nanomedicine. 10, 77–88.

Zhang, Y., Li, H., Huang, Y., Sun, L., Li, G., 2016. Peptide-induced bio-mineralization as a biomimetic means of detecting proteins in a mineralizing bio-context. Nano Res. 9, 1489–1496.

Chapter 8

Protein Assay Based on Protein–Small Molecule Interaction

Ya Cao and Juan Zhang
Center for Molecular Recognition and Biosensing, School of Life Sciences, Shanghai University, Shanghai, P. R. China

Chapter Outline

Benefiting from the efforts in drug discovery, small molecule ligands with high specificity and affinity have been developed for many important protein disease biomarkers. This provides another promising approach for the construction of nano-inspired biosensors for protein assay.

8.1 PROTEIN ASSAY BASED ON SMALL MOLECULE-TETHERED CHEMICAL PROBES

Small molecule-tethered chemical probes are generally prepared by simply conjugating the protein-specific small molecule ligand to a signaling moiety. The combination of specific binding and strong signaling has made these probes powerful tools in protein assay, as they allow simple, sensitive, and specific detection with high signal-to-noise ratios (Kubota and Hamachi, 2015; Pode et al., 2017; Zhuang et al., 2013). Recently, by coupling with recognition-driven supramolecular disassembly, these probes have been

Nano-inspired Biosensors for Protein Assay with Clinical Applications.
DOI: https://doi.org/10.1016/B978-0-12-815053-5.00008-8

utilized for fabricating nano-inspired biosensors to detect protein disease biomarkers.

8.1.1 Human Carbonic Anhydrases

Human carbonic anhydrases (hCAs) are a family of ubiquitous metalloenzymes that comprise 15 different isoforms and catalyze the reversible hydration of carbon dioxide to bicarbonate (Koutnik et al., 2017). Accumulating findings reveal that aberrant levels of hCAs are closely correlated to a range of diseases, including cancer, epilepsy, glaucoma, and osteoporosis (Bruno et al., 2017; Koukourakis et al., 2001; Liao et al., 2003). For this reason, a number of nano-inspired biosensors have been developed for hCAs detection by making use of small molecule-tethered chemical probes (Hou et al., 2015; Mizusawa et al., 2010; Molla et al., 2015; Takaoka et al., 2009; Wang et al., 2015a; Yoshii et al., 2014).

A typical design was proposed by Takaoka et al. (2009); (Fig. 8.1A). In the work, they designed a small molecule-tethered chemical probe composed of a ligand specific to human carbonic anhydrase I (hCAI) and a ^{19}F-containing group that gave signal in NMR spectroscopy. The basis of the biosensor was that the probes alone were NMR-silent because of their ability to self-assemble into supramolecular nanoparticles, but gave distinct ^{19}F signals in the presence of target hCAI through recognition-driven disassembly

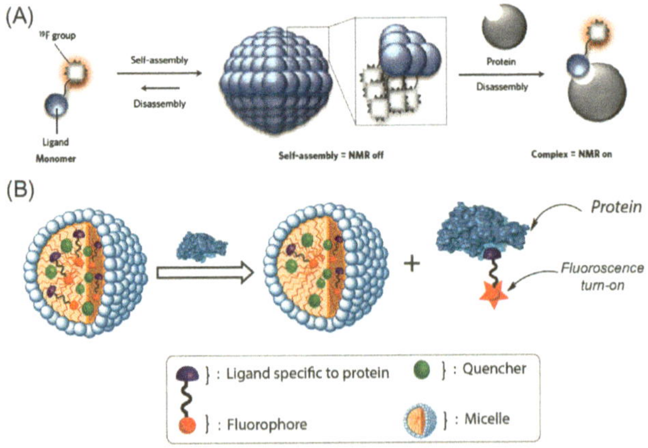

FIGURE 8.1 (A) Detection of hCAI based on self-assembling small molecule-tethered chemical probe. (B) Nano-inspired biosensor for hCAI detection based on supramolecular dissociation strategy. *(A) Recreated from Takaoka, Y., Sakamoto, T., Tsukiji, S., Narazaki, M., Matsuda, T., Tochio, H., et al., 2009. Self-assembling nanoprobes that display off/on 19F nuclear magnetic resonance signals for protein detection and imaging. Nat. Chem. 1, 557–561, with kind permission of Springer Nature. (B) Recreated from Wang, H., Zhuang, J., Raghupathi, K.R., Thayumanavan, S., 2015a. A supramolecular dissociation strategy for protein sensing. Chem. Commun. 51, 17265–17268, with kind permission of The Royal Society of Chemistry.*

of the nanoparticles. As the signal response was determined by specific protein–small molecule interaction, the biosensor has been successfully applied to not only detecting purified hCAI in test tubes but also visualizing the protein within live cells.

Based on a similar design, Hou et al. (2015) have proposed a nano-inspired fluorescent biosensor for detecting hCAs. The critical element of the biosensor was a near-infrared (near-IR) fluorogenic probe, where a small molecule ligand was conjugated to a novel γ-phenyl-substituted Cy5 fluorophore. In the absence of target hCAs, the probes were programmed to form a self-assembled nanoscale aggregate. However, if target protein was present, specific binding of the protein with small molecule ligand forced the disassembly of the nanoaggregate. In this way, the absence/presence of target hCAs was transformed into the assembly/disassembly of the probes, which resulted in switchable fluorescence emission. Taking human carbonic anhydrase II (hCAII) and transmembrane-type carbonic anhydrase (hCAIX) as evincive examples, the biosensor has been validated to permit specific detection and visualization of target proteins without washing operation.

Unlike the above two works, Wang et al. (2015a) have proposed a novel supramolecular dissociation strategy to develop a nano-inspired biosensor for hCA detection. As shown in Fig. 8.1B, in this work, a small molecule-tethered fluorogenic probe was designed to be noncovalently incorporated into micellar nanoassemblies equipped with quenchers. In this case, the fluorescence was in the "off" state due to the proximity of the fluorophore and the quencher. In contrast, upon the presence of target protein (hCAI), the specific protein–small molecule interaction reduced the hydrophobicity of the small molecule-tethered fluorogenic probe and therefore forced the dissociation of the probe away from the micelle. As a result, the proximity of fluorophore/quencher pair was significantly decreased, which turned the fluorescence to the "on" state. The biosensor obviates the requirement of inherent self-assembly features of small molecule-tethered probe, and is therefore simple in design and has great potential to open up new avenues in protein assay.

8.1.2 Biotin Receptor and Folate Receptor

Recent researches reveal that the biotin or folate-specific transporter systems are enhanced in many kinds of cancers, casing the overexpression of biotin receptor (BR) or folate receptor (FR) on cancer cells (Low et al., 2008; Ren et al., 2015). Considering the high affinities of BR and FR to the biotin and folic acid ($K_d = 10^{10}-10^{15}$ M) even after conjugation, small molecule-tethered chemical probe-based biosensors have been increasingly developed towards the two proteins (Azagarsamy et al., 2010; Fang et al., 2018; He et al., 2015; Li et al., 2017; Mizusawa et al., 2012).

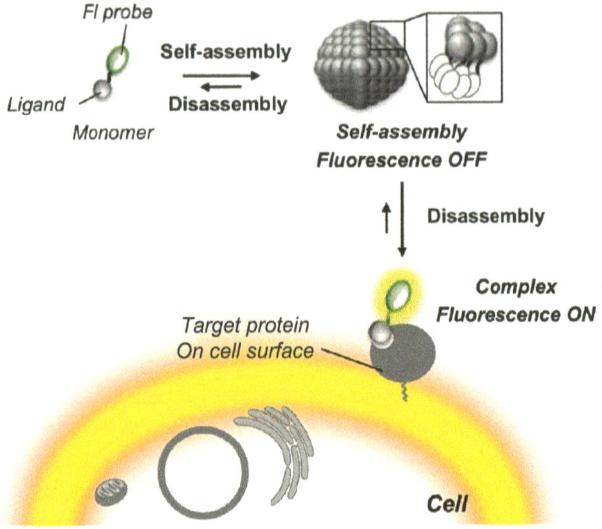

FIGURE 8.2 Nano-inspired biosensor targeted to FR on live cell surfaces based on self-assembling small molecule-tethered chemical probe. *Recreated from Mizusawa, K., Takaoka, Y., Hamachi, I., 2012. Specific cell surface protein imaging by extended self-assembling fluorescent turn-on nanoprobes. J. Am. Chem. Soc. 134, 13386–13395, with kind permission of American Chemical Society.*

In one such work, Mizusawa et al. (2012) have proposed a nano-inspired biosensor that fluorescently detected overexpressed FR on cancer cell surfaces (Fig. 8.2). The biosensor was based on a self-assembling small molecule-tethered fluorescent probe, which was designed to comprise a hydrophilic methotrexate group to serve as the ligand specific to FR, a fluorescein group to act as the signaling moiety and the hydrophobic core, and a linker group to control the hydrophilic/hydrophobic balance. With such a probe, the absence of target FR allowed the formation of self-assembled nanoparticles. In this state, only slight fluorescence emission could be observed as the fluorescein groups underwent self-quenching. However, once FR was introduced into the system, the methotrexate-tethered probe was specifically bound, causing the disassembly of the formed nanoparticles. As a consequence, switchable fluorescence emission took place, leading to the observation of strong fluorescence from cancer cell surfaces.

In another work, a nanostructured phthalocyanine assembly (NanoPcTB) has been explored to detect BR on living cell surfaces (Fig. 8.3; Li et al., 2017). The NanoPcTB self-assembled from a "one-for-all" building block that was synthesized with a versatile zinc(II) phthalocyanine derivative bearing biotin groups and possessed switchable photoactivities. By using avidin to simulate BR as the target, the NanoPcTB was proved to display selective protein-responsive partial disassembly, which generated increased

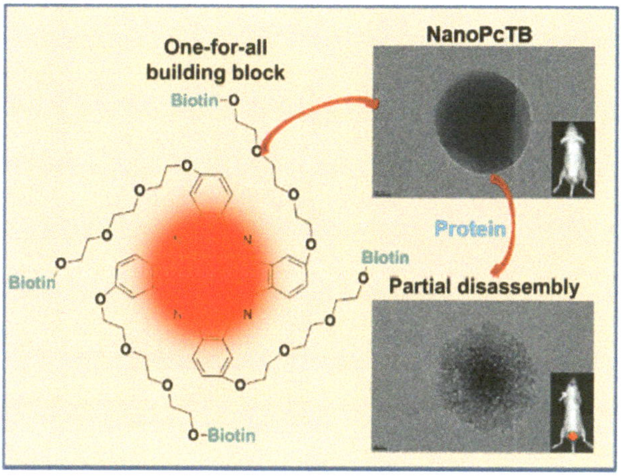

FIGURE 8.3 Nano-inspired biosensor for overexpressed BR on cancer cell surfaces based on a "one-for-all" nanomaterial. *Recreated from Li, X., Kim, C.Y., Lee, S., Lee, D., Chung, H.M., Kim, et al., 2017. Nanostructured phthalocyanine assemblies with protein-driven switchable photoactivities for biophotonic imaging and therapy. J. Am. Chem. Soc. 139, 10880–10886, with kind permission of American Chemical Society.*

fluorescence and reactive oxygen species, making NanoPcTB an ideal probe for BR. Further investigations revealed that the NanoPcTB was capable of distinguishing the BR-positive cancer cells (e.g., A549 and HeLa) from normal cells (e.g., WI38-VA13) that are low in BR expression, providing a powerful tool for evaluating the expression of BR on living cell surfaces.

8.2 PROTEIN ASSAY BASED ON SMALL MOLECULE-LINKED DNA

DNA is one of the most fascinating biopolymers and is well known as the carrier of genetic information (Teller and Willner, 2010). Over the last few decades, DNA has been found to be capable of displaying additional functions besides its central heredity features, such as acting as an enzyme mimetic or exhibiting specific binding affinity to biomolecules, and producing artificial DNA nanostructures that offer tremendous opportunities for signal amplification (Jung and Ellington, 2014; Lu et al., 2013; Wang et al., 2014; Zhang et al., 2012). Conjugating small molecule ligands to DNA strands can handily take advantage of these attractive features of DNA, and therefore paves a new way to construct nano-inspired biosensors towards protein disease biomarkers.

8.2.1 Folate Receptor

The pioneering work in this field was reported by Wu et al. in 2009 (Fig. 8.4). The basic biosensing hypothesis of this work was the terminal

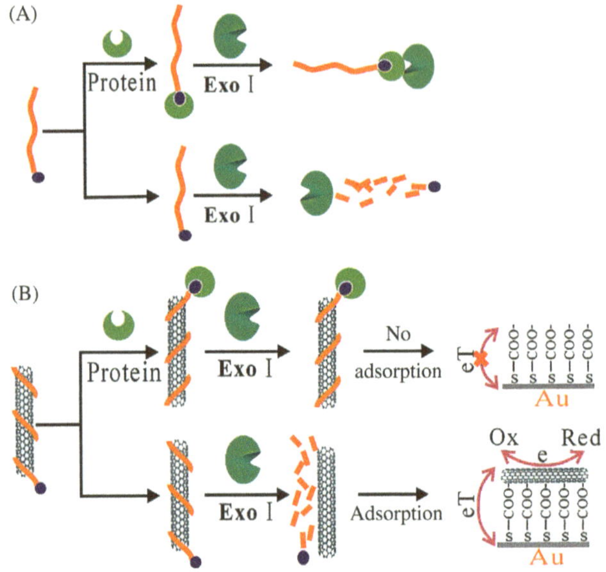

FIGURE 8.4 Electrochemical detection of FR based on terminal protection of small molecule-linked DNA and single-walled carbon nanotubes. (A) Terminal protection of small molecule-linked DNA. (B) A schematic view of the electrochemical method for detecting FR. *Recreated from Wu, Z., Zhen, Z., Jiang, J.H., Shen, G.L., Yu, R.Q., 2009. Terminal protection of small-molecule-linked DNA for sensitive electrochemical detection of protein binding via selective carbon nanotube assembly. J. Am. Chem. Soc. 131, 12325–12332, with kind permission of American Chemical Society.*

protection of small molecule-linked DNA, which relied on their finding that specific protein–small molecule interaction could bring dramatic steric hindrance that inhibited the enzyme (exonuclease I, Exo I) to approach the reaction site, consequently protecting the small molecule-linked DNA from Exo I-catalyzed degradation. By coupling with single-walled carbon nanotubes (SWNTs) and 16-mercaptohexadecanoic acid (MHA) modified gold electrode, they have successfully applied the terminal protection strategy to an electrochemical biosensor for FR detection. They found that after wrapping the small molecule (e.g., folate)-linked DNA on SWNTs through aromatic interactions, the DNA was still active for being degraded by Exo I, which produced "naked" SWNTs that could be recruited to the MHA-modified electrode to mediate efficient electron transfer, generating a strong redox current. However, in the presence of target FR, the specific binding of the protein to the folate ligand yielded significant steric hindrance, which served as the barrier to protect the DNA probe from digestion by Exo I. In this state, the SWNTs were kept wrapped by DNA probes and could not be recruited to the MHA-modified electrode, resulting in an ignorable redox current. By challenging with different concentrations of FR, the biosensor showed a desirable linear range (10 pM to 1.0 nM) and detection limit (3 pM) for FR detection, which was good enough for clinical applications.

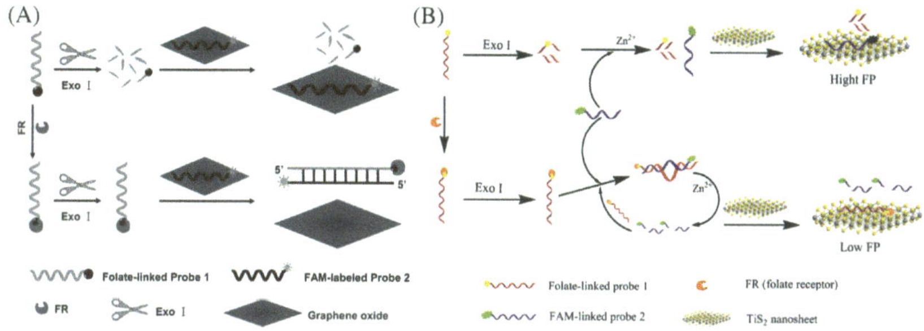

FIGURE 8.5 (A) Nano-inspired biosensor for FR detection based on terminal protection of small molecule-linked DNA and graphene oxide. (B) Detection of FR based on terminal protection of small molecule-linked DNA andTiS_2 nanosheet. *(A) Recreated from He, Y., Xing, X., Tang, H., Pang, D., 2013. Graphene oxide-based fluorescent biosensor for protein detection via terminal protection of small-molecule-linked DNA. Small 9, 2097–2101, with kind permission of Wiley-VCH. (B) Recreated from Li X., Ding X., Li Y., Wang L., Fan J., 2016. A TiS2 nanosheet enhanced fluorescence polarization biosensor for ultra-sensitive detection of biomolecules, Nanoscale 8, 9852–9860, with kind permission of The Royal Society of Chemistry.*

Apart from SWNTs, many other nanomaterials have been integrated with the terminal protection of small molecule-linked DNA to develop nano-inspired biosensors for FR detection (He and Jiao, 2015; He et al., 2013; Jiang et al., 2015; Li et al., 2016; Zhu et al., 2015). For example, graphene oxide, which is capable of distinguishing single-stranded DNA (ssDNA) from double-stranded DNA (dsDNA) and efficiently quenching fluorescence, has been utilized in such a biosensor proposed by Tang and coworkers (He et al., 2013). It can be seen in Fig. 8.5A that a 20-mer ssDNA (Probe 2) that was 3'-labeled with a FAM group was initially absorbed onto the surface of graphene oxide through π-π stacking, which facilitated the proximity between the FAM group and graphene oxide, yielding complete fluorescence quenching. Another 20-mer ssDNA (Probe 1) synthesized by tethering a folate group to its 3'-end was then used to adjust the Probe 2/GO composite in a selective manner. Specifically, in the absence of target FR, the Probe 1 was digested into mononucleotides by Exo I, hardly affecting the Probe 2/GO composite. However, once FR was added, the Probe 1 would be specifically bound and thus immune to Exo I digestion. In this state, the Probe 2/GO composite was fully disrupted because of the formation of dsDNA between Probe 1 and Probe 2, causing the FAM group to move far away from the GO surface. As a result, significant recovery of fluorescence emission could be observed, which in turn enabled selective detection of FR. Recently, TiS_2 nanosheet, a novel two-dimensional analog of GO, has also been involved in the construction of a biosensor towards FR (Fig. 8.5B; Li et al., 2016). The critical element of the biosensor pulls together the terminal protection of small molecule-linked DNA, the adsorption of dye-labeled ssDNA on TiS_2 surface via van der Waals interactions, and the Zn^{2+}-dependent DNAzyme.

Under optimized conditions, the biosensor allowed highly sensitive detection of FR down to 0.003 ng mL^{-1}, much lower than those reported in previous works.

The random oligonucleotides sequence of small molecule-linked DNA makes it readily engineered to initiate DNA amplification processes. Therefore, a variety of DNA amplification techniques have been incorporated with the terminal protection of small molecule-linked DNA in nano-inspired biosensors to achieve desirable detection sensitivity (Wang et al., 2017; Yang and Gao, 2014; Zhao et al., 2015a). Exonuclease III (Exo III), an enzyme that catalyzes the stepwise removal of mononucleotides from 3′-hydroxyl termini of dsDNA, has been proven to be powerful as a catalytic element for amplified detection of biomolecules (Zuo et al., 2010). By leveraging the advantages offered by Exo III-assisted DNA amplification, Yang and Gao (2014) proposed a novel colorimetric biosensor for the detection of FR. In the biosensing system, the presence of target FR was transformed into the preservation of folate-linked DNA through the terminal protection, which was subsequently designed to trigger a process of Exo III-catalyzed circular digestion of signaling ssDNA. Using gold nanoparticles (AuNPs) as the colorimetric reporter, the concentration of FR could be determined with a limit of 50 fM. In another nano-inspired biosensor developed by Zhao et al. (2015a), hybridization chain reaction (HCR), a newly-emerging enzyme-free DNA amplification technique, was utilized. As shown in Fig. 8.6, the folate-linked DNA that escaped from Exo I-catalyzed degradation only in the presence of target FR would be immobilized onto the electrode surface through partial hybridization and then triggered cascaded hybridization of DNA hairpins, producing long-range dsDNA strands. Afterward, copper nanoparticles (CuNPs) were synthesized on the electrode surface, which later released large amounts of Cu^{2+} ions to catalyze the oxidation of o-phenylenediamine (OPD) to 2, 3-diaminophenazine (DAP), yielding significant electrochemical responses. On the basis of the biosensor, quantitative detection of FR was established in a range from 0.01 ng mL^{-1} to 100 ng mL^{-1}. The detection limit was calculated to be 3 pg mL^{-1}, which was much lower than the previous reports.

In addition to Exo I, other nucleic acid tool enzymes such as Exo III, endonuclease Fok I and polymerase have also been involved in the terminal protection of small molecule-linked DNA (Shi et al., 2017; Wu et al., 2011; Zhao et al., 2015b; Zhen et al., 2012), which provides new opportunity for nanomaterial-assisted FR detection. A representative work in this field was reported by Zhao et al. (2015b) (Fig. 8.7). In their work, two DNA strands were involved in the detection procedure: the probe DNA whose 5'-terminus was modified with a folate group and the cDNA whose sequence was complementary to a segment of probe DNA, allowing the formation of a probe DNA/cDNA duplex. In the absence of target FR, the probe DNA/cDNA duplex could be recognized by Exo III, which subsequently catalyzed the

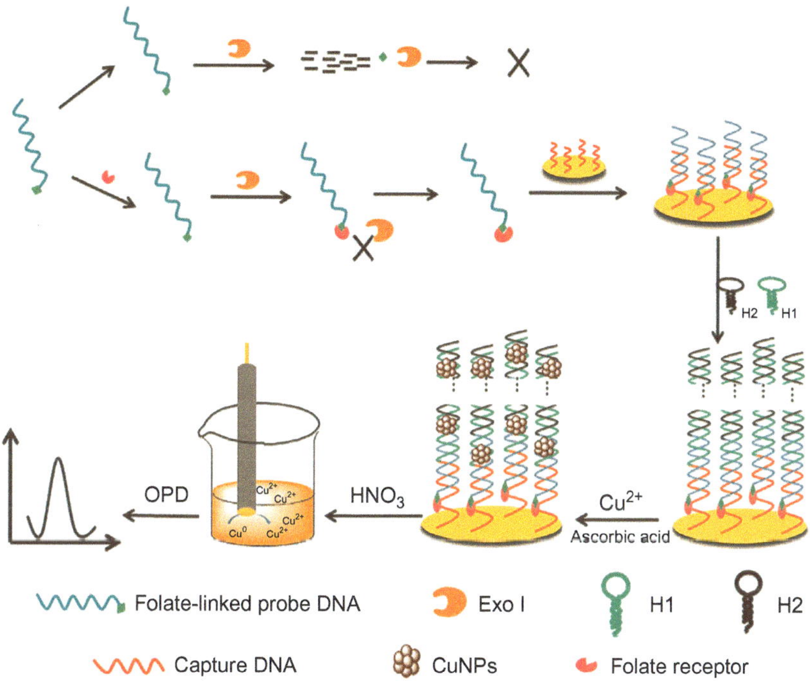

FIGURE 8.6 Electrochemical detection of FR based on terminal protection of small molecule linked DNA and hybridization chain reaction-assisted formation of copper nanoparticles. *Recreated from Zhao, J., Hu, S., Cao, Y., Zhang, B., Li, G., 2015a. Electrochemical detection of protein based on hybridization chain reaction-assisted formation of copper nanoparticles. Biosens. Bioelectron. 66, 327–331, with kind permission of Elsevier.*

degradation of cDNA from its 3′-terminus. However, if target FR was presented, specific binding of the protein with the folate group resulted in significant steric hindrance, preventing Exo III approaching the 3’-terminus of cDNA. In this case, the probe DNA/cDNA duplex could be retained, and then acted as a template for the formation of CuNPs. After adsorption onto the surface of magnetic graphene and separation with the aid of a magnet, the CuNPs were dissolved upon acid treatment and the obtained Cu^{2+} ions catalyzed the oxidation of OPD to generate DAP. By tracing the electrochemical response of DAP, a linear relationship between the response and the concentration of FR was observed in the range from 0.01 to 100 ng mL^{-1}. The detection limit was determined to be 7.8 pg mL^{-1}, which completely met the need for FR detection in pathological serum ($\sim$0.65 ng mL^{-1}).

Recent studies reveal that the steric hindrance arising from specific protein–small molecule interaction is not only limited to the terminal protection, but can be extended to novel biosensing strategies for protein assay relying on small molecule-linked DNA (Bell and Keyser, 2015; Fei et al., 2011;

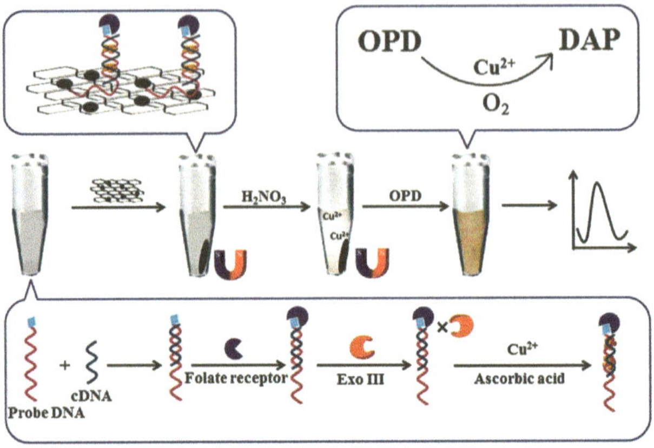

FIGURE 8.7 Nano-inspired electrochemical biosensor targeted to FR based on terminal protection of small molecule-linked DNA. *Recreated from Zhao, J., Lv, Y., Kang, M., Wang, K., Xiang, Y., 2015b. Electrochemical detection of protein by using magnetic graphene-based target enrichment and copper nanoparticles-assisted signal amplification. Analyst 140, 7818–7822, with kind permission of The Royal Society of Chemistry.*

Mahshid et al., 2015; Wang et al., 2016; Zhang et al., 2014). For instance, Mahshid et al. (2015) have developed a selective electrochemical biosensor that employed the steric hindrance effect to limit the hybridization of the small molecule-linked DNA with an electrode-attached complementary strand. Wang et al. (2016) have made use of the steric hindrance effect to alter the aptameric inhibition of Taq DNA polymerase and brought forth an effective sensing method. These novel biosensing strategies have also been applied to nanomaterial-assisted detection of FR. Fig. 8.8 illustrates the principle of one such work, in which a novel biosensing strategy termed as binding-regulated click ligation was utilized (Cao et al., 2016). In the detection procedure, three DNA probes were employed: the first one was P1 that held an azide (N_3) group located at its 5′-terminus; the second one was S-P2 that was designed to be complementary to P1 and 3′-labeled with a folate group to provide target binding; and the third one was P3 that was synthesized with an alkyne group at its 3′-end to enable ligation with P1 through Cu^+-catalyzed click chemistry. As depicted, P1 and S-P2 were originally hybridized into P1/S-P2 duplex, which brought the N_3 group close to the folate group. When ascorbate and copper ions (Cu^{2+}) were thereafter added, the P1/S-P2 duplex could not only serve as the template for the synthesis of fluorescent CuNPs through ascorbate-induced reduction of Cu^{2+}, but also be click ligated with P3 via Cu^+-catalyzed azide-alkyne cycloaddition. In this case, upon the introduction of GO into the detection system, the CuNPs were recruited onto the surface of GO due to the anchoring effect of the single-stranded domain of P1/S-P2/P3 complex, accomplishing the process of fluorescence quenching. Conversely, if target FR was presented in the detection

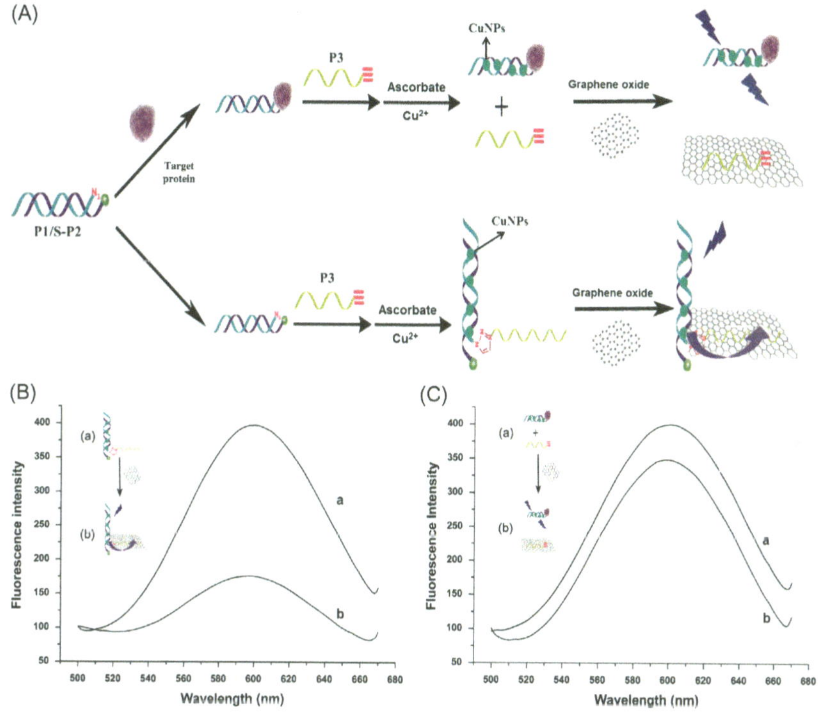

FIGURE 8.8 Nano-inspired biosensor for FR detection based on binding-regulated click ligation. (A) Representation of the fluorescence biosensor for detection of FR. (B) Fluorescence responses of the detection system containing 0 nM target without or with graphene oxide treatment. (C) Fluorescence responses of the detection system containing 100 nM target without or with graphene oxide treatment. *Recreated from Cao, Y., Han, P., Wang, Z., Chen, W., Shu, Y., Xiang, Y., 2016. Binding-regulated click ligation for selective detection of proteins. Biosens. Bioelectron. 78, 100–105, with kind permission of Elsevier.*

system, specific protein–small molecule interaction would take place, producing tremendous steric hindrance to block the click ligation between P3 and P1/S-P2 duplex. As a result, the CuNPs formed on the P1/S-P2 duplex stayed far away from the GO surface because of the weak and unstable adsorption between DNA duplex and GO, accordingly producing significant fluorescent response, which in turn enabled the selective detection of FR. Under optimized conditions, the proposed biosensor was demonstrated to allow detection of FR with desirable sensitivity and selectivity, which raises optimistic prospects in a wide range of applications.

8.2.2 Biotin Receptor

Nowadays, numerous nano-inspired biosensors that rely on small molecule-linked DNA have been developed for BR detection (Cao et al., 2017;

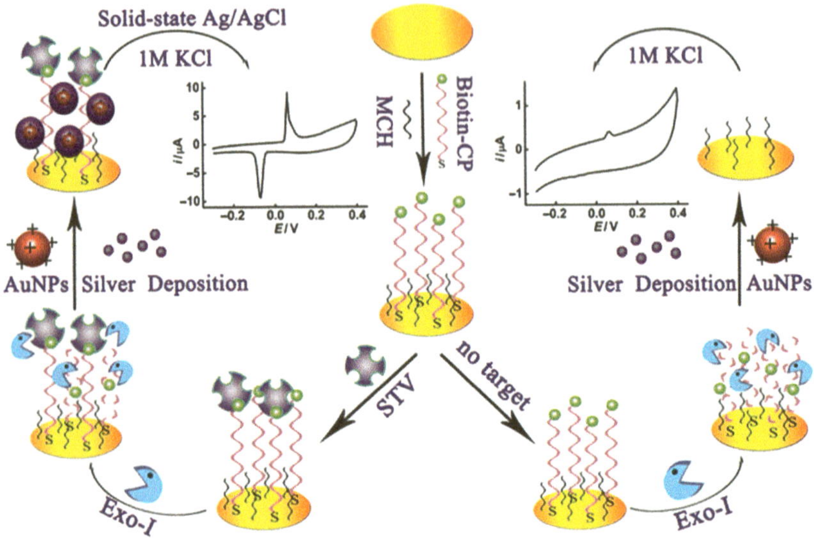

FIGURE 8.9 Nano-inspired biosensor for BR detection based on small molecule-linked DNA and a highly characteristic solid-state Ag/AgCl process. *Recreated from Wang, Q., Jiang, B., Xu, J., Xie, J., Xiang, Y., Yuan, R., et al., 2013. Amplified terminal protection assay of small molecule/protein interactions via a highly characteristic solid-state Ag/AgCl process. Biosens. Bioelectron. 43, 19–24, with kind permission of Elsevier.*

Chen et al., 2014; Chen et al., 2016a; Chen et al., 2016b; He and Jiao, 2016; Li et al., 2012; Wang et al., 2013; Wang et al., 2015b; Wei et al., 2018; Xiang et al., 2015; Zhang et al., 2015; Zhang et al., 2017). In these biosensors, terminal protection is the most commonly used sensing strategy, while avidin and streptavidin are usually used as the target to simulate BR.

An important branch of nano-inspired biosensors for BR detection is the development of nano-inspired electrochemical sensing methods, in which some nanomaterials that display unique electroactive properties play the role of electrochemical labels. A representative work in this vain was published by Wang et al. (2013). As shown in Fig. 8.9, the detection started with the binding of target streptavidin to a biotin-linked DNA, leading to the retention of DNA strands on the electrode surface, which was attributed to the terminal protection. Afterward, cetyltrimethylammonium bromide (CTAB)-functionalized and positively charged AuNPs were adsorbed onto the retained DNA strands through electrostatic interaction, which subsequently catalyzed the silver enhancement via seed-mediated growth process. Finally, a highly characteristic solid-state Ag/AgCl process of the silver deposited on the AuNPs was performed for signal output, enabling the detection limit of target streptavidin to be reduced down to 10 pM. Similar to silver-deposited AuNPs, QDs, which were reported to display prominent anodic stripping response, have also been employed as electrochemical labels for BR

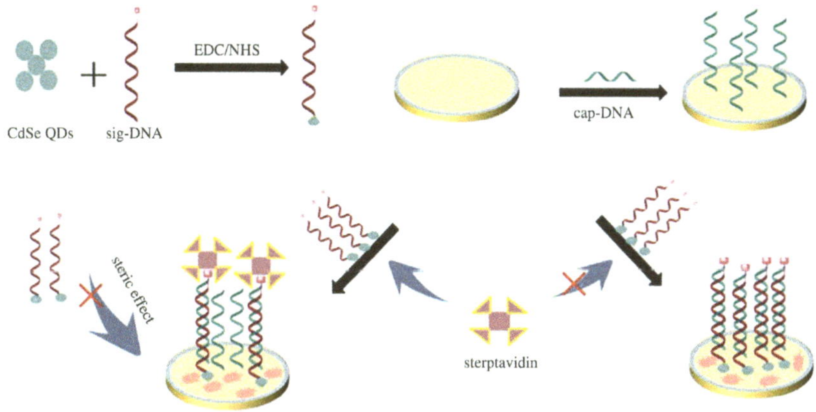

FIGURE 8.10 Highly sensitive electrochemical biosensor for BR detection based on small molecule-linked DNA and CdSe quantum dots. *Recreated from Wei, Y.P., Liu, X.P., Mao, C.J., Niu, H.L., Song, J.M., Jin, B.K., 2018. Highly sensitive electrochemical biosensor for streptavidin detection based on CdSe quantum dots. Biosens. Bioelectron. 103, 99–103, with kind permission of Elsevier.*

detection. Very recently, by coupling QDs with a steric hindrance hybridization sensing procedure, Wei et al. (2018) have fabricated an electrochemical biosensor that allowed highly sensitive detection of streptavidin (Fig. 8.10). In their design, a sensing DNA probe was labeled at its 3′-terminus with CdSe QDs and 5′-terminus with a biotin group, while a capturing DNA probe was modified at its 5′-end with thiol to self-assemble on gold electrode surface. In the absence of the target, the sensing DNA probe could be immobilized on the electrode surface via hybridization with the capture DNA probe, bringing large amounts of CdSe QDs close to the electrode. However, in the presence of streptavidin, the steric hindrance effect generated by specific streptavidin–biotin interaction limited the ability of the sensing DNA probe to hybridize with the capturing DNA probe, which consequently led to a large number of CdSe QDs far away from the electrode. In this way, the presence of target streptavidin was transformed into the decreased amount of CdSe QDs near the electrode surface, which could be detected by differential pulse anodic stripping voltammetry. Under optimal conditions, the biosensor allowed linear detection of target streptavidin form 1.96 pg mL^{-1} to 1.96 μg mL^{-1} with a desirable detection limit of 0.65 pg mL^{-1}.

Taking advantages of easy synthesis, low toxicity, and good biocompatibility, DNA-templated metal nanomaterials have been intensely investigated in the past few years, and have found widespread applications in BR detection because of their facile conjunction with small molecule-linked DNA (Cao et al., 2017; He and Jiao, 2016; Wang et al., 2015b; Zhang et al., 2017). In this area, Wang et al. (2015b) have recently developed an illustrative biosensor by making use of poly(thymine)(poly T)-templated CuNPs

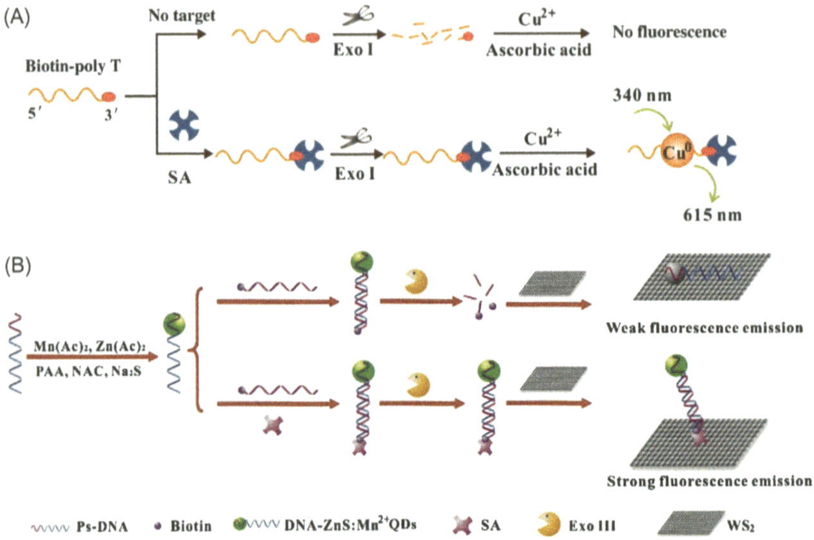

FIGURE 8.11 (A) Nano-inspired biosensor for BR detection based on terminal protection and poly T-templated CuNPs. (B) FRET-based BR sensing using DNA-ZnS: Mn^{2+} QDs as fluorescent labels. *(A) Recreated from Wang, H.B., Zhang, H.D., Chen, Y., Liu, Y.M., 2015b. A fluorescent biosensor for protein detection based on poly(thymine)-templated copper nanoparticles and terminal protection of small molecule-linked DNA. Biosens. Bioelectron. 74, 581–586, with kind permission of Elsevier. (B) Recreated from Zhang, C., Ding, C., Zhou, G., Xue, Q., Xian, Y., 2017. One-step synthesis of DNA functionalized cadmium-free quantum dots and its application in FRET-based protein sensing. Anal. Chim. Acta 957, 63–69, with kind permission of Elsevier.*

(Fig. 8.11A). A biotin-T30 probe was utilized not only as a recognition element towards BR (streptavidin as the model) but also as a template for fluorescent CuNPs formation. Due to the terminal protection of small molecule-linked DNA, the concentration of target streptavidin could be determined by measuring the fluorescence intensity of CuNPs at the maximum emission wavelength, which gave a linear correlation in the 0.5 to 100 nM concentration range with a detection limit of 0.1 nM. Similarly, Zhang et al. (2017) have proposed a FRET-based sensing method with one-step synthesized DNA-QDs as the fluorescent labels (Fig. 8.11B). In their work, DNA-Mn-doped ZnS (DNA-ZnS: Mn^{2+}) QDs were prepared by a one-pot route, in which the phosphorothiolate phosphate DNA (Ps-DNA) was used as the template. In the absence of the target streptavidin, the DNA-ZnS: Mn^{2+} QDs were hybridized with biotin-linked DNA to form DNA duplexes that were active for Exo III-catalyzed hydrolysis, which ultimately drove the DNA-ZnS:Mn^{2+} QDs in close proximity to WS_2, resulting in the fluorescence quenching through FRET. In contrast, once the target was introduced into the detection system, the terminal protection generated from specific protein–small molecule binding would impel the DNA-ZnS:Mn^{2+} QDs

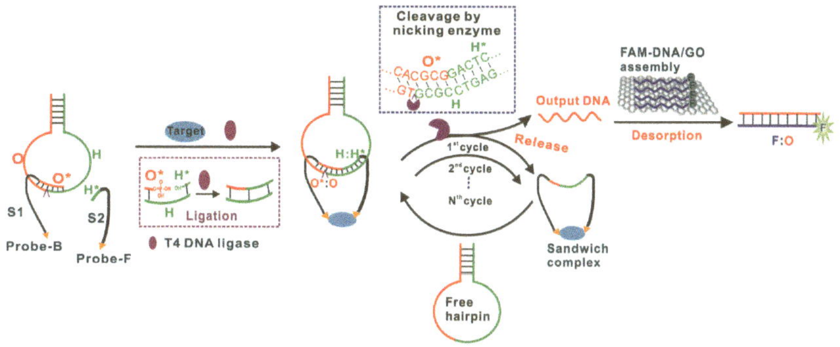

FIGURE 8.12 Nano-inspired biosensor for BR detection based on binding-induced DNA assembly. *Recreated from Chen J., Deng B., Wu P., Li F., Li X.F., Le X.C., et al., 2016a. Amplified binding-induced homogeneous assay through catalytic cycling of analyte for ultrasensitive protein detection, Chem. Commun. 52, 1816–1819, with kind permission of The Royal Society of Chemistry.*

apart from WS_2, making QDs retain their fluorescence, which in turn enabled quantitative detection of the target streptavidin down to 2.8 ng mL^{-1}.

Additionally, some new biosensing strategies that do not depend on the steric hindrance effect have gradually been utilized in nano-inspired biosensors for BR detection (Chen et al., 2016a; Li et al., 2012; Zhang et al., 2015). Fig. 8.12 illustrates a typical work published by Chen et al. (2016a). The essential design of the biosensor was the binding-induced DNA assembly, which converted the presence of target BR (streptavidin) into the ligation of biotin-linked DNA probes, forming a sandwich-type protein/DNA complex. With the aid of a nicking endonuclease that recognized specific sequences in dsDNA but cleaved only one of the strands at a fixed position, the sandwich-type protein/DNA complex participated in multiple cycles of hairpin cleavage, which generated large amounts of output DNA. As a result, the signaling DNA probe modified with a FAM group could be desorbed form the surface of GO, restoring the fluorescence emission of FAM. Based on this biosensor, streptavidin was sensitively detected with a detection limit of 1 pM.

8.3 CONCLUSION

By rational design and engineering of small molecule ligands-based probes, small molecule ligands have provided infinite opportunities for constructing nano-inspired biosensors for protein assay. However, it should be noted that the practical application of these biosensors is still restricted. One fatal problem is the limited target zone: as presented in this chapter, most of the biosensors are developed towards only three kinds of disease protein biomarkers: human carbonic anhydrase, folate receptor, and biotin receptor.

Another fatal issue is that most of these biosensors are performed and optimized in buffer solution rather than clinical samples. Certainly, considering the fast development in chemistry and biology, it is anticipated that nano-inspired biosensors using small molecule ligands as recognition elements will become much more powerful in clinical application in the future.

REFERENCES

Azagarsamy, M.A., Yesilyurt, V., Thayumanavan, S., 2010. Disassembly of dendritic micellar containers due to protein binding. J. Am. Chem. Soc. 132, 4550–4551.

Bell, N.A.W., Keyser, U.F., 2015. Specific protein detection using designed DNA carriers and nanopores. J. Am. Chem. Soc. 137, 2035–2041.

Bruno, E., Buemi, M.R., Di Fiore, A., De Luca, L., Ferro, S., Angeli, A., et al., 2017. Probing molecular interactions between human carbonic anhydrases (hCAs) and a novel class of benzenesulfonamides. J. Med. Chem. 60, 4316–4326.

Cao, J., Wang, W., Bo, B., Mao, X., Wang, K., Zhu, X., 2017. A dual-signal strategy for the solid detection of both small molecules and proteins based on magnetic separation and highly fluorescent copper nanoclusters. Biosens. Bioelectron. 90, 534–541.

Cao, Y., Han, P., Wang, Z., Chen, W., Shu, Y., Xiang, Y., 2016. Binding-regulated click ligation for selective detection of proteins. Biosens. Bioelectron. 78, 100–105.

Chen, C., Xiang, X., Liu, Y., Zhou, G., Ji, X., He, Z., 2014. Dual-color determination of protein via terminal protection of small-molecule-linked DNA and the enzymolysis of exonuclease III. Biosens. Bioelectron. 58, 205–208.

Chen, J., Deng, B., Wu, P., Li, F., Li, X.F., Le, X.C., et al., 2016a. Amplified binding-induced homogeneous assay through catalytic cycling of analyte for ultrasensitive protein detection. Chem. Commun. 52, 1816–1819.

Chen, J., Gao, C., Mallik, A.K., Qiu, H., 2016b. A WS2 nanosheet-based nanosensor for the ultrasensitive detection of small molecule–protein interaction via terminal protection of small molecule-linked DNA and Nt. BstNBI-assisted recycling amplification. J. Mater. Chem. B 4, 5161–5166.

Fang, F., Liu, J., Li, Y., Yang, J., Yang, J., 2018. A new colorimetric platform for protein detection based on recognition-induced cascade of polymeric nanoparticles disassembly. Macromol. Biosci. 18, 1700392.

Fei, Y.H., Liu, D., Wu, Z.S., Shen, G.L., Yu, R.Q., 2011. DNA-encoded signal conversion for sensitive microgravimetric detection of small molecule–protein interaction. Bioconjugate Chem. 22 (12), 2369–2376.

He, Q., Fan, X., Sun, S., Li, H., Pei, Y., Xu, Y., 2015. Highly selective turn-on detection of (strept)avidin based on self-assembled near-infrared fluorescent probes. RSC Adv. 5, 38571–38576.

He, Y., Jiao, B., 2015. High performance system for protein assays: synergistic effect of terminal protection strategy and graphene oxide platform. RSC Adv. 5, 101327–101332.

He, Y., Jiao, B., 2016. Detection of biotin-streptavidin interactions via terminal protection of small molecule linked DNA and the formation of fluorescent DNA-templated silver nanoclusters. Microchim. Acta 183, 3183–3189.

He, Y., Xing, X., Tang, H., Pang, D., 2013. Graphene oxide-based fluorescent biosensor for protein detection via terminal protection of small-molecule-linked DNA. Small. 9, 2097–2101.

Hou, T.C., Wu, Y.Y., Chiang, P.Y., Tan, K.T., 2015. Near-infrared fluorescence activation probes based on disassembly-induced emission cyanine dye. Chem. Sci. 6, 4643–4649.

Jiang, H., Xu, G., Sun, Y., Zheng, W., Zhu, X., Wang, B., et al., 2015. A "turn-on" silver nanocluster based fluorescent sensor for folate receptor detection and cancer cell imaging under visual analysis. Chem. Commun. 51, 11810–11813.

Jung, C., Ellington, A.D., 2014. Diagnostic applications of nucleic acid circuits. Acc. Chem. Res. 47, 1825–1835.

Koukourakis, M.I., Giatromanolaki, A., Sivridis, E., Simopoulos, K., Pastorek, J., Wykoff, C.C., et al., 2001. Hypoxia-regulated carbonic anhydrase-9 (CA9) relates to poor vascularization and resistance of squamous cell head and neck cancer to chemoradiotherapy. Clin. Cancer Res. 7, 3399–3403.

Koutnik, P., Shcherbakova, E.G., Gozem, S., Caglayan, M.G., Minami, T., Anzenbacher, P., 2017. Fluorescence-based assay for carbonic anhydrase inhibitors. Chem 2, 271–282.

Kubota, R., Hamachi, I., 2015. Protein recognition using synthetic small-molecular binders toward optical protein sensing in vitro and in live cells. Chem. Soc. Rev. 44, 4454–4471.

Li, F., Zhang, H., Lai, C., Li, X.F., Le, X.C., 2012. A molecular translator that acts by binding-induced DNA strand displacement for a homogeneous protein assay. Angew. Chem. Int. Ed. 51, 9317–9320.

Li, X., Ding, X., Li, Y., Wang, L., Fan, J., 2016. A TiS_2 nanosheet enhanced fluorescence polarization biosensor for ultra-sensitive detection of biomolecules. Nanoscale 8, 9852–9860.

Li, X., Kim, C.Y., Lee, S., Lee, D., Chung, H.M., Kim, G., et al., 2017. Nanostructured phthalocyanine assemblies with protein-driven switchable photoactivities for biophotonic imaging and therapy. J. Am. Chem. Soc. 139, 10880–10886.

Liao, S.Y., Ivanov, S., Ivanova, A., Ghosh, S., Cote, M.A., Keefe, K., et al., 2003. Expression of cell surface transmembrane carbonic anhydrase genes CA9 and CA12 in the human eye: overexpression of CA12 (CAXII) in glaucoma. J. Med. Genet. 40, 257–261.

Low, P.S., Henne, W.A., Doorneweerd, D.D., 2008. Discovery and development of folic-acid-based receptor targeting for imaging and therapy of cancer and inflammatory diseases. Acc. Chem. Res. 41, 120–129.

Lu, C.H., Willner, B., Willner, I., 2013. DNA nanotechnology: from sensing and DNA machines to drug-delivery systems. ACS Nano 7, 8320–8332.

Mahshid, S.S., Camire, S., Ricci, F., Vallee-Belisle, A., 2015. A highly selective electrochemical DNA-based sensor that employs steric hindrance effects to detect proteins directly in whole blood. J. Am. Chem. Soc. 137, 15596–15599.

Mizusawa, K., Ishida, Y., Takaoka, Y., Miyagawa, M., Tsukiji, S., Hamachi, I., 2010. Disassembly-driven turn-on fluorescent nanoprobes for selective protein detection. J. Am. Chem. Soc. 132, 7291–7293.

Mizusawa, K., Takaoka, Y., Hamachi, I., 2012. Specific cell surface protein imaging by extended self-assembling fluorescent turn-on nanoprobes. J. Am. Chem. Soc. 134, 13386–13395.

Molla, M.R., Prasad, P., Thayumanavan, S., 2015. Protein-induced supramolecular disassembly of amphiphilic polypeptide nanoassemblies. J. Am. Chem. Soc. 137 (23), 7286–7289.

Pode, Z., Peri-Naor, R., Georgeson, J.M., Ilani, T., Kiss, V., Unger, T., et al., 2017. Protein recognition by a pattern-generating fluorescent molecular probe. Nat. Nanotechnol. 12, 1161–1168.

Ren, W.X., Han, J., Uhm, S., Jang, Y.J., Kang, C., Kim, J.H., et al., 2015. Recent development of biotin conjugation in biological imaging, sensing, and target delivery. Chem. Commun. 51, 10403–10418.

Shi, H., Mao, X., Chen, X., Wang, Z., Wang, K., Zhu, X., 2017. The analysis of proteins and small molecules based on sterically tunable nucleic acid hyperbranched rolling circle amplification. Biosens. Bioelectron. 91, 136–142.

Takaoka, Y., Sakamoto, T., Tsukiji, S., Narazaki, M., Matsuda, T., Tochio, H., et al., 2009. Self-assembling nanoprobes that display off/on 19F nuclear magnetic resonance signals for protein detection and imaging. Nat. Chem. 1, 557–561.

Teller, C., Willner, I., 2010. Functional nucleic acid nanostructures and DNA machines. Curr. Opin. Biotech. 21, 376–391.

Wang, F., Lu, C.H., Willner, I., 2014. From cascaded catalytic nucleic acids to enzyme–DNA nanostructures: controlling reactivity, sensing, logic operations, and assembly of complex structures. Chem. Rev. 114, 2881–2941.

Wang, H., Zhuang, J., Raghupathi, K.R., Thayumanavan, S., 2015a. A supramolecular dissociation strategy for protein sensing. Chem. Commun. 51, 17265–17268.

Wang, H.B., Zhang, H.D., Chen, Y., Liu, Y.M., 2015b. A fluorescent biosensor for protein detection based on poly(thymine)-templated copper nanoparticles and terminal protection of small molecule-linked DNA. Biosens. Bioelectron. 74, 581–586.

Wang, Q., Jiang, B., Xu, J., Xie, J., Xiang, Y., Yuan, R., et al., 2013. Amplified terminal protection assay of small molecule/protein interactions via a highly characteristic solid-state Ag/AgCl process. Biosens. Bioelectron. 43, 19–24.

Wang, R., Xu, X., Li, P., Wang, Y., Jiang, W., 2017. Terminal protection-mediated autocatalytic cascade amplification coupled with graphene oxide fluorescence switch for sensitive and rapid detection of folate receptor. Talanta 174, 684–688.

Wang, Z., Li, Y., Han, P., Mao, X., Yin, Y., Cao, Y., 2016. Binding-responsive catalysis of Taq DNA polymerase for the sensitive and selective detection of cell-surface proteins. Chem. Commun. 52, 10684–10687.

Wei, Y.P., Liu, X.P., Mao, C.J., Niu, H.L., Song, J.M., Jin, B.K., 2018. Highly sensitive electrochemical biosensor for streptavidin detection based on CdSe quantum dots. Biosens. Bioelectron. 103, 99–103.

Wu, Z., Zhen, Z., Jiang, J.H., Shen, G.L., Yu, R.Q., 2009. Terminal protection of small-molecule-linked DNA for sensitive electrochemical detection of protein binding via selective carbon nanotube assembly. J. Am. Chem. Soc. 131, 12325–12332.

Wu, Z., Wang, H., Guo, M., Tang, L.J., Yu, R.Q., Jiang, J.H., 2011. Terminal protection of small molecule-linked DNA: a versatile biosensor platform for protein binding and gene yyping assay. Anal. Chem. 83, 3104–3111.

Xiang, X., Shi, J., Huang, F., Zheng, M., Deng, Q., Xu, J., 2015. MoS_2 nanosheet-based fluorescent biosensor for protein detection via terminal protection of small-molecule-linked DNA and exonuclease III-aided DNA recycling amplification. Biosens. Bioelectron. 74, 227–232.

Yang, X., Gao, Z., 2014. Gold nanoparticle-based exonuclease III signal amplification for highly sensitive colorimetric detection of folate receptor. Nanoscale 6, 3055–3058.

Yoshii, T., Mizusawa, K., Takaoka, Y., Hamachi, I., 2014. Intracellular protein-responsive supramolecules: protein sensing and in-cell construction of inhibitor assay system. J. Am. Chem. Soc. 136, 16635–16642.

Zhang, C., Ding, C., Zhou, G., Xue, Q., Xian, Y., 2017. One-step synthesis of DNA functionalized cadmium-free quantum dots and its application in FRET-based protein sensing. Anal. Chim. Acta 957, 63–69.

Zhang, H., Li, F., Dever, B., Li, X.F., Le, X.C., 2012. DNA-mediated homogeneous binding assays for nucleic acids and proteins. Chem. Rev. 113, 2812–2841.

Zhang, H., Lai, M., Zuehlke, A., Peng, H., Li, X.F., Le, X.C., 2015. Binding-induced DNA nanomachines triggered by proteins and nucleic acids. Angew. Chem. Int. Ed. 54, 14326–14330.

Zhang, Z., Hejesen, C., Kjelstrup, M.B., Birkedal, V., Gothelf, K.V., 2014. A DNA-mediated homogeneous binding assay for proteins and small molecules. J. Am. Chem. Soc. 136, 11115–11120.

Zhao, J., Hu, S., Cao, Y., Zhang, B., Li, G., 2015a. Electrochemical detection of protein based on hybridization chain reaction-assisted formation of copper nanoparticles. Biosens. Bioelectron. 66, 327–331.

Zhao, J., Lv, Y., Kang, M., Wang, K., Xiang, Y., 2015b. Electrochemical detection of protein by using magnetic graphene-based target enrichment and copper nanoparticles-assisted signal amplification. Analyst 140, 7818–7822.

Zhen, Z., Tang, L.J., Lin, J., Jiang, J.-H., Yu, R.Q., Xiong, X., et al., 2012. Endonucleolytic inhibition assay of DNA/Fok I transducer as a sensitive platform for homogeneous fluorescence detection of small molecule-protein interactions. Anal. Chem. 84, 5708–5715.

Zhu, Y., Wang, G., Sha, L., Qiu, Y., Jiang, H., Zhang, X., 2015. A ratiometric colorimetric detection of the folate receptor based on terminal protection of small-molecule-linked DNA. Analyst 140, 1260–1264.

Zhuang, Y.D., Chiang, P.Y., Wang, C.W., Tan, K.T., 2013. Environment-sensitive fluorescent turn-on probes targeting hydrophobic ligand-binding domains for selective protein detection. Angew. Chem. Int. Ed. 52 (31), 8124–8128.

Zuo, X., Xia, F., Xiao, Y., Plaxco, K.W., 2010. Sensitive and selective amplified fluorescence DNA detection based on exonuclease III-aided target recycling. J. Am. Chem. Soc. 132, 1816–1818.

Part III

Biosensing Technologies for Protein Assay

In this last part, we will mainly discuss how to convert the signals of molecular recognition into readable signals, that is, biosensing technologies, which mainly include various technical means of commonly used electrochemistry and spectrometry. It also includes some less common technologies that are also reported and applied. How to convert these readable signals into digital signals through certain electronic devices will not be discussed. In this part, electrochemical technology is relatively complicated because of its system and principle, so we will mainly introduce the electrochemical technology itself, and illustrate the application of electrochemical technology in different protein biosensing systems. In terms of spectrometric technology, the principle is relatively simple, but nanomaterials play a crucial role in the output of spectrometric signals, so we will mainly introduce the spectrometric signal output that nanomaterials participate in. For other relatively unpopular technical means, we only give brief introductions and examples. Finally, since signal amplification is critical in the analysis of low-abundance samples, we will present some commonly used signal amplification techniques in a separate chapter.

Chapter 9

Electrochemistry

Xiaoli Zhu[1] and Liu Shi[2]
[1]*Center for Molecular Recognition and Biosensing, School of Life Sciences, Shanghai University, Shanghai, P. R. China,* [2]*State Key Laboratory of Pharmaceutical Biotechnology, Department of Biochemistry, Nanjing University, Nanjing, P. R. China*

Chapter Outline

9.1 INTRODUCTION

Electrochemistry is a branch of physical chemistry that studies the relationship between electrical parameters and the identifiable chemical changes, with electricity considered as an outcome of a particular chemical change. In the past few decades, it has been pushing forward to accomplish an interdisciplinary application, in chemistry, biology, physics, electronics, and other related areas, which possesses a broad application prospect in the analysis of life, clinical testing, environmental protection, food inspection, and other fields (Buter et al., 2017; Dixit et al., 2016; Lojou and Bianco, 2006; Martin et al., 2017; Palecek et al., 2015). In recent years, with the development of nanotechnology, there has been a breakthrough in the research of electrochemistry (Ramanaviciene et al., 2006).

Based on electrochemical techniques, the electroanalytical method has been developed to qualitatively or quantitatively analyze target molecules through measuring the electrical and electrochemical parameters, where the test material is chosen as the recognition elements, electrode as the conversion components, and potential, current or conductance as the feature detection signal. Just as is shown in Fig. 9.1 (Ronkainen et al., 2010),

Nano-inspired Biosensors for Protein Assay with Clinical Applications.
DOI: https://doi.org/10.1016/B978-0-12-815053-5.00009-X

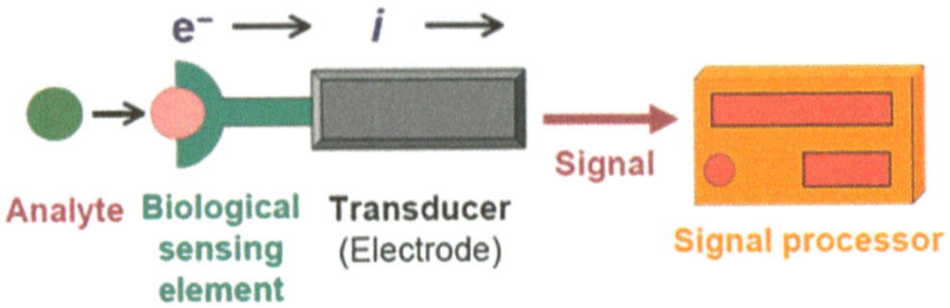

FIGURE 9.1 A schematic illustration of a biosensor with electrochemical transducer. *Reprinted from Ronkainen, N.J., Halsall, H.B., Heineman, W.R., 2010. Electrochemical biosensors, Chem. Soc. Rev. 39, 1747–1763.*

electrochemical sensors are mainly composed of two parts: the recognition system, which is the sensor; and the conversion system, which is the converter. The main function of the sensor is to selectively interact with the analytes to be tested, and to transmit the changes of chemical parameters in the system to the converter by means of reaction signals, while the main function of the converter is to recognize the response signal of the system, and then transform the response signal into voltage, current, or light intensity through the electrode, optical fiber, or a variety of sensing elements, followed by signal amplification output or direct output, so that the final output signal can be analyzed (Jacobs et al., 2010; Pei et al., 2014; Silwana et al., 2016; Yang et al., 2015a; Yasukawa et al., 2015; Zhu et al., 2015).

The microstructure on the surface of electrode can provide many kinds of utilizable potential field which are fit for effective separation and enrichment of the analyte under test, and the selectivity can be further improved by controlling the electrode potential (Zhu et al., 2015), but also the sensitivity of determination method and the selectivity of chemical reaction of the materials on the surface can be combined. Therefore electrochemical sensing is an ideal system which rolls up three functions including separation, enrichment, and selective determination into one, and possesses unique advantages in terms of improving selectivity and sensitivity (Liu et al., 2009). According to different detection signals, electrochemical sensors can be divided into potential, current, conductivity, resistance sensors, and so on, among which potential sensor and current sensor are of great importance for the application of analytical science. A potentiometric sensor converts biological recognition response signal to voltage signal which is directly proportional to the logarithm of the concentration of active substances produced or consumed in the process of the reaction, accordingly it is proportional to the logarithm of the concentration of the analytes under test (Bruner et al., 1991). A current sensor provides a driving force for electroactive electron transfer reaction through applying the potential of the working electrode, detecting changes in the current over time (Scherlis and Marzari, 2005), and the current directly measuring the rate of electron transfer reaction, reflects the speed of biological molecular recognition, that is, the current is proportional to the concentration of the analytes under test.

The research of electrochemical biosensors is mainly through the electrochemical analysis system, known as the electrochemical workstation, which is a commonly used measurement device in electrochemical research (Shan et al., 2009). The electrochemical workstation can be used for the control and detection of electrochemical parameters such as potentials, currents, etc. The workstation uses electrodes to input and output electrical signals. The commonly used electrochemical workstation is mainly a three-electrode system, which includes a working electrode, a counter electrode, and a reference electrode. The working electrode is the place where most electrochemical reactions occur and is the most direct sensor. Materials used to construct the working electrode usually include carbon, gold, platinum, mercury, etc. The counter electrode is usually a platinum electrode that possesses good conductivity. Therefore, it can constitute a circuit that forms a stable current with the working electrode. The standard reference electrode provides a stable potential reference for measuring the continuously changing potential of the working electrode (Manning and Purdy, 1970). Commonly used reference electrodes include saturated calomel, silver/silver chloride electrodes, etc. (Ramos Miguel et al., 2015; Sarkar et al., 2014; Stumpff and McGuigan, 2014; Wu and Su, 2014; Zhang et al., 2014).

9.2 ELECTROCHEMICAL TECHNIQUES

9.2.1 Voltammetry and Amperometry

Electrochemical measuring techniques of the research are based on the principle that the nature of the analytes under test (such as concentration) affects the redox reaction that is observed. Accordingly through the analysis of the input and output signals of modulation electrolytic cell, we can get related properties of the object under test, so as to realize the detection and analysis of the object (Hascup and Hascup, 2014; Kranz, 2014; Li et al., 2015). The techniques of electrochemical research referred to in this chapter include voltammetry/amperometry, impedance, conductometry, and potentiometry. Then the working principle of them is briefly introduced. Voltammetric and amperometric methods apply a potential to a working electrode versus a reference electrode to measure the current (Baciu et al., 2015; Dejmkova et al., 2011; Plotnikov et al., 2016).

The different combinations of parameters, including potential (E), current (I), charge (Q), and time (t), make for a long list of electrochemical techniques. In this section, we will provide a brief summary of the techniques that were used throughout this chapter. The current is generated from electrolysis at the working electrode where electrochemical oxidation or reduction occurs. The mass transfer rate of the reactant molecules from the bulk solution to the electrode interface limits the current. The potential rises at a given rate when the current is measured in a voltammetry biosensor, and the current usually realizes a peak or a plateau that is proportional to the analyte concentration (Davis, 1985). The amperometric biosensor detects the analyte by measuring the current with a constant applied potential at the working electrode, and comparing it to the reference electrode. The potential is stepped directly to the target value, followed by measurement of the current. Amperometric biosensors possess the advantage of selectivity because the oxidation or reduction potentials they use in the analysis vary in the analyte species (Wang et al., 2003). Different parameters including potential (E), current (I), charge (Q), and time (t) make up cross-combinations which enrich the electrochemical technique. In this section, we will briefly outline the techniques used in this chapter.

Electrochemical amperometry assays based on receptive nucleic acid aptamers have been developed with the principle of target-binding mediated conformational changes in aptamers. Recently, it has been applied to voltammetric detection of lysozyme which was assembled on a gold electrode with thiolated DNA aptamers (predecorated with electrostatically bound hexamine guanidine) (Cheng et al., 2007). In the presence of the cationic lysozyme target (isoelectric point 11), the electroactive hexamine was competitively shut down and the resulting current signal was correspondingly reduced to 500 ng mL^{-1} in a calibratable manner (Fig. 9.2). Amperometric protein biosensors are also established by means of imprinted polymers in the presence of solution phase redox probes. Viswanathan et al. (Viswanathan et al., 2012) made use of a gold electrode immobilized with polyphenol MIP to detect the ovarian cancer marker CA-125, taking advantage of the principle that the reduced diffusive route of a solution added redox probe was reduced in a linear calibrating manner through target binding.

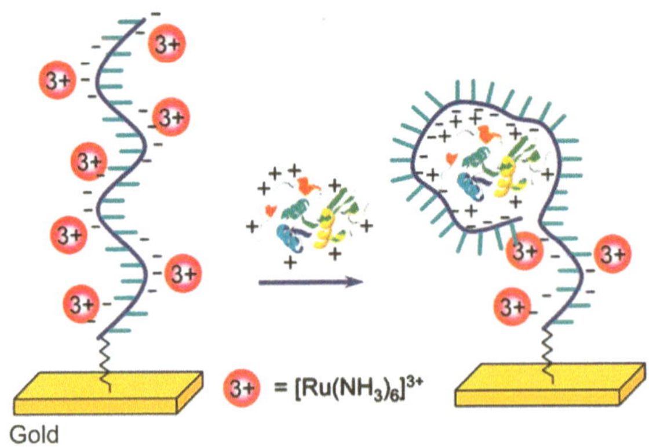

FIGURE 9.2 Scheme of voltammetric detection for lysozyme utilizing antilysozyme DNA aptamers immobilized on gold electrodes. *Reprinted from Cheng, A.K.H., Ge B., Yu, H.-Z., 2007. Aptamer-based biosensors for label-free voltammetric detection of lysozyme, Anal. Chem. 79, 5158–5164.*

9.2.2 Impedance

The impedance method (EIS) is an electrochemical technique for measuring the response of a system to a perturbation in a steady state by means of perturbing the electrode by a small amplitude voltage or current. The impedance method reflects the nature of the electrode surface and the electron transfer rate on the surface of the modified electrode, thereby judging the assembly of molecules on the electrode surface and verifying the experimental process. Perturbation of the system by a small amplitude sinusoidal AC excitation signal can reduce the interference, which is favorable for the study of modified electrodes, and can also establish the linear relationship between the dynamic response of the electrode system and the disturbance signal. Having obtained the impedance spectroscopy, the resistance and capacitance components of the circuit can be determined after measuring the in-phase and out-of-phase current responses (Li et al., 2013c).

Mount et al. have proposed a method for the detection of two wound infection protein biomarkers with electrochemical impedance spectroscopy techniques by means of triggering Receptor-1 Expressed on Myeloid cells (TREM-1) and Matrix Metallo Peptidase 9 (MMP-9) at antibody-modified screen-printed electrodes. 10 mL of mock wound fluid realizes the operation in the assay where both proteins can be detected down to ng mL^{-1} levels, which exceeds the clinical threshold required to indicate infection (0.8 ng mL^{-1} for TREM-1141 and 1000 ng mL^{-1} for MMP-9142). Similarly, Davis and his coworkers have established a highly effective method to detect the acute phase protein, C-reactive protein (CRP), using both constrained peptide (affimer) and antibody-modified electrodes (Johnson et al., 2012).

9.2.3 Conductometry

The conductivity of the sample solution or nanowire is monitored by the conductometry, because the composition of the solution/medium changes during the chemical reaction. Conductometric measurements of biosensors usually contain the enzyme whose charge product leads to changes in ionic strength and thus the increase in conductivity. Biosensors of environmental monitoring and clinical analysis usually adopted conductometry as the technique of signal output. A tyrosinase biosensor that employed the conductometric method was established to measure the ppb amounts of pollutants diuron and atrazine and its metabolites. As for the establishment of label-free conductometric protein biosensors, the transmission of charge through conductive media (such as nanowires) is necessary for the identification and binding of the target.

A biosensor for the detection of cancer biomarker CA-125 has been established by Mulchandani et al. (Bangar et al., 2009)which is based on a 3-micron gap between the gold electrode and the antibody-modified single

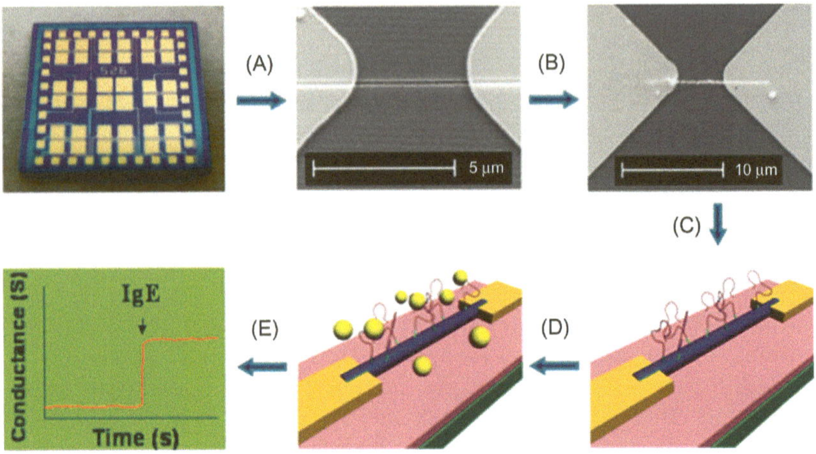

FIGURE 9.3 Images and schematic representations of conductometric protein sensor utilizing polyaniline nanowire. *Reprinted from Luo, X., Lee, I., Huang, J., Yun, M., Cui, X.T., 2011. Ultrasensitive protein detection using an aptamer-functionalized single polyaniline nanowire, Chem. Commun. 47, 6368–6370.*

polypyrrole nanowires. These nanowires are capable of presenting the conductance change of CA-125 concentration with orders of magnitude of the target. More recently as shown in Fig. 9.3, Cui et al. have also proposed a conductometric method for ultrasensitive detection of IgE based on the synthesis of polyaniline nanowires between gold microelectrode pairs on silicon (Luo et al., 2011).

9.2.4 Potentiometry

Potentiometric sensors measure the potential of an electrochemical cell under the condition that the drawing current is negligible. Glass pH electrode and ion selective electrode, such as K^+, Ca_2^+, Na^+, and Cl^- are typical examples of potentiometric sensors that commonly include an electrochemical cell and two-reference electrode in order to measure the potential of the membrane which selectively reacts with the target ions. These chemical sensors can be successfully alerted into biosensors by means of modifying biosensing element, such as an enzyme that catalyzes a reaction forming ions.

A potentiometric sensor was developed by Anderson et al. (May et al., 2005) for the detection of vascular endothelial growth factor (VEGF, a cancer marker) that utilized a monolayer of human umbilical vein endothelial cells which was attached to the cellulose triacetate surface of a potassium-selective ISE. Initially, the ion transport was blocked by the cells, but when VEGF existed, the permeability was tremendously increased which led to the signal soaring. In a recent example, a potentiometric sensor was developed on a molecularly imprinted alkane thiol surface to detect the cancer biomarkers CEA (Wang et al., 2010). After the template target was removed, highly specific footprint cavities in the single-layer matrix were considered to be created. Subsequently a detectable change of surface potential was brought forth after the target binding on the surface (Wang et al., 2008).

9.3 ELECTROCHEMICAL BIOSENSORS

9.3.1 Introduction

The detection of clinically relevant proteins is paving the way for understanding their biological and physiological functions and thus developing clinical diagnostics. Clinical analysis is no longer confined to the clinical laboratories only (Hood et al., 2004). Instead, it is usually performed in some settings, including both hospital care settings and nonhospital care settings (D'Orazio, 2003), which call for simple and convenient equipment. Electrochemical biosensors are well suited for these new applications (Urdea et al., 2006).

An electrochemical biosensor is an integrated device that includes a class of biological elements, such as enzymes, antibodies, nucleic acids, microorganisms, and cells, which performs highly specific response for the target through molecular recognition and conformational changes and provides analytical information, either qualitative or quantitative, by means of an electrochemical transducer.

Electronic communication between the transducer and biomolecules is the basis of electrochemical biosensors (Cao et al., 2014; Kim et al., 2015). As shown in Fig. 9.4, biometric components (antibodies, fitness, polypeptides, etc.) and signal transduction elements (optics, electrochemistry, thermodynamics, etc.) constitute the two most basic components of a typical electrochemical biosensor. The recognition elements, such as primary antibodies are immobilized onto a supporting surface and a labeled secondary antibody recognizes target antigens. The following electronic signal, such as voltammetric/amperometric, impedance, conductometric, and potentiometric signal, which is proportional to the concentration of the target, is generated after the transformation of redox-active labels. With unique advantages such as specificity, sensitivity, quick analysis, economical and practical properties, suitability for detection and analysis of target in complex systems, and low concentration samples (Lazcka et al., 2007), electrochemical assay is very attractive for protein detection. As a high and new technology where multiple disciplines integrate with each other, compared with the traditional analysis method, it increasingly plays an important role in the protein assay (Shen et al., 2014). And with the continuous development of science and technology, the electrochemical biosensor tends to high automation, integration, and miniaturization, therefore it has a prospect of broad application in the fields of pharmaceuticals and health care, biological engineering, environmental monitoring, food, and military medicine, and other fields.

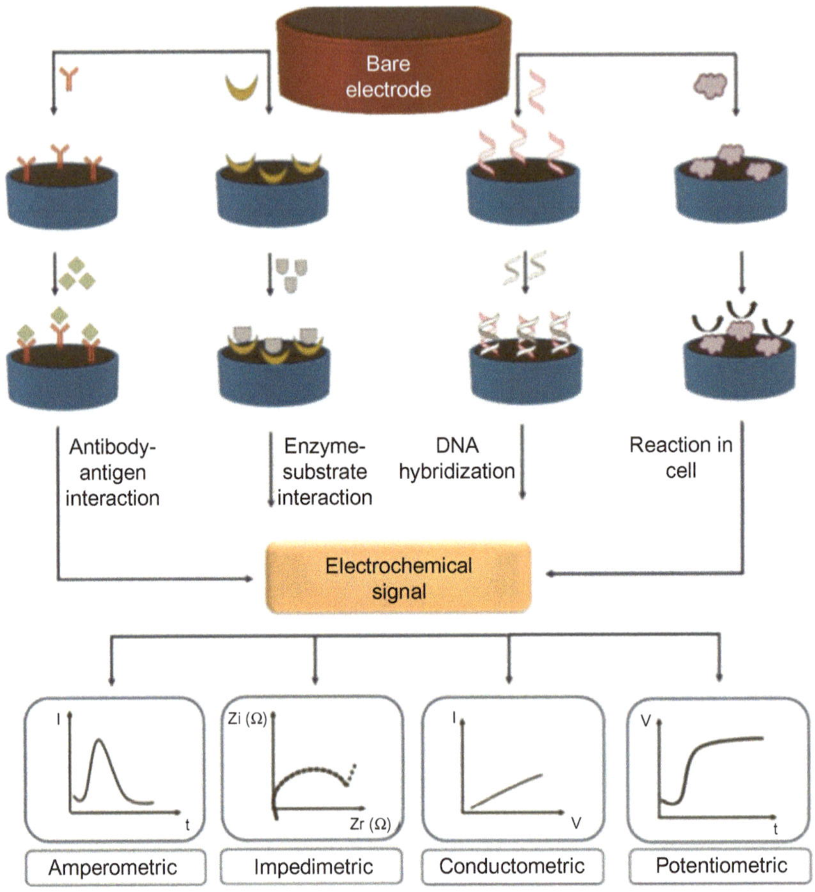

FIGURE 9.4 Scheme of an electrochemical biosensor. *Reprinted from Bahadır, E. B., Sezgintürk, M. K., 2015. Electrochemical biosensors for hormone analyses, Biosens. Bioelectron. 68, 62–71.*

9.3.2 Category of Electrochemical Biosensors

Electrochemical biosensors contain a biological recognition element (enzymes, proteins, antibodies, nucleic acids, cells, etc.) that could specifically bind to the target, and generate an electrical signal that is proportional to the concentration of the analyte under test after reactions. Therefore, on account of different biological recognition processes, the electrochemical biosensors can be divided into two main categories, biocatalytic sensors and affinity devices. The main component of biocatalytic devices mainly contains enzymes, cells, or tissue slices. They recognize the target analyte, perform redox reactions, and subsequently produce electroactive species that will be represented in the form of electrochemical signals. Enzymatic sensors occupy half of the mountain in terms of biocatalytic sensors, and the common targets include glucose, lactose, phosphatase, and xanthine. Affinity sensors, otherwise, rely on a selective binding interaction by virtue of affinity between the analyte and a biological element, such as an antibody, nucleic acid, or a receptor. DNA hybridization-based biosensors, immunosensors, aptamer-based biosensors, peptide-based sensors with electrochemical protein assay will be discussed as examples of affinity sensors (Ronkainen et al., 2010).

9.3.2.1 *Biocatalytic Sensors*

Biocatalytic reactions play important roles in the process of electrochemical detection of proteins, and enzymes are selected as the main component in many biosensing devices due to their high biocatalytic activity, specificity, and reusability. Biorecognition and signal amplification are the main function of enzymes in biocatalytic sensors, which broaden and deepen the range of biological detection without expensive instrumentation and trivial operations. As macromolecular biological catalysts, enzymes usually have a high degree of specificity for their substrates. Enzyme is the recognition element, a catalyst in life activities, which determinates that they can be used as the biomarker and drug target for the treatment and therapy of disease. The active sites of the enzymes have exquisite specificity for certain molecules, known as their substrates, which have some features similar to those of the antigen-binding fragment of antibodies, endowing these catalysts sensing functions by nature. In addition, they are commonly conjugated to a secondary recognition element; therefore signal amplification is achieved by circular reactions. Biocatalytic biosensor research began in the 1960s, and blood glucose monitoring was the original study, and the most widely studied and used of all enzyme recognition element-based biosensors as well. The object under test diffuses into and reacts with the enzyme active site, producing or consuming some active substances, and the extent of reaction is measured by amperometric or potentiometric methods to reflect the concentration of the target. Changes in quantity resulting from the enzymatic reaction are transformed into signal changes, and the numerical value is related to the substrate concentration (Harper and Anderson, 2010).

Clinically relevant enzymes which are vital to the body's various processes are chosen as detection targets by scientists. For example, many efforts have been made for the assay of α-glucosidase activity and the inhibitor screening, as is shown in Fig. 9.5. Zhang et al. proposed an electrochemical method through one-step displacement reaction. Compared to dopamine (DA)/silver nanoparticles (AgNPs), 4-aminophenyl- α-D-glucopyranoside (pAPG)/magnetic nanoparticles (MNPs) have high affinity to pyrene boric acid (PBA) which was immobilized on the surface of graphite electrode (GE), and when α-glucosidase exists, they can be specifically catalyzed into MNPs/pAP which has no binding capacity with PBA. Therefore, the activity of both isolated and membrane-bound enzymes can be well evaluated by means of this proposed method (Zhang et al., 2015).

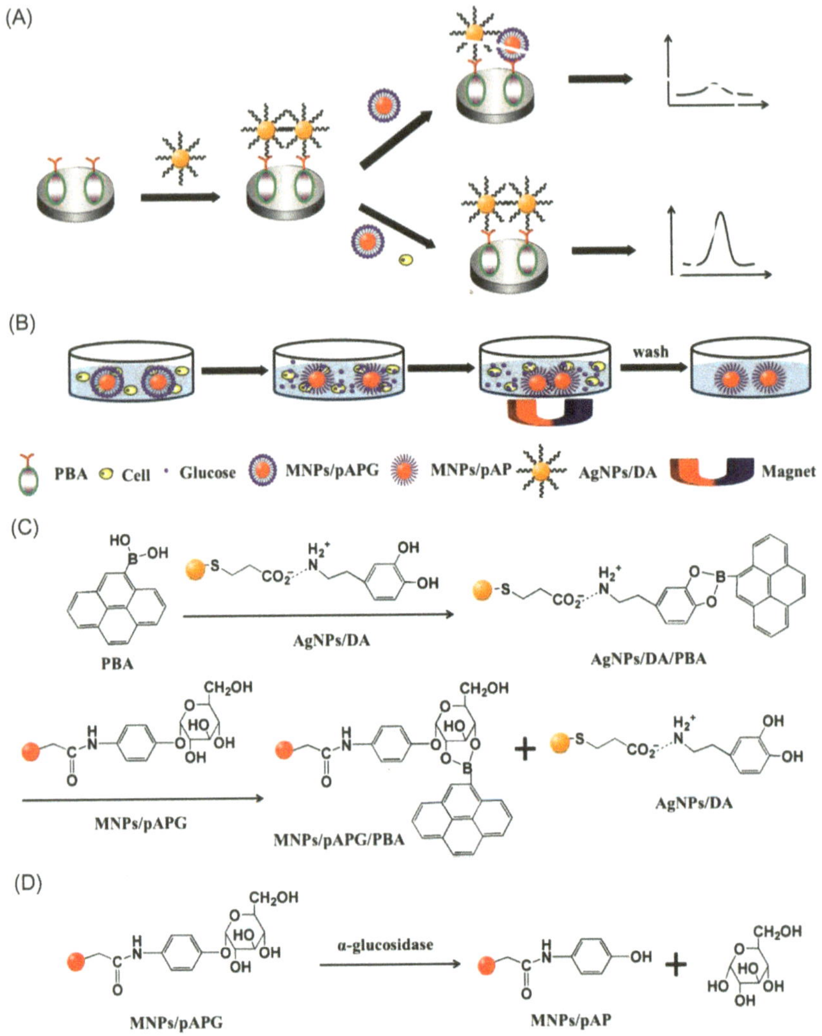

FIGURE 9.5 Scheme of the detection of α-glucosidase activity by means of MNPs/pAPG with PBA/GE. MNPs: magnetic nanoparticles; pAPG: 4-aminophenyl--d-glucopyranoside. *Reprinted from Zhang, J., Liu, Y., Wang, X., Chen, Y., Li G., 2015. Electrochemical assay of α-glucosidase activity and the inhibitor screening in cell medium, Biosens. Bioelectron. 74, 666–672.*

Sphingosine kinase 1 (SphK1) is selected as a potential marker for the diagnosis and treatment of cancer because of its phosphorylation product, a phosphosphingosine (S1P), regulates cell proliferation and death, which has attracted the attention of many researchers. For example, Gao et al. have reported a simple, sensitive, and effective electrochemical method for SphK1 assay assisted by liposome which can simulate the catalytic environment. As is shown in Fig. 9.6, sphingosine, a substrate of SphK1, was catalyzed on

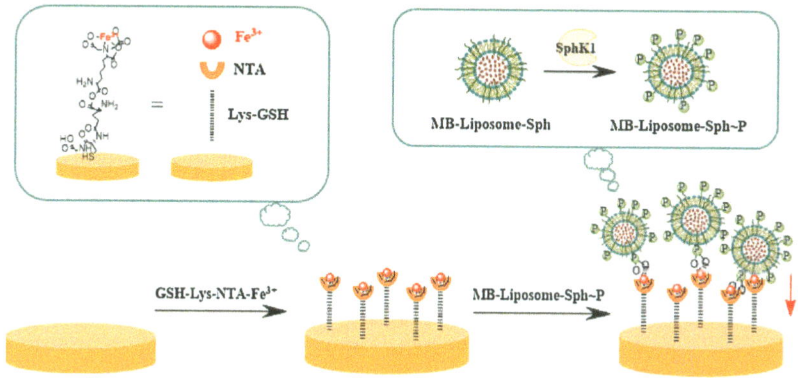

FIGURE 9.6 Schematic illustration of the electrochemical assay for lipid kinase activity based on liposome. *Reprinted from Gao, T., Gu, S., Mu, C., Zhang, M., Yang, J., Liu, P., et al., 2017. Electrochemical assay of lipid kinase activity facilitated by liposomes, Electrochim. Acta 252, 362–367.*

the surface of the liposome by kinase, which generated S1P. In this condition, Fe(III) can bind to phosphorylation sites with high affinity, thus dragging the phosphorylated liposome closer to Fe(III)-modified electrode surface. The electrically active MB molecules encapsulated in the liposomes can directly transmit the electrical signals to the electrode surface for the purpose of detecting kinase activity (Gao et al., 2017).

Enzymes can not only play the role of target, but take part in other protein assays as the significant part of the signal amplification system. Hatad et al. have developed an enzymatic biosensor utilizing the fructose base amino acid oxidase (FAOD) and the hexaammineruthenium chloride as electronic media. The response current generated by the sensor system has a significant correlation with glycosylated albumin, which can be used to determine the content of glycosylated albumin in the protease digestion sample (1.3 μL) within 1 min. Moreover, the enzymatic biosensor possesses high sensitivity, good stability, and can maintain good features at 37°C storage for 3 months (Hatada et al., 2017).

Another example that reflects the function of signal amplification is terminal deoxynucleotidyl transferase (TdT), which could mediate the DNA elongation without the template. T4 polynucleotide kinase (T4PNK) can remove phosphate groups of DNA 3'-end, which can also be regarded as DNA dephosphorylation that plays a vital role in many cellular processes. Combining the necessity of detection and the function of catalysis, Wu et al. reported a surface-extended DNA nanotail strategy for simple and ultrasensitive detection of DNA 3'-phosphatases by terminal deoxynucleotidyl transferase (TdT) mediated signal amplification. The substrate DNA probe labeled with thiols at their 5'-terminals and phosphoryls at 3'-terminals is modified on the electrode surface. When T4PNK exists, the phosphate at the 3'-terminals is hydrolyzed into hydroxyl, and then nucleotides will be added

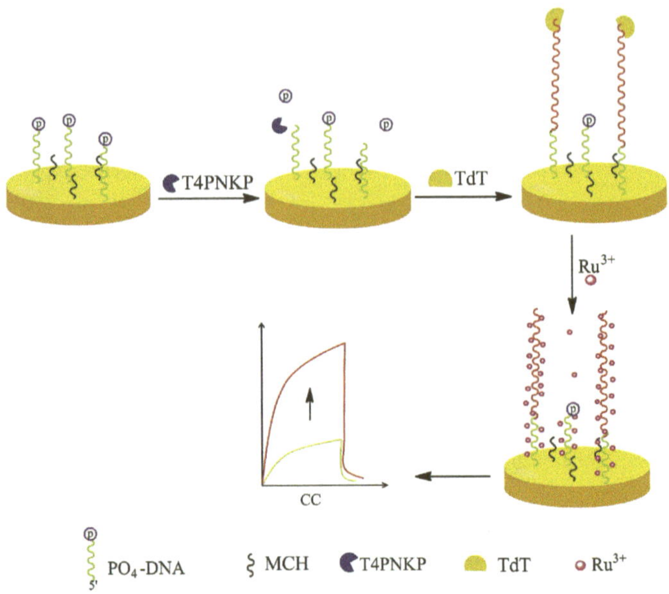

FIGURE 9.7 Scheme of surface-extended DNA nanotail strategy for the detection of T4PNKP. *Reprinted from Wu, D., Li, C., Hu, X., Mao, X., Li, G., 2016. Electrochemical detection of DNA 3′-phosphatases based on surface-extended DNA nanotail strategy, Anal. Chim. Acta 924, 29–34.*

to the 3'-terminal hydroxyl of the substrate DNA that was catalyzed by terminal deoxyribonucleotide transferase, thus forming long DNA chains of hundreds of nucleotides. The newly formed long single-stranded DNA nanotail on the electrode surface can adsorb amounts of hexaammineruthenium chloride ($[Ru(NH_3)_6]^{3+}$) by electrostatic effect, generating amplified signal, so as to realize the DNA 3'-phosphatase detection that is simple, sensitive, and convenient (Wu et al., 2016; Fig. 9.7).

9.3.2.2 *Affinity Biosensors*

Affinity sensors allow the selective and strong binding of biologically recognized substances such as antibodies (Abs), membrane receptors, or oligonucleotides to produce measurable electrical signals (Lenain et al., 2015; Schoukroun-Barnes and White, 2015; Schultz, 2015). The size and shape of the complementary binding site of the biorecognition substance have an important influence on the biorecognition capability of the affinity sensor. The high affinity and specificity of biomolecules for their ligands make these sensors very sensitive and selective, such as DNA hybridization or antibody–antigen (Ab-Ag) complex binding processes. Affinity-type electrochemical biosensors have attracted widespread attention because of their advantages such as fast response, ease of use, low cost, and miniaturization. In recent years, the development of molecular biotechnology, nanomaterials, analytical chemistry, etc. has continued to develop (Crack et al., 2014). The affinity electrochemical biosensors have been rapidly developed in various fields of research and are capable of acting on a variety of biologically active substances (cells, bacteria, nucleic acids). The rapid and accurate detection of inorganic ions, peptides, etc. have a very active research frontier in clinical diagnostics (Sarkar et al., 2013; Wang and Anzai, 2015).

9.3.2.2.1 DNA Hybridization-Based Electrochemical Biosensors

Sequence-specificity and transcription factors play an important role in genome replication, transcription, recombination, and repair, as well as the promoter's various biological activities and the specific binding of proteins to DNA. It is of great importance in basic biology, as well as in clinical diagnosis and drug development. At present, it is still a challenging task to analyze and identify related proteins that bind to specific nucleic acid sequences at the molecular level. For example, in the process of gene expression, the level of gene expression is not only controlled by the amount of the relevant polymerase, but also by the relevant promoter sequence, regulating protein–protein or protein–nucleic acid binding and initiation efficiency. Therefore, a basic mechanism which can comprehensively analyze DNA–protein interactions is very important in cell activities.

Electrochemical biosensors based on DNA hybridization, which use DNA or RNA nucleic acid fragments, known as aptamers, recognizing and binding to three-dimensional protein surfaces, are quite widely used among modern protein detection and analysis technologies. Since the recent expansion of the knowledge of proteins and the technologies of manipulating them, nucleic acids have been becoming increasingly important as biometric identification reagents in sensors. DNA affinity probes are commonly used in medical diagnostics to detect cancer, viral infections, and genetic diseases.

The realization of genome-related functions is accomplished by the interaction between various proteins and DNA elements. At present, the main problem lies in how to analyze specifically the sequence of related DNA binding with proteins in a convenient and cost-effective way. In response, Li et al. have introduced an electrochemical assay, utilizing the motion behavior between DNAzyme and transcription factor (FT). To some extent, it has proven that there is a greatly increased specificity and sensitivity of interactions between transcription factors and nucleic acids, which may open up new opportunities for an in-depth analysis of the sequence specificity of DNA binding proteins and the study of nucleotide polymorphisms in known protein binding sites. He and his team screened the binding sequence of HNF-4α protein in vitro, and analyzed the related protein binding sequence with related mutations and binding affinity. In addition, they also analyzed three protein binding factors, NF-κB, SP6 RNA polymerase, and HNF-4α, in a rapid, sensitive, and selective way (Li et al., 2017; Fig. 9.8).

Related studies have shown that the accumulation and misbinding of oxidatively-damaged 8-oxo-7, 8-dihydroguanine triphosphate (8-oxo-dGTP)

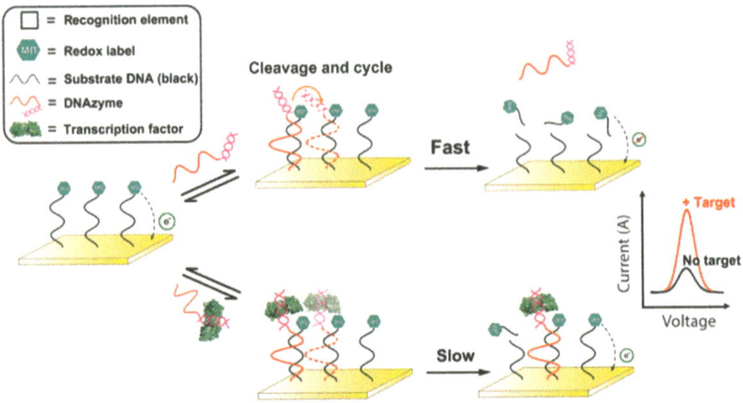

FIGURE 9.8 Schematic diagram of proposed principle of transcription factor (TF) detection. *Reprinted from Li, C., Tao, Y., Yang, Y., Xiang, Y., Li, G., 2017. In vitro analysis of DNA–protein interactions in gene transcription using DNAzyme-based electrochemical assay, Anal. Chem. 89, 5003–5007.*

in genomic DNA can lead to severe cell dysfunction. MutTH Homolog 1 (MTH1) as a protease may be of benefit to prevent the mistaken incorporation of 8-oxo-dGTP. In response, Gao et al. have introduced an electrochemical assay in this work for the assay of MTH1 activity in a simple and convenient way. In his work, a mismatch-based ("8-oxoG: A" mismatch) DNA chain elongation strategy (MB-DCE) is firstly proposed to reveal the misincorporation efficiency of 8-oxo-dGTP, which can be established to reveal the activity of MTH1 through MB-DCE by preventing 8-oxo-dGTP misincorporation. As a proof of principle, this method can be used to detect the activity of MTH1 in different breast cancer cell lines, and to some extent provided a feasible way for future biomedical research and clinical diagnosis of breast cancer-related diseases (Gao et al., 2016; Fig. 9.9).

Yang et al. proposed an ultrahigh sensitivity target protein analysis method by double amplification without enzyme assistance, which is relatively simple and inexpensive, having great potential for future clinical diagnosis and biomedical research. The specific aptamer captures the target protein to complete the recognition, and then the DNA strand replacement reaction is mediated by a targeted click chemistry reaction (first round of recovery). The protruding ends of the electrode surface aptamers can hybridize with RP-DNA and initiate a hybridization chain reaction (second round of recovery). Therefore, a large amount of electrochemical hexamethylammonium chloride ($[Ru(NH_3)_6]^{3+}$) can be inserted into double-stranded DNA to generate an amplified electrochemical signal. A wide linear dynamic range from 100 fM to 10 nM and a detection limit of 30 fM (S/N = 3) can be obtained by taking thrombin as a model analyte (Yang et al., 2015b; Fig. 9.10).

Zhao et al. proposed an electrochemical method based on the small molecule-assisted metal nanoparticles formation to sensitively and

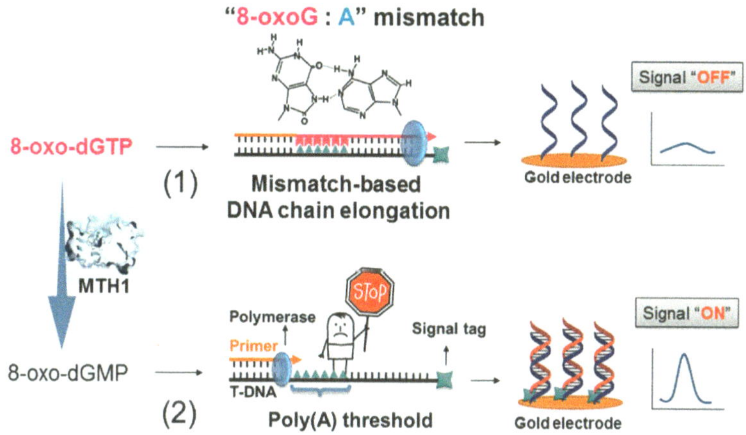

FIGURE 9.9 The schematic is a principle of an electrochemical assay method for detecting MTH1 activity based on the mismatched DNA strand elongation (MB-DCE) strategy. *Reprinted from Gao, T., Gu, S., Liu, F., Li, L., Wang, Z., Yang, J., et al., 2016. Investigation of MTH1 activity via mismatch-based DNA chain elongation, Anal. Chim. Acta 905, 66–71.*

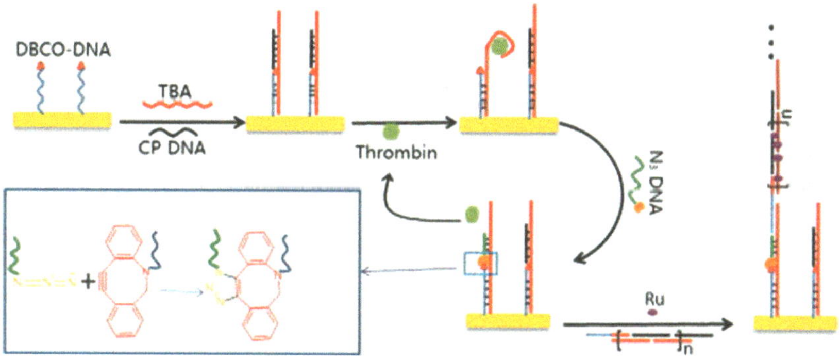

FIGURE 9.10 Schematic diagram of the principle of protein detection in this plan. *Reprinted from Yang, D., Ning, L., Gao, T., Ye, Z., Li, G., 2015b. Enzyme-free dual amplification strategy for protein assay by coupling toehold-mediated DNA strand displacement reaction with hybridization chain reaction, Electrochem. Commun. 58, 33–36.*

specifically detect proteins by means of hybridization chain reaction (HCR). When the target protein is present, the DNA probe immobilized on the surface of the electrode hybridizes with the captured DNA, triggering a surface hybridization chain reaction (HCR). Using long double-stranded DNA oligomers generated by HCR as templates, copper nanoparticles can be formed on the electrode surface. In addition, the release of active oxygen during the acid dissolution process of copper nanoparticles can catalyze the oxidation of o-phenylenediamine, which acts as an electrical signal amplifier. Since the folate receptor (FR) binds to folic acid, the probe DNA is protected from the degradation catalyzed by exonuclease I. The method proposed in the article

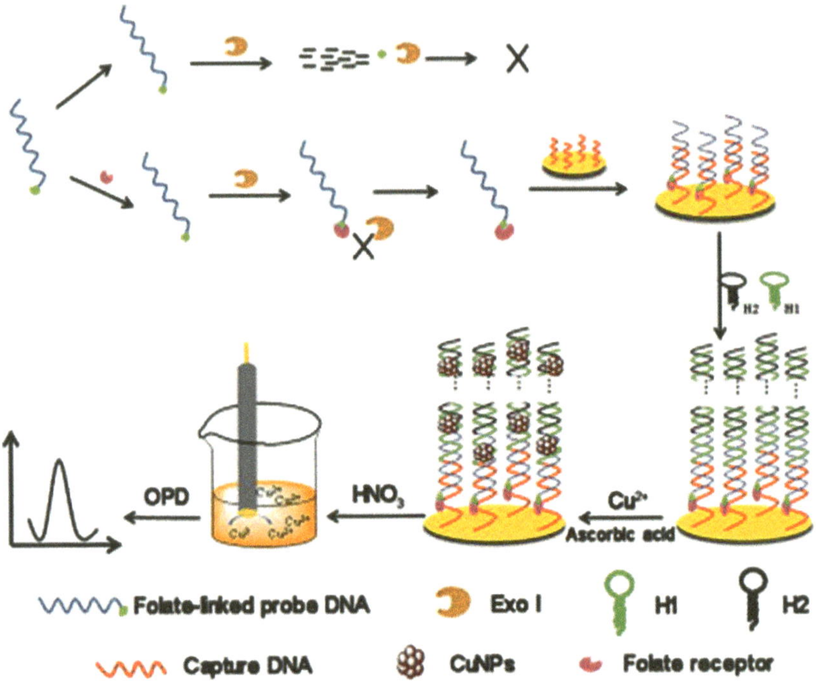

FIGURE 9.11 Hybridization chain reaction assisted formation of copper nanoparticle protein electrochemical detection schematic. *Reprinted from Zhao, J., Hu, S., Cao, Y., Zhang, B., Li, G., 2015. Electrochemical detection of protein based on hybridization chain reaction-assisted formation of copper nanoparticles, Biosens. Bioelectron. 66, 327–331.*

can specifically detect the target protein in both buffer and complex serum samples, which is promising for the development of the technique for protein detections since many other proteins can be assayed by changing the corresponding small molecule (Zhao et al., 2015; Fig. 9.11).

9.3.2.2.2 Immunosensors

The immunosensor is designed by utilizing the principle that an antigen and an antibody specifically combine to cause an electrochemical change in an organism-specific immune response. The complementary region of Ab and the antibody Ag have a high degree of specificity and affinity. The electrochemical sensor responds to the binding reaction and stores, outputs, and amplifies the electrochemical signal in response.

The complementary region of Ab and the antibody Ag have a high degree of specificity and affinity, which can be developed for a wide range of substances. Immunoassays and immunosensors have been developed for quantitative and qualitative applications, which are well-known in analytical methods because of their extremely low detection limits, and have been widely used for the detection of bacteria, viruses, drugs, hormones, and pesticides (Ronkainen-Matsuno et al., 2002). Examples for the application of immunosensors include monitoring of food safety related to severe allergies (e.g., peanuts), detection of environmental pollutants (e.g., herbicides and pesticides in water and soil), detection of biomedical substances (e.g., warfarin), and monitoring of biological warfare agents such as toxins, bacteria, viruses, and spores. Nowadays, with the in-depth research and the abundance of antibodies, the development of affinity biosensors has entered a blowout period (Ronkainen-Matsuno et al., 2002). Palomar et al. reported an impedimetric immunosensor that was dependent on the deposition of carbon nanotube (CNT) which can significantly improve immunosensor performance by controlling thicknesses for enhanced electroactive surface areas. In order to form an ideal immunosensor device, poly(cyclodrolyl)-nitrotriacetic acid (polypyrrole-NTA) was electro-coated to form a Cu(II) complex with NTA function, the CNT precipitate functionalized, and then the biotin-modified cholera toxin B was fixed with the NTA(II) complex. Cyclic voltammetry and electrochemical impedance spectroscopy (EIS) are used to ensure the formation of the immunosensor and the subsequent binding of the analyte antibody anticholera toxin. After majorization, the resulting impedimetric cholera sensor shows great reproducibility, increased sensitivities, and satisfied detection limit (Palomar et al., 2017; Fig. 9.12).

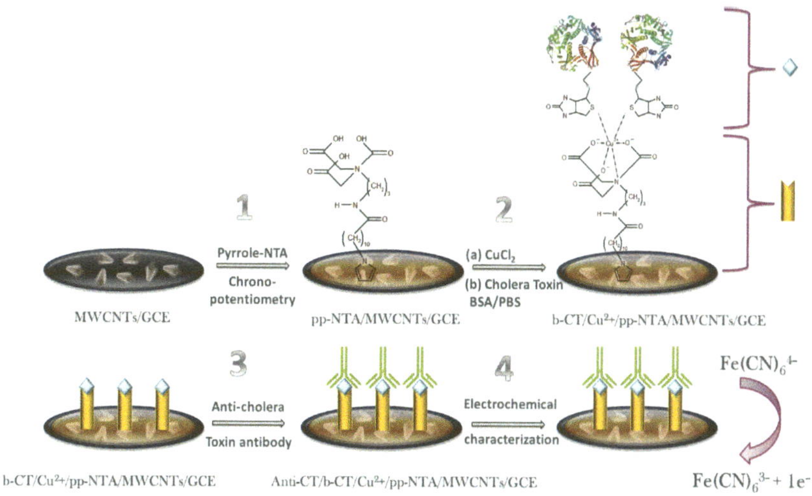

FIGURE 9.12 Formation and immunosensor schematic operation in this work: (1) polymerizing pyrrole -NTA electrically; (2) b-CTB poly (pyrrole -NTA) / Cu^{2+} binding of the ligand; (3) b-CTB Immune recognition with anti-CT antibodies; (4) identifying events using EIS and cyclic voltammetry. *Reprinted from Palomar, Q., Gondran, C., Holzinger, M., Marks, R., Cosnier S., 2017. Controlled carbon nanotube layers for impedimetric immunosensors: high performance label free detection and quantification of anti-cholera toxin antibody, Biosens. Bioelectron. 97, 177–183.*

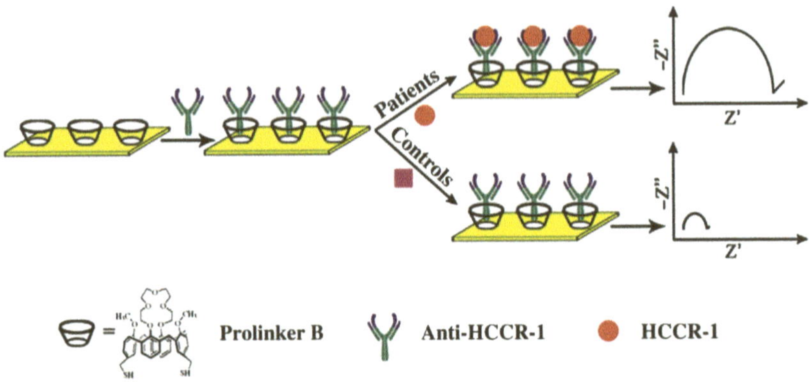

FIGURE 9.13 Direct unmarked human cervical cancer in serum oncoprotein-1 (HCCR-1) impedimetric detection. *Reprinted from Chen, D., Shen, M., Cao, Y., Bo, B., Chen, Z., Shu, Y., et al., 2013. Electrochemical identification of hepatocellular carcinoma based on the assay of human cervical cancer oncoprotein-1 in serum, Electrochem. Commun. 27, 38–41.*

Chen et al. developed an advanced electrochemical immunosensor method that was established for the early detection of overexpressed human cervical cancer oncoprotein-1 (HCCR-1) in HCC. This immunosensor is very valuable for early HCC identification, considering that there is no effective treatment for patients with advanced HCC at the present. This method can easily identify HCCs, based on the comparison of EIS response differences between HCC patients and healthy subjects, with calixarene derivatives immobilized on the gold electrode surface, and the EIS of HCCR-1 in serum is measured, which can be captured by the HCCR-1 antibody loaded on the electrode surface in serum. There is no doubt that this method is simple and quick and has considerable potential for early diagnosis of HCC in the future (Chen et al., 2013; Fig. 9.13).

Nowadays, more and more nanomaterials are being applied to the design of redox nanoprobe in EIS for doping detection. As is shown in Fig. 9.14, functionalized with polyamidoamine and AuNPs, with the assistance of surfactant tetraoctylammonium bromide as a booster, the inner redox activity of the C60-based nanomaterials was greatly raised, which was explored as a new approach for the use of carbon nanomaterials as redox nanoprobes in the field of electrochemical immunoassay (Han et al., 2015).

At present, relevant studies have shown that Golgi protein 73 (GP73) is closely related to the occurrence of biliary tract cancer. It has become an important biomarker for the study of biliary tract cancer. Considering that the current electrochemical research is subject to the conductivity of the material and other related factors, which no doubt will influence electrochemical signals, Yu et al. developed an electrochemical immunosensor for the determination of GP73 with the excellent electrical conductivity of reduced graphene oxide (rGO) and electrochemical signals with outstanding

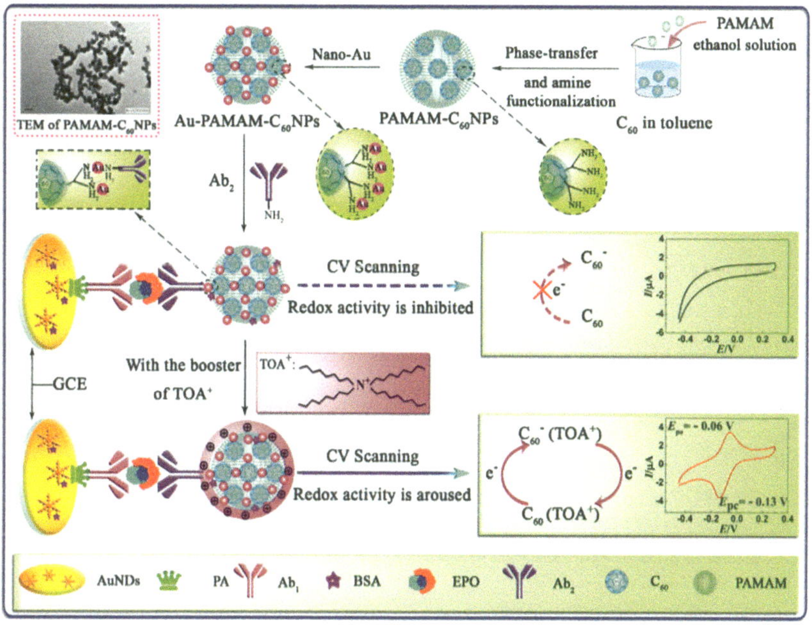

FIGURE 9.14 Preparation of immunosensors and its possible mechanism of electrochemical reaction with C60-based nanomaterials to construct redox nanoprobes. *Reprinted from Han, J., Zhuo, Y., Chai, Y.-Q., Xiang, Y., Yuan, R., 2015. New type of redox nanoprobe: C60-based nanomaterial and its application in electrochemical immunoassay for doping Detection, Anal. Chem. 87, 1669–1675.*

quantum dots, which show specificity and sensitivity in the direct detection of serum. The results can clearly predict the effect of surgical resection of primary tumors, revealing great potential for sensors applications in the future (Fig. 9.15; Yu et al., 2014).

9.3.2.2.3 Aptamer-Based Electrochemical Biosensors

Aptamers are synthetic oligonucleotides that have undergone iterative rounds of in vitro selection for binding with high affinity to specific analytes of choice, which have become increasingly important molecular tools for diagnostics and therapeutics. A sensitive yet simple method to utilize aptamers as recognition entities for the development of biosensors is to transduce the signal electrochemically (You et al., 2003; Jenison et al., 1994).

Nowadays, aptamer technology has received great attention in the scientific community and industry. Over several decades, aptamers (DNA or RNA) have been discovered that bind tightly to a wide range of targets (e.g., proteins, peptides, amino acids, drugs, metal ions, or even whole cells), especially with the rapid, automated, and selective development of technologies (Tombelli et al., 2007).

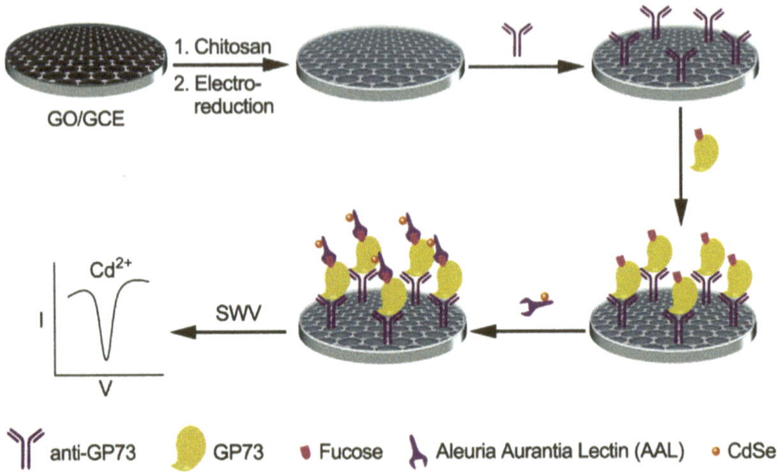

FIGURE 9.15 The proposed mechanism for BTC assay. *Reprinted from Yu, Y., Gao, T., Li, H., Ye, Z., Chen, Z., Li, G., 2014. A novel electrochemical immunosensor for Golgi Protein 73 assay, Electrochem. Commun. 42, 6–8.*

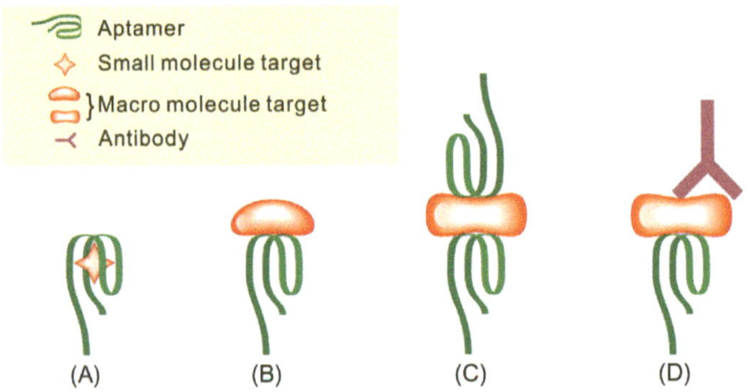

FIGURE 9.16 Several aptamer-assisted identification forms. (A) Small molecule target buried in the aptamer structure; (B) single-site format; (C) sandwich structure with two aptamer sites for dual site binding; and (D) sandwich structure with aptamer and antibody binding (Song et al., 2008). *Reprinted from Song, S., Wang, L., Li, J., Fan, C., Zhao, J., 2008. Aptamer-based biosensors, TrAC, Trends Anal. Chem. 27, 108–117.*

Compared to biosensors using natural receptors such as antibodies and enzymes, aptamer-based biosensors show several unprecedented advantages as follows (Song et al., 2008; Fig. 9.16): (1) high specificity and affinity, ranging from small molecules to large proteins and even cells; (2) easy to synthesize and purify, and highly chemically stability, compared with protein-based antibodies or enzymes; and (3) significant conformational changes make it easy to design novel biosensors with high sensitivity and selectivity.

With a consistent understanding of the conformation and ligand binding properties of nucleic aptamers, a series of bioassay methods that rely on

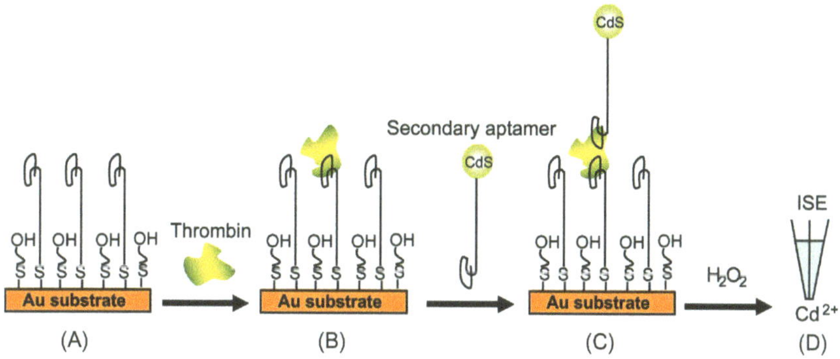

FIGURE 9.17 Aptamer-based potentiometric measurements of proteins using ion-selective microelectrodes. (A) A mixed monolayer of thiolated aptamers on a gold substrate is formed. (B) Thrombin is added and combined with the aptamer. (C) Combined with CdS-labeled aptamers. (D) Dissolve the CdS label, then use a solid contact Cd^{2+} selective microelectrode for detection (Numnuam et al., 2008). *Reprinted from Numnuam, A., Chumbimuni-Torres, K.Y., Xiang, Y., Bash, R., Thavarungkul, P., Kanatharana, P., et al., 2008. Aptamer-based potentiometric measurements of proteins using ion-selective microelectrodes, Anal. Chem. 80, 707–712.*

aptamer sensors were produced. Recent advances in aptamer-based electrochemical sensors are described. Most of the designs can be classified into two types of configurations: single-site binding; and dual-site binding.

For example, as is shown in Fig. 9.17, two aptamers, of which one is immobilized onto gold substrate through Au-S bonds, and the other is primarily conjugated with CdS particles, cooperated with the target forming an aptamer–protein–aptamer sandwich structure. Dissolved CdS was measured electrochemically for the detection of the protein target thrombin.

Antibody and aptamer can be used simultaneously in one system. Zhu et al. proposed a concept by introducing a magneto-controlled moveable architecture (MCMA), on which anti-CEA antibody was functionalized to capture and separate the target cancer cells magnetically. Horseradish peroxidase-labeled mucin-1 aptamer (HRP-apt) was also prepared to recognize the target cells. While being pulled onto a working electrode under magnetic field, the HRP on the cell surface could present an electrocatalytic signal (Fig. 9.18; Zhu et al., 2014).

9.3.2.2.4 Peptide-Based Electrochemical Sensors

Peptides such as enzymes, antibodies, drug receptors, and transmembrane proteins are a clever choice to mimic molecular recognition mechanisms that occur in biomolecules. Multiple interactions can be formed within the macromolecules which have high specificity or high catalytic activity. A multitude of work has been done to comprehend the mechanism with the view of generating artificial macrostructures for binding and/or catalysis. One of the most representative is olfactory receptor protein, which is difficult to purify from natural sources and

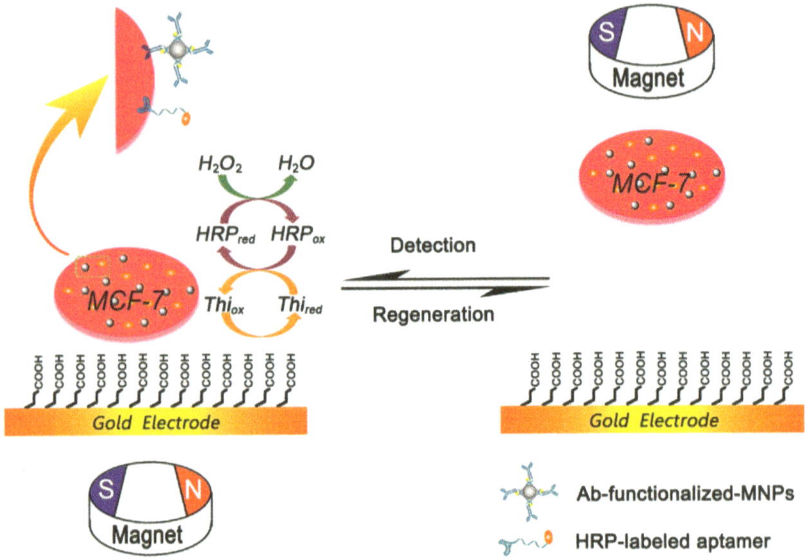

FIGURE 9.18 Reusable electrochemical biosensor for detecting human breast cancer cells with MCMA. *Reprinted from Zhu, X., Feng, C., Ye, Z., Chen, Y., Li, G., 2014. Fabrication of magneto-controlled moveable architecture to develop reusable electrochemical biosensors, Sci. Rep. 4, 4169.*

impossible to synthesize (Zarzo, 2007). Considering their widely application, the design of short, peptide-based, artificial receptors capable of highly specific recognition is necessary (Skerra, 2007). Peptide-based sensors have been developed in recent years according to different strategies.

The peptamer which is constructed by several different well-prepared peptide motifs have great targeting and catalytic abilities. In reference, the peptamer to an integrin can be controlled by a synthetic ligand and the affinity of the peptamer with a catalytic cofactor. Cupric ion will be changed by the conformational rearrangement of the peptamer induced by both integrin and the synthetic ligand, which leads to greatly contrasted efficiency of catalysis in the presence/absence of integrin. This simple and sensitive peptamer model may aid in the realization of the structural reconfiguration of proteins, and also may provide a new strategy to assess protein activity under pathological conditions such as cancer (Li et al., 2016a; Fig. 9.19).

Hampered by the limited choice of signal readout methods for electrical signals, although protein-binding peptides are considered as effective artificial affinity reagents for protein assays, their applications are limited. A new signal readout method for protein-binding peptides utilizes the chemistry characteristic of host–guest of cucurbituril. Protein-binding peptides

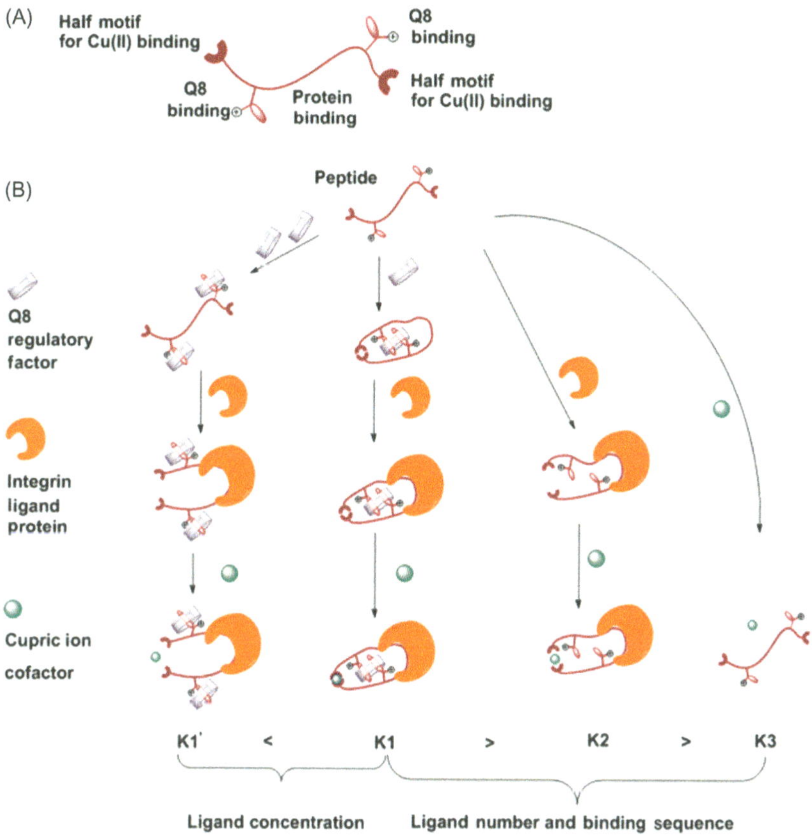

FIGURE 9.19 Detection of tumor invasive biomarker using a peptamer. (A) Construction of peptamer. (B) Regulation of the peptamer conformation and affinity with the cofactor ion based on the interaction with the ligand. *Reprinted from Li, H., Li, W., Liu, F., Wang, Z., Li, G., Karamanos, Y.,2016a. Detection of tumor invasive biomarker using a peptamer of signal conversion and signal amplification, Anal. Chem. 88, 3662–3668.*

can be noncovalently coupled to electrochemical reporter molecules via supramolecule formation between cucurbituril, electrochemical reporters, and peptides. The protein-binding peptide is first self-assembled at the sensing layer, capturing the target protein, with a protein boundary conforming, and then the protein-free peptide is coupled with the electrochemical reporter, resulting in a signal readout that is inversely proportional to the amount of target protein captured. The versatility and target specificity of the proposed methods in the determination of two target proteins: tumor necrosis factor-alpha and amyloid beta 1-42 soluble oligomers have been successfully demonstrated, as have tumor necrosis factor-α in HL-60 cells (Li et al., 2013b; Fig. 9.20).

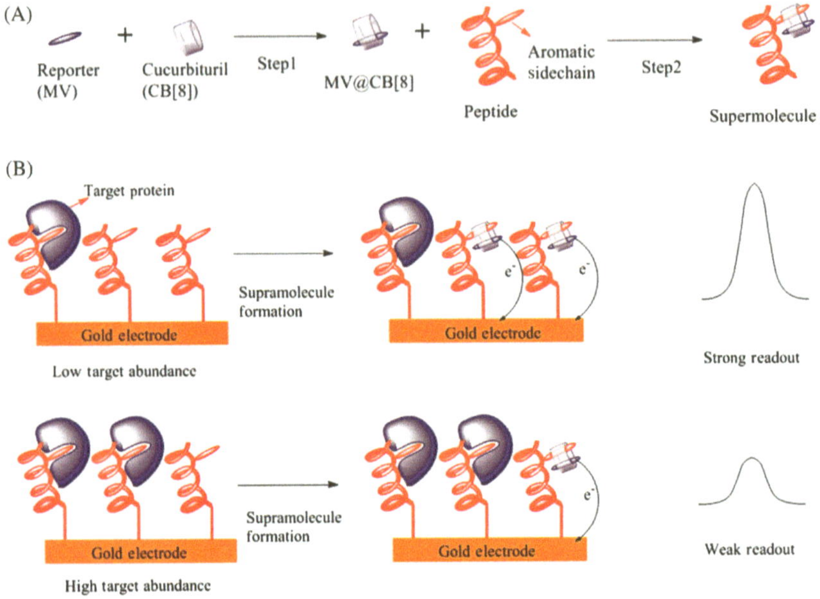

FIGURE 9.20 A general way to assay protein by coupling peptide with signal reporter via supermolecule formation. (A) Combining peptide with reporter through supermolecule effect, and (B) assay for protein detection. *Reprinted from Li, H., Xie, H., Cao, Y., Ding, X., Yin, Y., Li, G., 2013b. A general way to assay protein by coupling peptide with signal reporter via supermolecule formation, Anal. Chem. 85, 1047–1052.*

9.4 SUMMARY AND PROSPECTS

Simplicity, sensitivity, and accuracy of electrochemical biosensors will endow electrochemistry with paramount status in protein assays. Biocatalytic and affinity biosensors for electrochemical detection play an increasingly important role in diagnostics and disease monitoring due to their unparalleled sensitivity and selectivity. Recent developments in nanotechnology and materials science will offer building blocks for the super kingdom of the electrochemical biosensors. The biggest limitations of biosensors are the shelf life and stability of the biorecognition element, as well as nonspecific binding. Therefore, more and more strategies are desired to deal with these challenges.

REFERENCES

Baciu, A., Ardelean, M., Pop, A., Pode, R., Manea, F., 2015. Simultaneous voltammetric/amperometric determination of sulfide and nitrite in water at BDD electrode. Sensors (Basel) 15, 14526–14538.

Bangar, M.A., Shirale, D.J., Chen, W., Myung, N.V., Mulchandani, A., 2009. Single conducting polymer nanowire chemiresistive label-free immunosensor for cancer biomarker. Anal. Chem. 81, 2168–2175.

Bahadır, E.B., Sezgintürk, M.K., 2015. Electrochemical biosensors for hormone analyses, Biosens. Bioelectron. 68, 62–71.

Bruner, L.H., Miller, K.R., Owicki, J.C., Parce, J.W., Muir, V.C., 1991. Testing ocular irritancy in vitro with the silicon microphysiometer. Toxicol. In Vitro 5, 277–284.

Buter, L., Frensemeier, L.M., Vogel, M., Karst, U., 2017. Dual reductive/oxidative electrochemistry/liquid chromatography/mass spectrometry: Towards peptide and protein modification, separation and identification. J. Chromatogr. A. 1479, 153–160.

Cao, Y., Chen, D., Chen, W., Yu, J., Chen, Z., Li, G., 2014. Aptamer-based homogeneous protein detection using cucurbit[7]uril functionalized electrode. Anal. Chim. Acta 812, 45–49.

Chen, D., Shen, M., Cao, Y., Bo, B., Chen, Z., Shu, Y., et al., 2013. Electrochemical identification of hepatocellular carcinoma based on the assay of human cervical cancer oncoprotein-1 in serum. Electrochem. Commun. 27, 38–41.

Cheng, A.K.H., Ge, B., Yu, H.-Z., 2007. Aptamer-based biosensors for label-free voltammetric detection of lysozyme. Anal. Chem. 79, 5158–5164.

Crack, J.C., Green, J., Thomson, A.J., Le Brun, N.E., 2014. Iron-sulfur clusters as biological sensors: the chemistry of reactions with molecular oxygen and nitric oxide. Acc. Chem. Res. 47, 3196–3205.

D'Orazio, P., 2003. Biosensors in clinical chemistry. Clin. Chim. Acta 334, 41–69.

Davis, G., 1985. Electrochemical techniques for the development of amperometric biosensors. Biosensors 1, 161–178.

Dejmkova, H., Stoica, A.I., Barek, J., Zima, J., 2011. Voltammetric and amperometric determination of 2,4-dinitrophenol metabolites. Talanta 85, 2594–2598.

Dixit, C.K., Kadimisetty, K., Otieno, B.A., Tang, C., Malla, S., Krause, C.E., et al., 2016. Electrochemistry-based approaches to low cost, high sensitivity, automated, multiplexed protein immunoassays for cancer diagnostics. Analyst 141, 536–547.

Gao, T., Gu, S., Liu, F., Li, L., Wang, Z., Yang, J., et al., 2016. Investigation of MTH1 activity via mismatch-based DNA chain elongation. Anal. Chim. Acta 905, 66–71.

Gao, T., Gu, S., Mu, C., Zhang, M., Yang, J., Liu, P., et al., 2017. Electrochemical assay of lipid kinase activity facilitated by liposomes. Electrochim. Acta 252, 362–367.

Han, J., Zhuo, Y., Chai, Y.-Q., Xiang, Y., Yuan, R., 2015. New type of redox nanoprobe: C60-based nanomaterial and its application in electrochemical immunoassay for doping Detection. Anal. Chem. 87, 1669–1675.

Harper, A., Anderson, M.R., 2010. Electrochemical glucose sensors—developments using electrostatic assembly and carbon nanotubes for biosensor construction. Sensors 10, 8248–8274.

Hascup, K.N., Hascup, E.R., 2014. Electrochemical techniques for subsecond neurotransmitter detection in live rodents. Comp. Med. 64, 249–255.

Hatada, M., Tsugawa, W., Kamio, E., Loew, N., Klonoff, D.C., Sode, K., 2017. Development of a screen-printed carbon electrode based disposable enzyme sensor strip for the measurement of glycated albumin. Biosens. Bioelectron. 88, 167–173.

Hood, L., Heath, J.R., Phelps, M.E., Lin, B., 2004. Systems biology and new technologies enable predictive and preventative medicine. Science 306, 640–643.

Jacobs, C.B., Peairs, M.J., Venton, B.J., 2010. Review: carbon nanotube based electrochemical sensors for biomolecules. Anal. Chim. Acta 662, 105–127.

Jenison, R., Gill, S., Pardi, A., Polisky, B., 1994. High-resolution molecular discrimination by RNA. Science 263, 1425–1429.

Johnson, A., Song, Q., Ko Ferrigno, P., Bueno, P.R., Davis, J.J., 2012. Sensitive affimer and antibody based impedimetric label-free assays for C-reactive protein. Anal. Chem. 84, 6553–6560.

Kim, J., Banks, A., Cheng, H., Xie, Z., Xu, S., Jang, K.I., et al., 2015. Epidermal electronics with advanced capabilities in near-field communication. Small 11, 906–912 (Weinheim an der Bergstrasse, Germany).

Kranz, C., 2014. Recent advancements in nanoelectrodes and nanopipettes used in combined scanning electrochemical microscopy techniques. Analyst 139, 336–352.

Lazcka, O., Del Campo, F.J., Munoz, F.X., 2007. Pathogen detection: a perspective of traditional methods and biosensors. Biosens. Bioelectron. 22, 1205–1217.

Lenain, P., De Saeger, S., Mattiasson, B., Hedstrom, M., 2015. Affinity sensor based on immobilized molecular imprinted synthetic recognition elements. Biosens. Bioelectron. 69, 34–39.

Li, H., Xie, H., Cao, Y., Ding, X., Yin, Y., Li, G., 2013b. A general way to assay protein by coupling peptide with signal reporter via supermolecule formation. Anal. Chem. 85, 1047–1052.

Li, H., Xie, H., Yang, N., Huang, Y., Sun, L., Li, G., 2013c. Design of a bi-functional peptide for protein assays: observation of cortactin expression in human placenta. Chem. Commun. 49, 5387–5389.

Li, Y., Zelakiewicz, B.S., Allison, T.C., Tong, Y.J., 2015. Measuring level alignment at the metal-molecule interface by in situ electrochemical C NMR. ChemPhysChem 16, 747–751.

Li, H., Li, W., Liu, F., Wang, Z., Li, G., Karamanos, Y., 2016a. Detection of tumor invasive biomarker using a peptamer of signal conversion and signal amplification. Anal. Chem. 88, 3662–3668.

Li, C., Tao, Y., Yang, Y., Xiang, Y., Li, G., 2017. In vitro analysis of dna–protein interactions in gene transcription using DNAzyme-based electrochemical assay. Anal. Chem. 89, 5003–5007.

Liu, Z., Du, J., Qiu, C., Huang, L., Ma, H., Shen, D., et al., 2009. Electrochemical sensor for detection of p-nitrophenol based on nanoporous gold. Electrochem. Commun. 11, 1365–1368.

Lojou, É., Bianco, P., 2006. Application of the electrochemical concepts and techniques to amperometric biosensor devices. J. Electroceram. 16, 79–91.

Luo, X., Lee, I., Huang, J., Yun, M., Cui, X.T., 2011. Ultrasensitive protein detection using an aptamer-functionalized single polyaniline nanowire. Chem. Commun. 47, 6368–6370.

Manning, C.W., Purdy, W.C., 1970. Reference electrode for electrochemical studies in dimethylformamide. Anal. Chim. Acta 51, 124–126.

Martin, L.L., Kubeil, C., Simonov, A.N., Kuznetsov, V.L., Corbin, C.J., Auchus, R.J., et al., 2017. Electrochemistry of cytochrome P450 17alpha-hydroxylase/17,20-lyase (P450c17). Mol. Cell. Endocrinol. 441, 62–67.

May, K.M.L., Vogt, A., Bachas, L.G., Anderson, K.W., 2005. Vascular endothelial growth factor as a biomarker for the early detection of cancer using a whole cell-based biosensor. Anal. Bioanal. Chem. 382, 1010–1016.

Numnuam, A., Chumbimuni-Torres, K.Y., Xiang, Y., Bash, R., Thavarungkul, P., Kanatharana, P., et al., 2008. Aptamer-based potentiometric measurements of proteins using ion-selective microelectrodes. Anal. Chem. 80, 707–712.

Palecek, E., Tkac, J., Bartosik, M., Bertok, T., Ostatna, V., Palecek, J., 2015. Electrochemistry of nonconjugated proteins and glycoproteins. Toward sensors for biomedicine and glycomics. Chem. Rev. 115, 2045–2108.

Palomar, Q., Gondran, C., Holzinger, M., Marks, R., Cosnier, S., 2017. Controlled carbon nanotube layers for impedimetric immunosensors: high performance label free detection and quantification of anti-cholera toxin antibody. Biosens. Bioelectron. 97, 177–183.

Pei, X., Kang, W., Yue, W., Bange, A., Heineman, W.R., Papautsky, I., 2014. Disposable copper-based electrochemical sensor for anodic stripping voltammetry. Anal. Chem. 86, 4893–4900.

Plotnikov, E., Korotkova, E., Voronova, O., Sazhina, N., Petrova, E., Artamonov, A., et al., 2016. Comparative investigation of antioxidant activity of human serum blood by amperometric, voltammetric and chemiluminescent methods. Arch. Med. Sci. 12, 1071–1076. AMS.

Ramanaviciene, A., Ramanavicius, A., Finkelsteinas, A., 2006. Basic electrochemistry meets nanotechnology: electrochemical preparation of artificial receptors based on nanostructured conducting polymer, polypyrrole. J. Chem. Educ. 83, 1212–1214.

Ramos Miguel, A., Ramos Macias, A., Viera Artiles, J., Perez Zaballos, M.T., 2015. The effect of reference electrode position in cochlear implants. J Int Adv Otol 11, 222–228.

Ronkainen-Matsuno, N.J., Thomas, J.H., Halsall, H.B., Heineman, W.R., 2002. Electrochemical immunoassay moving into the fast lane. TrAC, Trends Anal. Chem. 21, 213–225.

Ronkainen, N.J., Halsall, H.B., Heineman, W.R., 2010. Electrochemical biosensors. Chem. Soc. Rev. 39, 1747–1763.

Sarkar, T., Gao, Y., Mulchandani, A., 2013. Carbon nanotubes-based label-free affinity sensors for environmental monitoring. Appl. Biochem. Biotechnol. 170, 1011–1025.

Sarkar, S., Mathwig, K., Kang, S., Nieuwenhuis, A.F., Lemay, S.G., 2014. Redox cycling without reference electrodes. Analyst 139, 6052–6057.

Scherlis, D.A., Marzari, N., 2005. π-Stacking in thiophene oligomers as the driving force for electroactive materials and devices. J. Am. Chem. Soc. 127, 3207–3212.

Schoukroun-Barnes, L.R., White, R.J., 2015. Rationally designing aptamer sequences with reduced affinity for controlled sensor performance. Sensors (Basel) 15, 7754–7767.

Schultz, J.S., 2015. Thirty-fifth anniversary of the optical affinity sensor for glucose: a personal retrospective. J. Diabetes. Sci. Technol. 9, 153–155.

Shan, C., Yang, H., Song, J., Han, D., Ivaska, A., Niu, L., 2009. Direct electrochemistry of glucose oxidase and biosensing for glucose based on graphene. Anal. Chem. 81, 2378–2382.

Shen, J., Li, Y., Gu, H., Xia, F., Zuo, X., 2014. Recent development of sandwich assay based on the nanobiotechnologies for proteins, nucleic acids, small molecules, and ions. Chem. Rev. 114, 7631–7677.

Silwana, B., Van Der Horst, C., Iwuoha, E., Somerset, V., 2016. A brief review on recent developments of electrochemical sensors in environmental application for PGMs. J. Environ. Sci. Health, Part A: Toxic/Hazard. Subst. Environ. Eng. 51, 1233–1247.

Skerra, A., 2007. Alternative non-antibody scaffolds for molecular recognition. Curr. Opin. Biotechnol. 18, 295–304.

Song, S., Wang, L., Li, J., Fan, C., Zhao, J., 2008. Aptamer-based biosensors. TrAC, Trends Anal. Chem. 27, 108–117.

Stumpff, F., McGuigan, J.A., 2014. Measuring Ca(2 +) binding to short chain fatty acids and gluconate with a Ca(2 +) electrode: role of the reference electrode. Anal. Biochem. 459, 46–52.

Tombelli, S., Minunni, M., Mascini, M., 2007. Aptamers-based assays for diagnostics, environmental and food analysis. Biomol. Eng. 24, 191–200.

Urdea, M., Penny, L.A., Olmsted, S.S., Giovanni, M.Y., Kaspar, P., Shepherd, A., et al., 2006. Requirements for high impact diagnostics in the developing world. Nature 444, 73–79.

Viswanathan, S., Rani, C., Ribeiro, S., Delerue-Matos, C., 2012. Molecular imprinted nanoelectrodes for ultra sensitive detection of ovarian cancer marker. Biosens. Bioelectron. 33, 179–183.

Wang, B., Anzai, J.I., 2015. Recent progress in electrochemical HbA1c sensors: a review. Mater 8, 1187–1203 (Basel, Switzerland).

Wang, J., Musameh, M., Lin, Y., 2003. Solubilization of carbon nanotubes by nafion toward the preparation of amperometric biosensors. J. Am. Chem. Soc. 125, 2408–2409.

Wang, Y., Zhang, Z., Jain, V., Yi, J., Mueller, S., Sokolov, J., et al., 2010. Potentiometric sensors based on surface molecular imprinting: detection of cancer biomarkers and viruses. Sens. Actuators, B: Chemical 146, 381–387.

Wang, Y., Zhou, Y., Sokolov, J., Rigas, B., Levon, K., Rafailovich, M., 2008. A potentiometric protein sensor built with surface molecular imprinting method. Biosens. Bioelectron. 24, 162–166.

Wu, Z., Su, S., 2014. A dynamic selection method for reference electrode in SSVEP-based BCI. PLoS. One 9, e104248.

Wu, D., Li, C., Hu, X., Mao, X., Li, G., 2016. Electrochemical detection of DNA 3′-phosphatases based on surface-extended DNA nanotail strategy. Anal. Chim. Acta 924, 29–34.

Yang, C., Denno, M.E., Pyakurel, P., Venton, B.J., 2015a. Recent trends in carbon nanomaterial-based electrochemical sensors for biomolecules: a review. Anal. Chim. Acta 887, 17–37.

Yang, D., Ning, L., Gao, T., Ye, Z., Li, G., 2015b. Enzyme-free dual amplification strategy for protein assay by coupling toehold-mediated DNA strand displacement reaction with hybridization chain reaction. Electrochem. Commun. 58, 33–36.

Yasukawa, T., Kiba, Y., Mizutani, F., 2015. A dual electrochemical sensor based on a test-strip assay for the quantitative determination of albumin and creatinine. Anal. Sci. 31, 583–589.

You, K.M., Lee, S.H., Im, A., Lee, S.B., 2003. Aptamers as functional nucleic acids: in vitro selection and biotechnological applications. Biotechnol. Bioprocess Eng. 8, 64–75.

Yu, Y., Gao, T., Li, H., Ye, Z., Chen, Z., Li, G., 2014. A novel electrochemical immunosensor for Golgi Protein 73 assay. Electrochem. Commun. 42, 6–8.

Zarzo, M., 2007. The sense of smell: molecular basis of odorant recognition. Biol Rev. 82, 455–479.

Zhang, F., Liu, J., Ivanov, I., Hatzell, M.C., Yang, W., Ahn, Y., et al., 2014. Reference and counter electrode positions affect electrochemical characterization of bioanodes in different bioelectrochemical systems. Biotechnol. Bioeng. 111, 1931–1939.

Zhang, J., Liu, Y., Wang, X., Chen, Y., Li, G., 2015. Electrochemical assay of α-glucosidase activity and the inhibitor screening in cell medium. Biosens. Bioelectron. 74, 666–672.

Zhao, J., Hu, S., Cao, Y., Zhang, B., Li, G., 2015. Electrochemical detection of protein based on hybridization chain reaction-assisted formation of copper nanoparticles. Biosens. Bioelectron. 66, 327–331.

Zhu, X., Feng, C., Ye, Z., Chen, Y., Li, G., 2014. Fabrication of magneto-controlled moveable architecture to develop reusable electrochemical biosensors. Sci. Rep. 4, 4169.

Zhu, C., Yang, G., Li, H., Du, D., Lin, Y., 2015. Electrochemical sensors and biosensors based on nanomaterials and nanostructures. Anal. Chem. 87, 230–249.

FURTHER READING

Kang, Y., Feng, K.-J., Chen, J.-W., Jiang, J.-H., Shen, G.-L., Yu, R.-Q., 2008. Electrochemical detection of thrombin by sandwich approach using antibody and aptamer. Bioelectrochemistry 73, 76–81.

Li, C., Wang, Z., Gao, T., Duan, A., Li, G., 2013a. Fabrication of hand-in-hand nanostructure for one-step protein detection. Chem. Commun. 49, 3760–3762.

Li, H., Huang, Y., Zhang, B., Yang, D., Zhu, X., Li, G., 2014. A new method to assay protease based on amyloid misfolding: application to prostate cancer diagnosis using a panel of proteases biomarkers. Theranostics 4, 701–707.

Li, W., Li, H., Wu, S., Feng, C., Li, G., 2016b. Highly sensitive protein detection based on DNAzyme cycling activated surface assembly of peptide decorated nanoparticles. Electrochem. Commun. 71, 84–88.

Yu, Y., Li, H., Zhang, B., Pan, X., Zhu, X., Ding, Y., et al., 2015. Peptide network for detection of tissue-remodeling enzyme in the prognosis of hepatocellular carcinoma. ACS Appl. Mater. Int. 7, 4401–4405.

Chapter 10

Spectrometry

Xiaoli Zhu and Tao Gao
Center for Molecular Recognition and Biosensing, School of Life Sciences, Shanghai University, Shanghai, P. R. China

Chapter Outline

10.1 COLORIMETRY

Colorimetric assays have gained great interest because of their inherent advantages, including simple operation, quick response, adaptable sensitivity, and long linear range of the quantitative assay that is based on spectrometry.

Nano-inspired Biosensors for Protein Assay with Clinical Applications.
DOI: https://doi.org/10.1016/B978-0-12-815053-5.00010-6

Due to the advantages of low cost, simple operation, and no need for complicated apparatus, colorimetric biosensors are especially suitable for clinical and point-of-care diagnosis. There are mainly two types of colorimetric biosensors. One type is the colorimetric biosensors based on enzyme-catalyzed organic chromogenic substrates such as ABTS, o-phenylenediamine, and TMB to form colored products, the other type is the colorimetric biosensors based on localized surface plasmon resonance (LSPR) of noble metal nanoparticles, including gold nanoparticles (AuNPs), silver nanoparticles (AgNPs), and etc.

Here, LSPR is mainly discussed. The enzyme-linked colorimetric assays will be depicted in Chapter 4, Other Nanomaterials, in detail. LSPR is a phenomenon where when two particles are in close proximity, a strong interparticle plasmon coupling and an associated perturbation in the LSPR band occurs, leading to a red-shift in the absorbance peak. An obvious color change of the colloid solution is easily achieved by altering the interparticle distance, morphology, nanoparticle size, as well as the local dielectric environment, which is suitable for the fabrication of biosensors because of the label-free, sensitive, robust, and facile detection (Aldewachi et al., 2017; Piriya et al., 2017; Tang and Li, 2017). A simple explanation of the principle of the LSPR-based colorimetric sensor is shown in Fig. 10.1 (Zhao et al., 2008).

Interparticle cross-linking aggregation is a mechanism in which metal nanoparticles are brought together through the formation of linkages between the individual particles. This occurs either by using cross-linkers that have two binding sites that link two AuNPs to each other, or by the direct interaction (without cross-linkers) such as DNA hybridization and antigen–antibody interaction. Utilizing this concept, for example, a colorimetric assay for protein was developed based on peptide-decorated gold nanoparticles (AuNPs). As illustrated in Fig. 10.2, the peptides modified on AuNPs acted as the specific binary recognition elements and also as the cross-linkers of AuNPs. These peptide-decorated AuNPs can form bulky aggregates by the introduction of cucurbit[8] uril, due to the selective accommodation of two N-terminal aromatic residues of peptides into the hole of cucurbit[8] uril. However, in the presence of the target protein, the N-terminal aromatic residue of the peptide was occupied, and the specific binding between target protein and the peptide could inhibit cucurbit[8] uril-induced aggregation of AuNPs. With vascular endothelial growth factor receptor 1 (Flt-1) as an example, a detection limit of 0.2 nM was achieved, which was comparable with traditional methods (Wei et al., 2015).

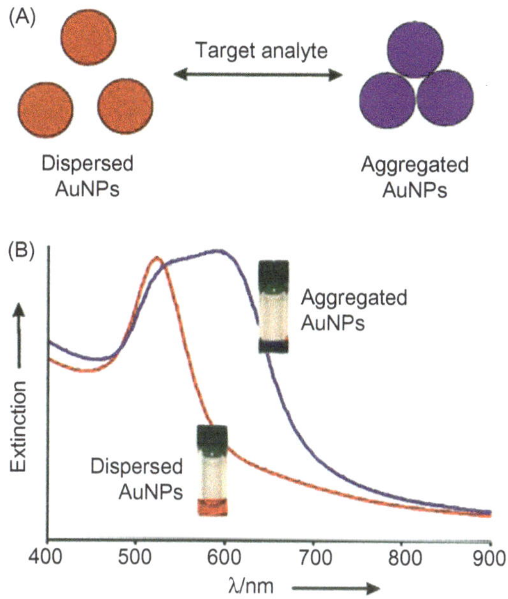

FIGURE 10.1 (A) General principle of the LSPR-based colorimetric assay based on AuNPs aggregation and dispersion. (B) Typical surface plasmon absorption bands for dispersed (red) and aggregated (blue) AuNPs with a diameter of 13 nm. *Reprinted from Zhao, W., Brook, M.A., Li, Y.F., 2008. Design of gold nanoparticle-based colorimetric biosensing assays. Chembiochem. 9, 2363–2371.*

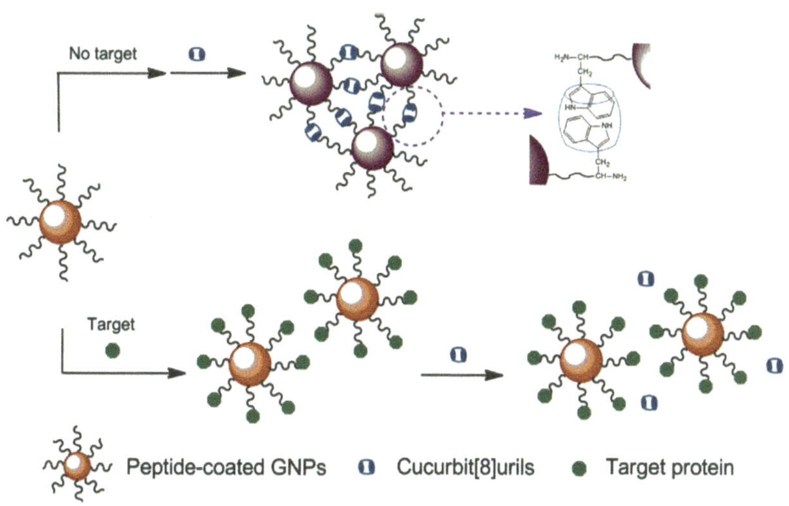

FIGURE 10.2 Scheme illustration of a peptide-decorated gold nanoparticles-based colorimetric assay. *Reprinted from Wei, L., Wang, X., Li, C., Li, X., Yin, Y., Li, G., 2015. Colorimetric assay for protein detection based on "nano-pumpkin" induced aggregation of peptide-decorated gold nanoparticles. Biosens. Bioelectron. 71, 348–352.*

10.2 FLUORESCENCE

10.2.1 Introduction

Fluorescence is an optical phenomenon where the absorption of photons at a certain wavelength typically results in the emission of photons at a longer wavelength. The loss in energy between the absorbed and emitted photons is the result of vibrational relaxation, and this difference is referred to as a Stokes shift. Förster or fluorescence resonance energy transfer (FRET) is a phenomenon where energy transfers from a fluorescent donor to a fluorescent acceptor when the emission spectrum of the fluorescent donor overlaps with the excitation spectrum of the fluorescent acceptor, as long as they are close enough with a distance typically less than 10 nm (Fig. 10.3) (Sapsford et al., 2006). Fluorescence methods have been widely applied for the detection of clinical biomarkers, since they have shown lots of unique advantages, such as nondestructive operation, fast response, convenience of optical signal transduction, and availability of multiplex detection.

Due to the quantum confinement effect, a lot of nanomaterials can directly generate fluorescent emissions with high quantum yield, excellent photostability, and long fluorescence lifetime. In addition, nanomaterials can

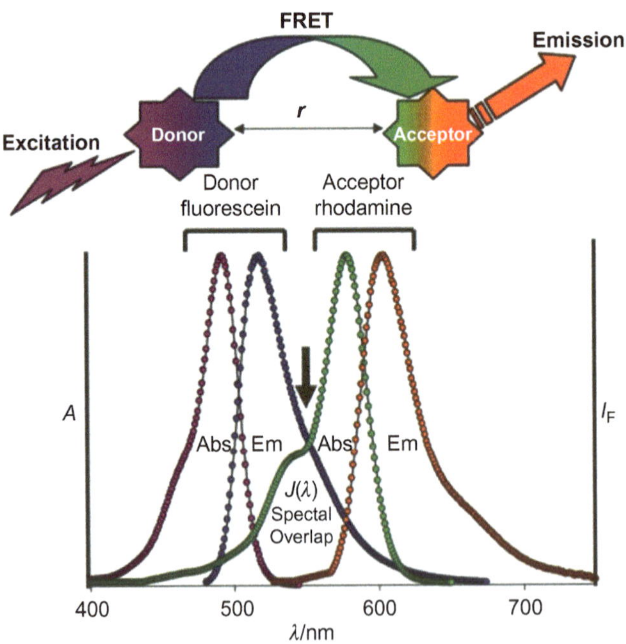

FIGURE 10.3 The principle of fluorescence resonance energy transfer. *Reprinted from Sapsford, K.E., Berti, L., Medintz, I.L., 2006. Materials for fluorescence resonance energy transfer analysis: beyond traditional donor-acceptor combinations. Angewandte Chemie 45(28), 4562–4589.*

also be used as efficient fluorescence quenchers for their excellent quenching capabilities. Moreover, by making use of their large surface-to-volume area, a huge numbers of fluorescent dyes or recognition elements can be modified, which may greatly enhance sensitivity of the fluorescence assays. Furthermore, some metal nanomaterials can be used to enhance the fluorescence of fluorophores located near their surface, which is called metal-enhanced fluorescence (MEF). Obviously, the introduction of nanomaterials not only improves the performance of fluorescence assay for clinical biomarkers, but also provides a lot of novel strategies to design fluorescent sensors (Bartelmess et al., 2015; Chinen et al., 2015; Tian et al., 2016; Wegner and Hildebrandt, 2015).

10.2.2 Fluorescent Nanomaterials

Semiconductor quantum dots (QDs) are a kind of fluorescent nanomaterial that have broad absorption spectra and narrow emission peaks, and both of which can be tuned by changing their size, composition, and shape. Quantitative detection of multiple tumor markers using QDs with different colors has been achieved by the development of an immunochromatographic test strip (Wang et al., 2015). By conjugating QDs with capture antibodies, the test strip can be used to simultaneously detect alpha fetoprotein (AFP) and carcinoembryonic antigen (CEA), with only one test line and one control line. Both analytes were measured with satisfactory sensitivity and specificity. Carbon dots (CDs), another kind of fluorescent nanomaterials, have drawn a lot of attention for their simplicity of synthesis and abundance of the raw material in nature. For example, Miao et al. have described a facile-green strategy to synthesize carbon dots (CDs) with tomato juice served as the carbon source, which was used for the label-free detection of CEA (Fig. 10.4). With rich carboxyl groups on their surface, CDs can absorb a lot of single-strand CEA aptamer through π-π stacking interactions, leading to effective fluorescence quenching. In the presence of CEA, the aptamer strongly binds to CEA, and the conformation change of aptamer causes the dissociation of CDs, resulting in immediate recovery of the fluorescence. Through the developed method, CEA can be quantified in a range from 1 ng mL^{-1} to 0.5 ng mL^{-1}, with the detection limit of 0.3 ng mL^{-1} (Miao et al., 2016). Upconversion-based fluorescent nanoparticles (UCNPs) can convert near-infrared (NIR) light into visible light, which makes them suitable for in vivo assays, because of good biological tissue penetration of NIR light (Grebenik et al., 2016; Hildebrandt et al., 2017; Li et al., 2017b; Sun and Lei, 2017). In another study by Li et al., UCNPs were used for ultrasensitive detection of CEA based on FRET between UCNPs and palladium nanoparticles (PdNPs) (Fig. 10.5). These UCNPs were modified with CEA aptamer that could interact with PdNPs and pull it close to UCNPs, leading to efficient quenching of fluorescence. In the presence of CEA, the

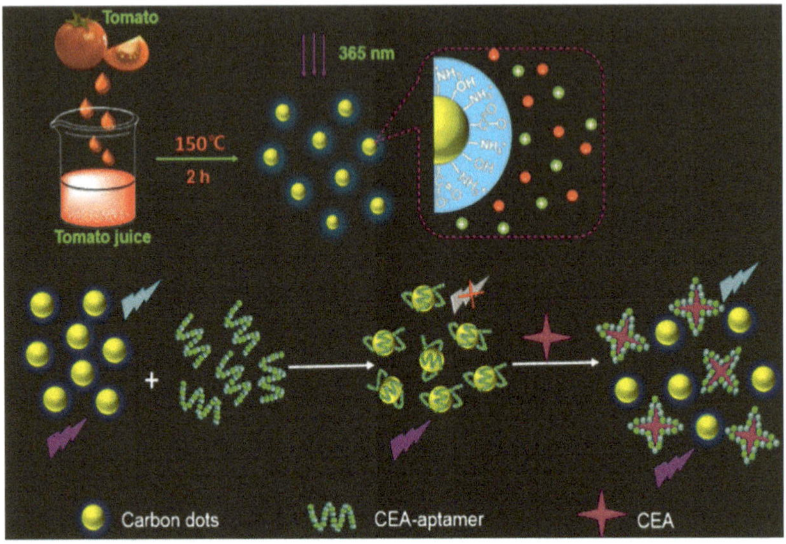

FIGURE 10.4 Schematic illustration of CD preparations and its use for CEA detection. *Reprinted from Miao, H., Wang, L., Zhuo, Y., Zhou, Z., Yang, X. 2016. Label-free fluorimetric detection of CEA using carbon dots derived from tomato juice. Biosens. Bioelectron, 86, 83–89.*

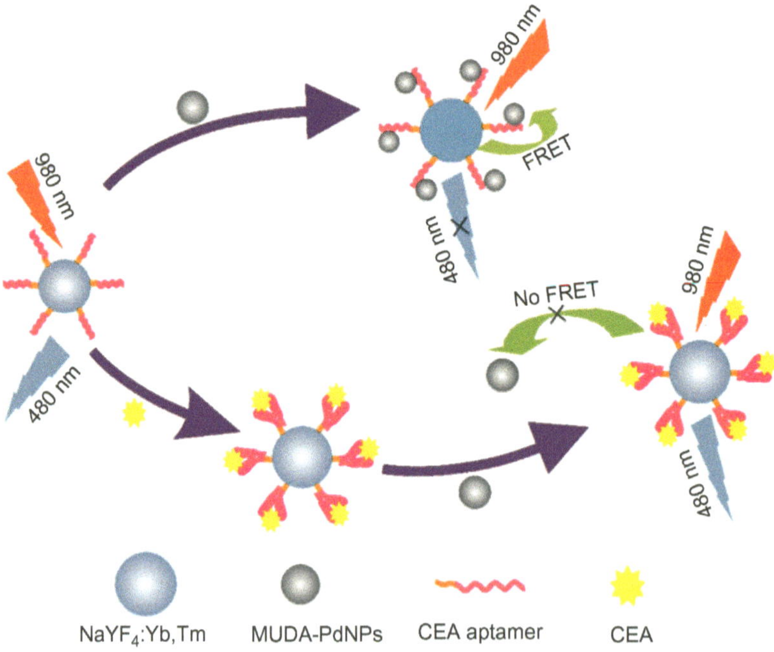

FIGURE 10.5 Schematic illustration of the CEA biosensor based on FRET from CEA aptamer-attached UCPs to PdNPs. *Reprinted from Li, H., Shi, L., Sun, D. E., Li, P., Liu, Z. 2016. Fluorescence resonance energy transfer biosensor between upconverting nanoparticles and palladium nanoparticles for ultrasensitive CEA detection. Biosens. Bioelectron, 86, 791–798.*

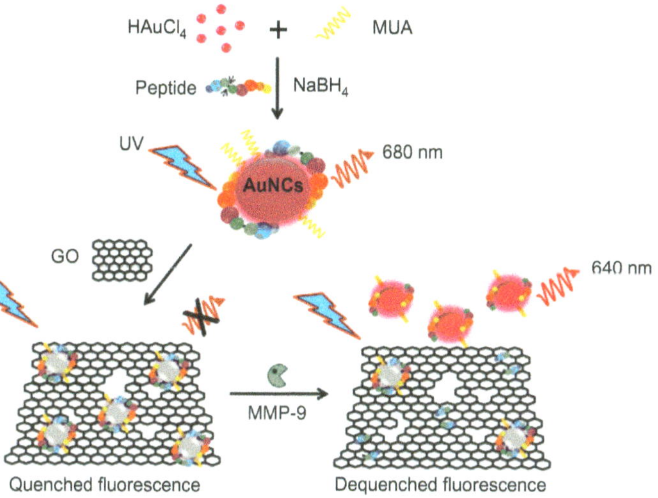

FIGURE 10.6 Scheme illustration of gold nanocluster based fluorescence assay for the detection of metalloproteinase-9. *Reprinted from Nguyen, P.D., Cong, V.T., Baek, C., Min, J. 2017. Fabrication of peptide stabilized fluorescent gold nanocluster/graphene oxide nanocomplex and its application in turn-on detection of metalloproteinase-9. Biosens. Bioelectron 89, 666–672.*

strong interaction between CEA and CEA aptamer weakened the interaction between the aptamer and PdNPs, causing the recovery of fluorescence. This method has a linear range from 2 pg mL^{-1} to 100 pg mL^{-1} with the detection limit of 0.8 pg mL^{-1} (Li et al., 2016c).

Metal nanoclusters are another kinds of nanomaterials that show size-dependent fluorescent properties. They have shown good biocompatibility when they are synthesized by using DNA, protein, and peptide as the scaffolds. Metal nanoclusters have thus been widely used for construction of fluorescent biosensors. For example, a gold nanocluster has been used for fluorescence assay of cancer-related enzyme matrix metalloproteinase-9 (Fig. 10.6). With peptides and mercaptoundecanoic acid working as cotemplating ligands, fluorescent gold nanoclusters were synthesized in a facile and one-step way. By designing a metalloproteinase-9 cleavage site in the peptide, the peptide acted as both a stabilizer and a targeting ligand for the enzyme detection. Graphene oxide was employed as an efficient quencher in this assay. Once the peptide was cut by the enzyme, the nanoclusters released from the graphene oxide, resulting in the recovery of fluorescence.

10.2.3 Nanomaterials as Fluorescence Quenchers

Some nanomaterials don't exhibit fluorescence, but they can quench a wide range of fluorophores. In combination with other excellent properties,

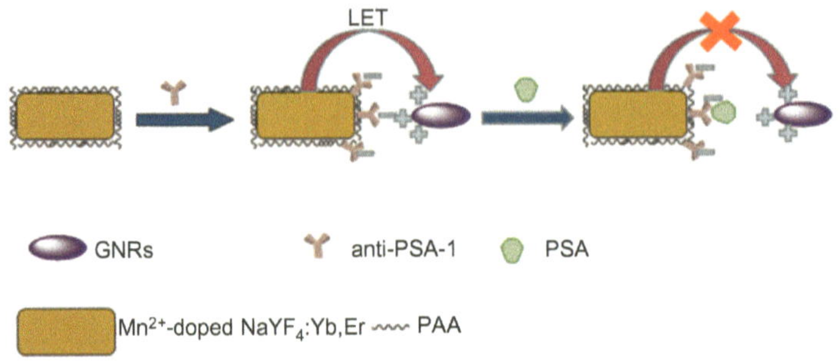

FIGURE 10.7 Schematic illustration of the LET assays based on upconversion nanorods and GNRs for the detection of PSA. *Reprinted from Zhang, J., Wang, S., Gao, N., Feng, D., Wang, L., Chen, H., 2015. Luminescence energy transfer detection of PSA in red region based on Mn(2+)-enhanced NaYF4:Yb, Er upconversion nanorods. Biosens. Bioelectron. 72, 282–287.*

nanomaterials hold great potential to design different types of fluorescence biosensors. Among which, gold nanoparticles (AuNPs) are the most widely used quenchers in the fluorescence-based protein assays. AuNPs have the advantages of easy synthesis, functionalization, controllable particle size, as well as good stability in bioassays. In the meantime, graphene oxide (GO) has become another promising quencher because of its good solubility, biocompatibility, and the simplicity of functionalization (Shi et al., 2015; Tian et al., 2016).

Taking advantage of the fluorescence quenching ability of gold nanorods (GNRs), Zhang et al. have described a luminescence energy transfer, antibody-based detection of prostate specific antigen (PSA) (Fig. 10.7). NaYF4: Yb, Er upconversion nanorods were functionalized with PSA antibodies and GNRs, and were used to construct the sensor for PSA detection. After attachment of UCNPs to GNRs, the fluorescence was quenched. While in the presence of PSA antigens, the greater affinity binding between UCNPs and PSA separated upconversion nanorods from GNRs, thereby recovering the UC signal. The analytical method exhibited a linear range from 0.1172 to 18.75 ng mL^{-1} and a detection limit of 0.1129 ng mL^{-1} (Zhang et al., 2015).

10.2.4 Nanomaterials as Fluorophores Carriers

Some nanomaterials neither exhibit fluorescence nor quench fluorescence. However, they can be used as the scaffolds in many fluorescence assays. Modification or encapsulation of large amounts of fluorescent dyes or QDs in a nanoparticle can improve their water solubility, reduce toxicity of fluorophores, and greatly enhance fluorescence signal (Deng et al., 2013).

By using the high surface-to-volume ratio and biocompatibility of silica microspheres, He et al. have developed a smart DNA walker biosensor for label-free detection of carcinoembryonic antigen (CEA) (Fig. 10.8). Coupled with exonuclease III-assisted target recycling amplification, the presence of

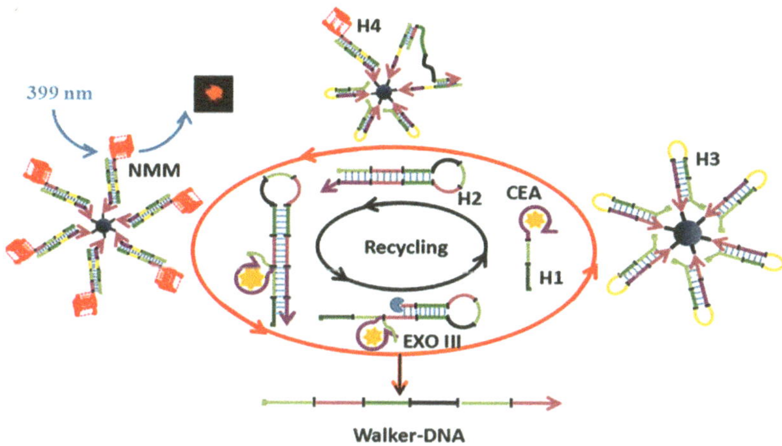

FIGURE 10.8 A silica microspheres based DNA-walker assay for the fluorescent detection of CEA. *Reprinted from He, M., Wang, K., Wang, W.J., Yu, Y.L., Wang, J.H., 2017. A smart DNA machine for carcinoembryonic antigen detection by exonuclease III-assisted target recycling and DNA walker cascade amplification. Anal. Chem. 89, 9292–9298.*

CEA could result in the generation of quantities of walker DNA strands, which could autonomously travel on the substrate-modified silica microspheres and trigger the G-quadruplex to bind to the microsphere's surface. An ultrasensitive fluorescent signal can be produced by the G-quadruplex with the help of N-methylmesoporphyrin IX (NMM). This fluorescence-based nanosensor has shown a favorable specificity and achieves a low detection limit of 1.2 pg mL^{-1}, with a linear detection range from 10 pg mL^{-1} to 100 ng mL^{-1} (He et al., 2017).

Silica nanoparticles (SiNPs) doped with fluorescent dyes have also found extensive use in fluorescence-based protein assay. For example, bimodal, magnetically encoded fluorescent SiNPs ($[CdTe/Fe_3O_4]@SiO_2$) were shown to achieve magnetic separation, capture, and fluorescent detection of three antigens (cancer antigen 125, AFP, and CEA) with detection limits of 20 kU L^{-1}, 10 ng mL^{-1}, and 5 ng mL^{-1}, respectively. This strategy was proposed to be applicable to multicomponent separation and analysis of biomolecules in a facile, rapid, and economical way (Song et al., 2014).

10.2.5 Metal-Enhanced Fluorescence

Metal-enhanced fluorescence (MEF) is another strategy to improve the sensitivity of fluorescence assays. When the fluorophores and the metal surface are close to a proper distance, interactions between dipole moments of fluorophores and the surface plasmon field of metal nanoparticles can result in fluorescence enhancement (Deng et al., 2013).

For example, Yang et al. have described a MEF strategy based on AuNPs and Ag nanoclusters for specific detection of CEA (Fig. 10.9). In the

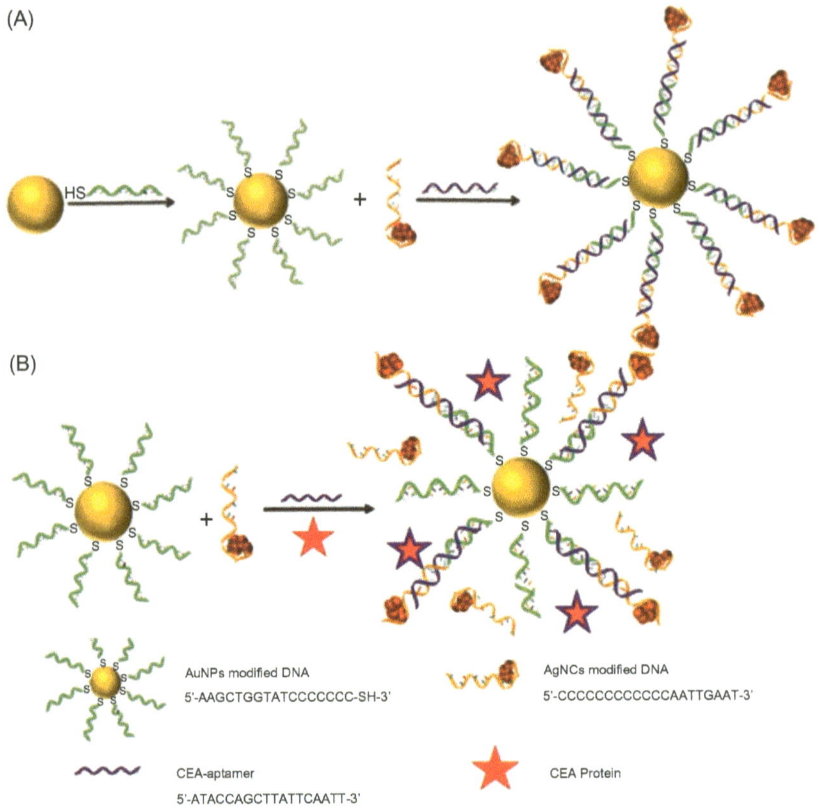

FIGURE 10.9 Scheme illustration of (A) the construction of MEF sensor based on AuNPs and Ag nanoclusters, and (B) their application for CEA detection. *Reprinted from Yang, X., Zhuo, Y., Zhu, S., Luo, Y., Feng, Y., Xu, Y., 2015. Selectively assaying CEA based on a creative strategy of gold nanoparticles enhancing silver nanoclusters' fluorescence. Biosens. Bioelectron. 64, 345–351.*

strategy, silver nanoclusters provided the original fluorescence signal, and AuNPs acted as the fluorescence enhancer. CEA aptamer was used to link AuNPs and Ag nanoclusters, which facilitated MEF. However, the specific binding of CEA and the aptamer separated AuNPs and Ag nanoclusters, which destroyed MEF, leading to decreased fluorescence. This sensor can specifically detect CEA with a linear range from 0.01 ng mL^{-1} to 1 ng mL^{-1}, with a detection limit of 3 pg mL^{-1} (Yang et al., 2015).

10.3 CHEMILUMINESCENCE

10.3.1 Introduction

Chemiluminescence (CL) is the luminescence produced by chemical reactions that induce the transition of an electron from its ground state to an excited electronic state. When the excited molecule decays to the electronic ground state, CL emission at different wavelengths occurs, from ultraviolet-visible to infrared radiation. Similar to FRET, if there is an overlapping of CL emission spectrum with the absorption spectrum of a fluorescent acceptor, CL resonance energy transfer (CRET) takes place with energy transfer from a chemiluminescent donor to a fluorophore acceptor (Tiwari and Dhoble, 2017).

Compared to absorbance and fluorescence assays, CL assays have lower background signal, leading to higher sensitivity. However, most of the CL-based reactions suffer from low quantum efficiency, and hence produce weak luminescence. This may restrict their applications in analytical assays. In recent years, the introduction of nanomaterials has brought new capabilities into CL assays, due to their excellent optical, electronic, and catalytic properties, which can be used to improve the performance of CL assays as catalyzers and fluorescence acceptors (Ehsani et al., 2017; Xu et al., 2017).

10.3.2 Nanozyme-Based Chemiluminescence

It has been demonstrated that some nanomaterials exhibit catalytic activities. Therefore, they are called nanozymes. More information on nanozyme can be found in Chapter 4, Other Nanomaterials. Here because of their unique properties, they have been used as catalysts to build new and efficient chemiluminescence-based biosensors for protein assay.

For example, Li et al. have developed a label-free chemiluminescent immunosensor based on dual functional cupric oxide nanorods (CuONRs) as peroxidase mimics (Fig. 10.10). CuONRs was first synthesized and deposited onto an epoxy-activated glass-slide. After which, capture antibodies were immobilized through a streptavidin bridge. In the presence of antigens, CL substrate was excluded from the surface due to the formation of immunocomplexes, leading to decrease of CL intensity. CuONRs acted as catalyzers and showed excellent enhancement of CL intensity. By using CEA as a model antigen, this sensor achieved a wide linear range from 0.1 to 60 ng mL^{-1}, and a low detection limit to 0.05 ng mL^{-1}. This work has constructed a novel, rapidly, and cost-efficient chemiluminescent immunosensing platform for protein assays (Li et al., 2017a).

10.3.3 Nanomaterials in Chemiluminescence Resonance Energy Transfer

Chemiluminescence Resonance Energy Transfer (CRET) takes place with energy transfer from a chemiluminescent donor to a fluorophore acceptor, which is similar to fluorescence resonance energy transfer (FRET). However, the occurrence of CRET doesn't need an external excitation source, thus obtaining a very low background signal. In recent years, a lot of

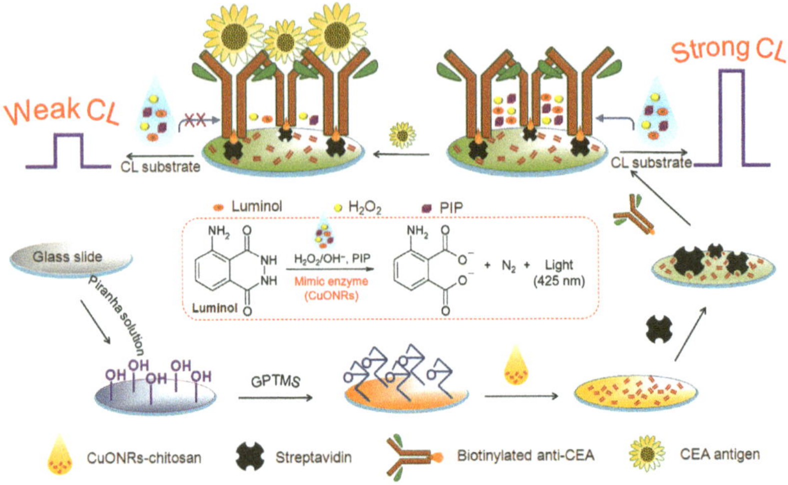

FIGURE 10.10 Schematic illustration of a chemiluminescent immunosensor based on dual functional cupric oxide nanorods as peroxidase mimics for label-free detection of CEA. *Reprinted from Li, J., Cao, Y., Hinman, S.S., Mckeating, K.S., Guan, Y., Hu, X., et al., 2017. Efficient label-free chemiluminescent immunosensor based on dual functional cupric oxide nanorods as peroxidase mimics. Biosens. Bioelectron. 100, 304–311.*

nanomaterials with unique properties have been used as the energy acceptor, in order to improve the performance of chemiluminescent sensors for protein assay (Zheng et al., 2017; Zhou et al., 2015).

Utilizing the excellent quenching ability of GO, Liu et al. have described a novel detection method for CEA, on the basis of proximity hybridization-regulated CRET (Fig. 10.11). The oxidation of TCPO in the presence of H_2O_2 and energy transfer between excited TCPO and Cy5 caused strong chemiluminescent emission of Cy5. In the absence of targets, Cy5-labeled ssDNA was absorbed on GO, leading to the quench of chemiluminescence. However, the introduction of CEA induced proximity hybridization to occur to form a proximate complex, which could specifically bind with Cy5-labeled ssDNA. Combined with nicking endonuclease Nt. BbvCI for in situ recycling, a huge chemiluminescence signal was obtained (Liu et al., 2016).

10.4 ELECTROCHEMILUMINESCENCE

10.4.1 Introduction

Electrochemiluminescence (ECL) is a light emission process in which species generated at the electrode surface undergo exergonic electron transfer reaction to form excited states that emit light. As illustrated in Fig. 10.12, the mechanism of ECL can be divided into two pathways, the annihilation pathway and the coreactant pathway. In the annihilation pathway, only a

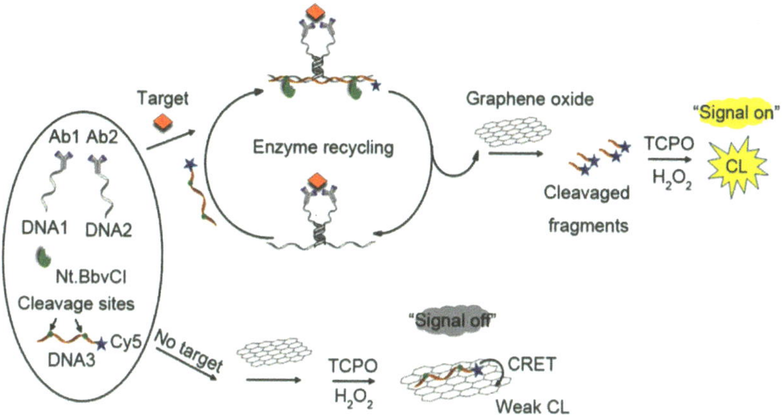

FIGURE 10.11 Schematic illustration of proximity hybridization-regulated CRET for homogeneous detection of CEA. *Reprinted from Liu, M., Wu, J., Yang, K., Zong, C., Lei, J., Ju, H., 2016. Proximity hybridization-regulated chemiluminescence resonance energy transfer for homogeneous immunoassay. Talanta 154, 455–460.*

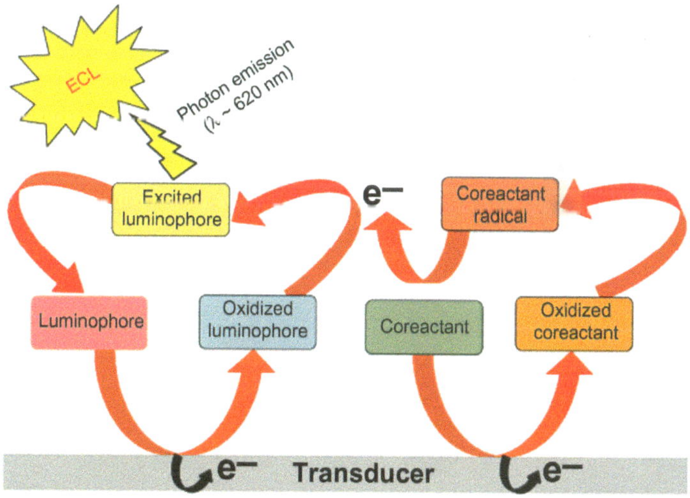

FIGURE 10.12 Scheme illustration of the light emission process at the electrode surface in electrochemiluminescence. *Reprinted from Rizwan, M., Mohdnaim, N.F., Ahmed, M.U., 2018. Trends and advances in electrochemiluminescence nanobiosensors. Sensors 18, 166.*

single emitter is needed to generate ECL, while in the coreactant pathway, both emitter and coreactant are involved (Rizwan et al., 2018).

ECL assay can be regarded as a combination of a spectrometric assay and an electrochemical assay, thus holding the advantages of both two methods. Besides, because excitation light is not needed in ECL, this method exhibits nearly zero background fluorescence. Therefore, ECL has now become a powerful analytical strategy for clinical biomarkers. In recent years, in order to improve the performance of ECL biosensors, different kinds of nanomaterials have been employed as ECL labels, ECL emitters, and the supporting elements to modify the electrode surfaces (Chen et al., 2018; Rizwan et al., 2018).

10.4.2 Nanomaterials as Electrochemiluminescence Labels

Due to their biocompatibility and photostability, graphene quantum dots (GQDs) have been widely used as ECL labels to fabricate ECL biosensors. For example, Nie et al. have described an ECL immunosensor for CEA detection with Au nanoparticles-decorated graphene quantum dots (GQDs@AuNP) as ECL labels (Fig. 10.13). Poly(5-formylindole)/reduced graphene oxide nanocomposite (P5FIn/erGO) was used as an effective matrix, which not only facilitated the ion transport during the redox reactions but also provided larger surface area for the immobilization of Ab1. In the presence of CEA, a sandwich complex was formed with Ab1 immobilized on P5FIn/erGO, CEA, and GQDs labeled Ab2. This ECL sensor exhibited a broad linear range from 0.1 pg mL^{-1} to 10 ng mL^{-1}, and a low detection limit of 3.78 fg mL^{-1} (Nie et al., 2018).

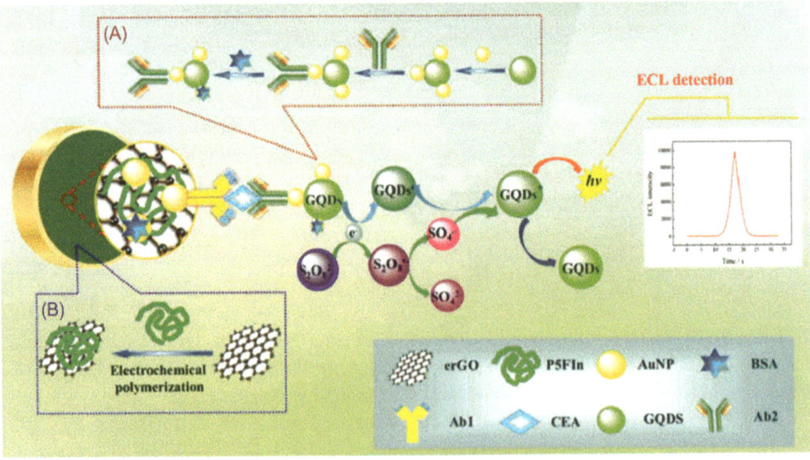

FIGURE 10.13 Scheme illustration of GQDs-based electrochemiluminescence immunosensor for CEA detection. The inset shows (A) the modification of GQDs and (B) the modification of the working electrode. *Reprinted from Nie, G., Wang, Y., Tang, Y., Zhao, D., Guo, Q., 2018. A graphene quantum dots based electrochemiluminescence immunosensor for carcinoembryonic antigen detection using poly(5-formylindole)/reduced graphene oxide nanocomposite. Biosens. Bioelectron. 101, 123–128.*

10.4.3 Nanomaterials-Enhanced Electrochemiluminescence

Modifications of electrode surface with metal nanomaterials not only enhance electron transfer of the electrode, but also provide an effective matrix for the immobilization of recognition elements. As illustrated in Fig. 10.14, the nanocomposite of polydopamine and Ag nanoparticles (PDA-AgNPs) was used to modify electrode surfaces to improve the performance of ECL sensor. By coupling with carbon quantum dots (CQDs)-attached graphene oxide (PEI-GO), this ECL sensor was successful used for CEA detection with a linear range from 5 pg mL^{-1} to 500 ng mL^{-1}, and a detection limit of 1.67 pg mL^{-1} (Li et al., 2017c).

In another study, NCs niobate-Au nanoparticles@bismuth sulfide ($KNbO_3$-AuNPs@Bi_2S_3) was used by Li et al. to modify glassy carbon electrode (GCE) for the detection of prostate-specific antigen (PSA) (Fig. 10.15). $KNbO_3$-AuNPs@Bi_2S_3 enhanced electron transfer of the nanocomposite at the surface of the electrode, and both the sensitivity and the stability of the ECL nanobiosensor have been improved. For the detection of PSA, a wide detection range from 0.005 ng mL^{-1} to 5 ng mL^{-1}, and a low detection limit of 3 pg mL^{-1} were achieved (Li et al., 2015).

Moreover, multifunctionalized flower-like Au@BSA nanoparticles was also utilized to create a novel ECL nanobiosensor (Fig. 10.16). In this case, Au@BSA not only offered good electrical conductivity and excellent biocompatibility, but also provided a large surface area for the immobilization of luminol molecules. Au@BSA acted as signal probes and recognition probes at the same time. With CEA as targets, the ECL nanobiosensor

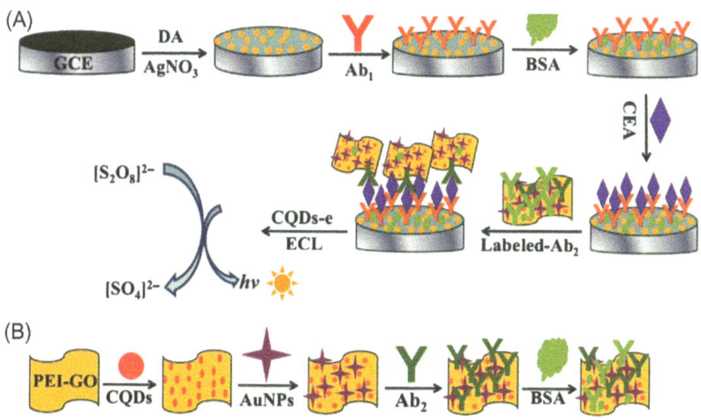

FIGURE 10.14 Scheme illustration of (A) the working principle of ECL immunosensor, and (B) the fabrication process of AuNPs, CQDs, and Ab2 immobilized PEI-GO matrix. *Reprinted from Li, N.L., Jia, L.P., Ma, R.N., Jia, W.L., Lu, Y.Y., Shi, S.S., et al., 2017. A novel sandwiched electrochemiluminescence immunosensor for the detection of carcinoembryonic antigen based on carbon quantum dots and signal amplification. Biosens. Bioelectron. 89, 453–460.*

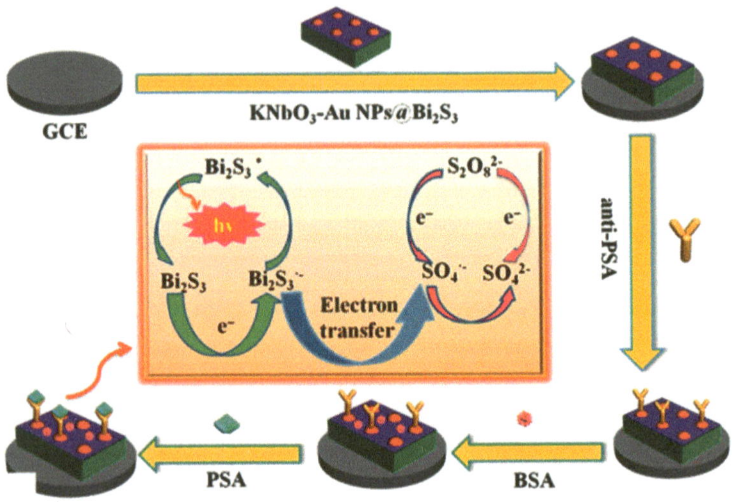

FIGURE 10.15 Scheme illustration of the fabrication of an ECL nanobiosensor by incorporating the nanocomposite of $KNbO_3$-AuNPs@Bi_2S_3. *Reprinted from Li, J., Ma, H., Wu, D., Li, X., Zhao, Y., Zhang, Y., et al., 2015. A label-free electrochemiluminescence immunosensor based on KNbO3-Au nanoparticles@Bi2S3 for the detection of prostate specific antigen. Biosens. Bioelectron. 74, 104–112.*

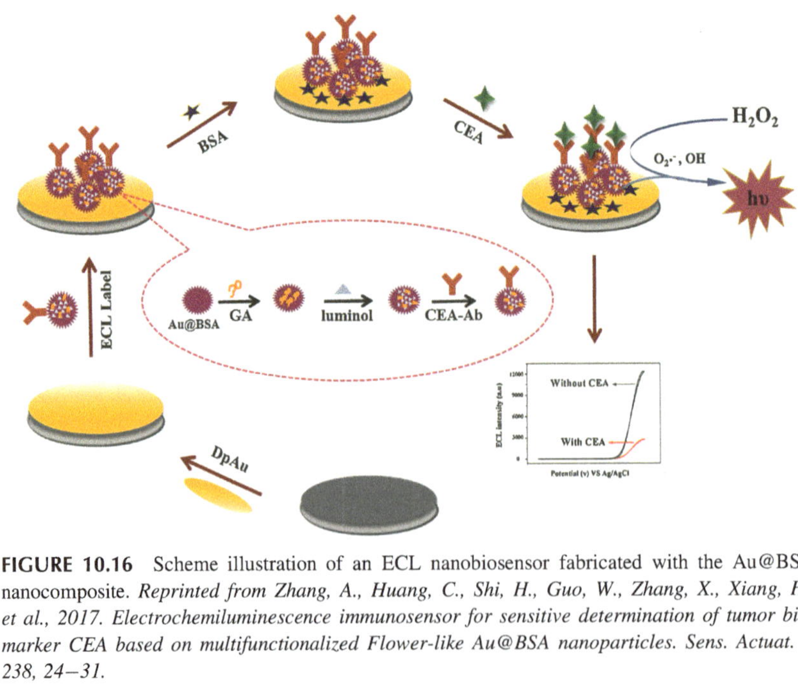

FIGURE 10.16 Scheme illustration of an ECL nanobiosensor fabricated with the Au@BSA nanocomposite. *Reprinted from Zhang, A., Huang, C., Shi, H., Guo, W., Zhang, X., Xiang, H., et al., 2017. Electrochemiluminescence immunosensor for sensitive determination of tumor biomarker CEA based on multifunctionalized Flower-like Au@BSA nanoparticles. Sens. Actuat. B 238, 24–31.*

achieved a good stability, excellent reproducibility, and favorable selectivity with a wide detection range from 0.001 ng mL^{-1} to 200 ng mL^{-1}, and a low limit of detection of 0.0003 ng mL^{-1} (Zhang et al., 2017).

10.5 SURFACE PLASMON RESONANCE ASSAY

Surface plasmon resonance (SPR) is a phenomenon where the electrons in the metal surface layer are excited by photons of incident light with a certain angle of incidence, and then propagate parallel to the metal surface (Fig. 10.17; Zeng et al., 2017). With a constant light source wavelength and a metal thin surface, the certain angle that triggers SPR is dependent on the refractive index of the material near the metal surface. Therefore, a small change in the reflective index of the sensing medium will hinder the occurrence of SPR, which makes it possible for analytes detection. In SPR assay, the amounts of analytes are determined by monitoring the reflected light intensity or tracking the resonance angle shifts, which makes it a real-time and label-free detection method. So far, various types of SPR biosensors have been developed for the detection of clinical relevant biomarkers. Many nanomaterials are also utilized to improve the performance of SPR biosensors (Hiep et al., 2015; Lisi et al., 2016).

Nanomaterials are ideal materials for signal amplification in SPR. For example, a recent study described the utilization of multiwalled carbon nanotubes (MWCNTs) as signal amplification tags for the construction of a SPR biosensor (Fig. 10.18). Due to the large surface area, a lot of secondary antibodies were modified onto MWCNTs to enhance the capture ability. For tau protein

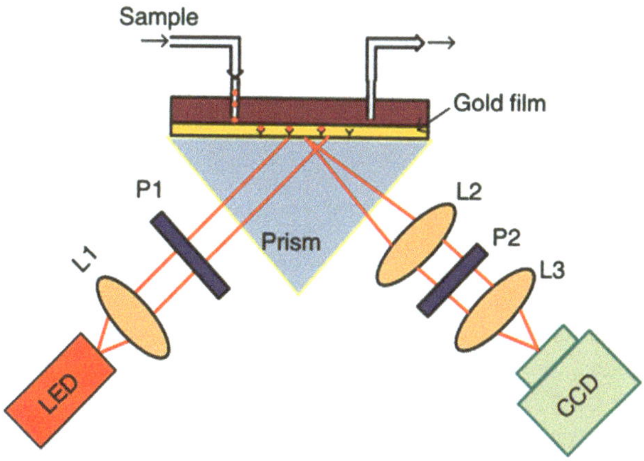

FIGURE 10.17 Scheme illustration of surface plasmon resonance assay. *Reprinted from Zeng, Y., Hu, R., Wang, L., Gu, D., He, J., Wu, S. Y., et al. 2017. Recent advances in surface plasmon resonance imaging: detection speed, sensitivity, and portability. Nanophotonics, 6(5), 1017–1030.*

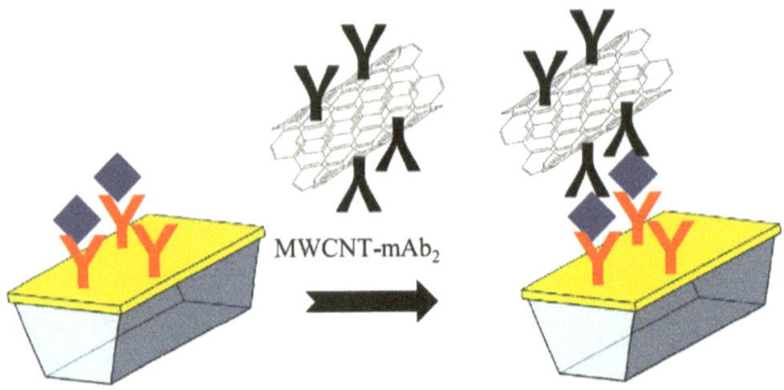

FIGURE 10.18 Scheme illustration of the multiwalled carbon nanotubes-based SPR sensor. *Reprinted from Lisi, S., Scarano, S., Fedeli, S., Pascale, E., Cicchi, S., Ravelet, C., et al., 2016. Toward sensitive immuno-based detection of tau protein by surface plasmon resonance coupled to carbon nanostructures as signal amplifiers. Biosens. Bioelectron. 93, 289–292.*

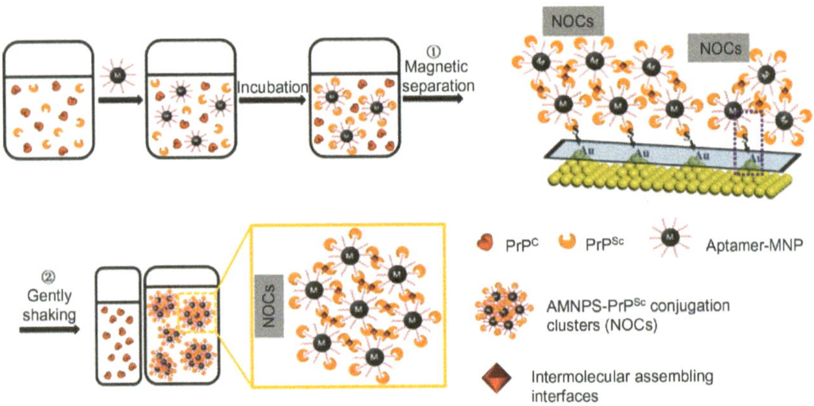

FIGURE 10.19 Scheme illustration of a SPR sensor fabricated with the magnetic conjugated clusters. *Reprinted from Lou, Z., Han, H., Zhou, M., Wan, J., Sun, Q., Zhou, X., et al., 2017. Fabrication of magnetic conjugation clusters via intermolecular assembling for ultrasensitive surface plasmon resonance (SPR) detection in wide-range concentration. Anal. Chem. 89, 13472–13479.*

detection, the signal of this MWCNTs method was 10^2-fold higher compared to that of direct assay without signal amplification tags (Lisi et al., 2016).

Magnetic nanomaterials have a lot of advantages, such as large surface for modification, low cost, and most importantly, ability to be manipulated by a magnetic field. These fascinating properties have been explored to fabricate SPR biosensors. Lou et al. have developed an aptamer − Fe_3O_4 nanoparticles (AMNPs)-based SPR sensor for ultrasensitive detection of prion disease-associated isoform (PrPSc) (Fig. 10.19). In the presence of PrPSc, PrPSc conjugating magnetic nanoparticle clusters were generated in an external magnetic field, relying on the high affinity and the intermolecular

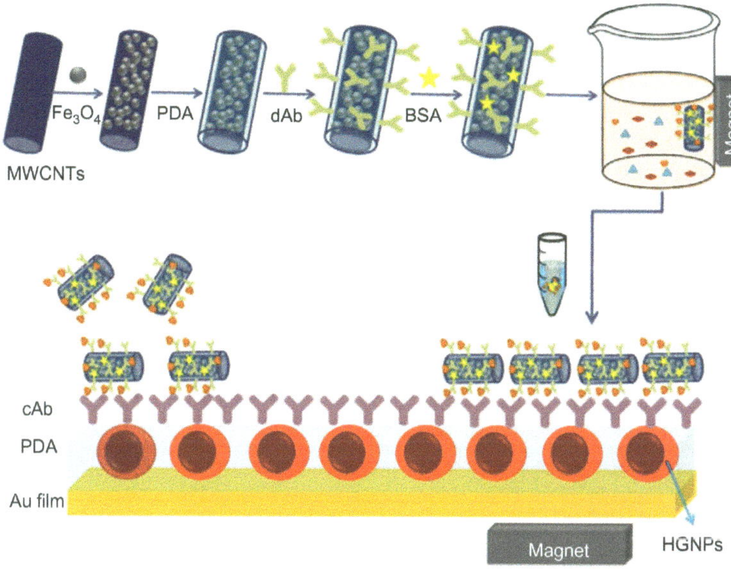

FIGURE 10.20 The fabrication and detection process of magnetic field-assisted SPR biosensor fabricated with hollow gold nanoparticles. *Reprinted from Wu, Q., Sun, Y., Zhang, D., Li, S., Zhang, Y., Ma, P., et al., 2017. Ultrasensitive magnetic field-assisted surface plasmon resonance immunoassay for human cardiac troponin I. Biosens. Bioelectron. 96, 288–293.*

assembly among PrPSc. Due to the formation of large magnetic nanoparticle clusters, a 215-fold increase of the SPR signal was obtained with a lower detection limit of $0.1\ pg \cdot mL^{-1}$, and a wide quantitation range from 1×10^{-4} ng mL^{-1} to 1×10^{5} ng mL^{-1} (Lou et al., 2017).

Recently, as shown in Fig. 10.20, hollow gold nanoparticles (HGNPs) were utilized to coat the gold sensing film in a SPR biosensor for the detection of human cardiac troponin I (cTnI). A remarkable amplification of SPR signal was observed, because of the electronic coupling of the surface plasmon waves originating from the HGNPs and the gold film. After further modification of polydopamine (PDA), a large number of capture antibodies (cAb) were immobilized. Assisted with PDA-wrapped magnetic multiwalled carbon nanotubes (MMWCNTs PDA) that were decorated with detection antibodies (dAb), separation and enrichment of cTnI in sample were successfully accomplished. As a result, significant enhancement of sensitivity was achieved with a detection limit of 1.25 ng mL^{-1} (Wu et al., 2017).

10.6 SURFACE-ENHANCED RAMAN SCATTERING

Compared to other spectroscopy, Raman scattering spectroscopy has unique advantages, such as the frequency shifts to provide "fingerprint" information of an analyte's chemical structure and narrow peaks for multicomponent

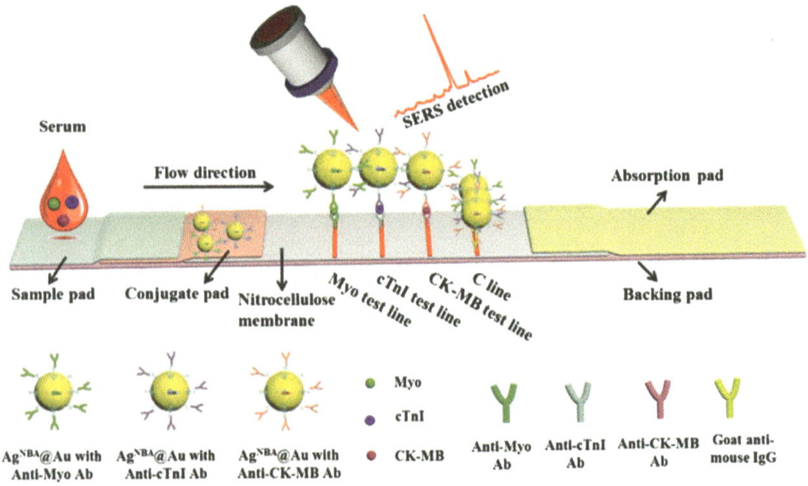

FIGURE 10.21 Scheme illustration of a multiplex lateral flow assay with core–shell SERS nanotags. *Reprinted from Zhang, D., Huang, L., Liu, B., Ni, H., Sun, L., Su, E., et al., 2018. Quantitative and ultrasensitive detection of multiplex cardiac biomarkers in lateral flow assay with core-shell SERS nanotags. Biosens. Bioelectron. 106, 204–211.*

analysis. However, Raman spectroscopy suffers from low sensitivity, making it helpless in the diagnosis of clinical biomarkers. Fortunately, rough metal surfaces or nanostructures have been found to have an enhancement to the Raman signal of molecules adsorbed or close to them, which is called surface-enhanced Raman scattering (SERS). As the enhancement of Raman signal can reach 10^{11}-fold, SERS has been widely used to develop sensitive and high-throughput molecular assays. Nanomaterials were also utilized to improve the performance of SERS assay from different aspects (Cialla-May et al., 2017; Wang et al., 2017).

In addition to enhancing the Raman signal of molecules near the surface of nanomaterials, nanomaterials can also be utilized to embed Raman dyes and carry recognition elements. Moreover, their advantages of stability, water solubility, and low cost are suitable for the design of SERS assays. Silver core and gold shell nanoparticles were used to embed Raman dyes for development of a lateral flow assay (Fig. 10.21). Due to the large amount of embedded Raman dyes, a great enhancement of Raman signal was observed. Moreover, due to the fabrication of three test lines with different antibodies, three cardiac biomarkers, i.e., Myo, cTnI, and CK-MB, were detected simultaneously (Zhang et al., 2018).

Magnetic beads-based SERS assay was also developed for the diagnosis of prostate cancer. As illustrated in Fig. 10.22, magnetic beads acted as both recognition elements, after decoration with antibodies, and magnetic separation elements. By coupling two types of antibody-conjugated AuNPs that were modified with two different Raman reporter nanotags, a sandwich

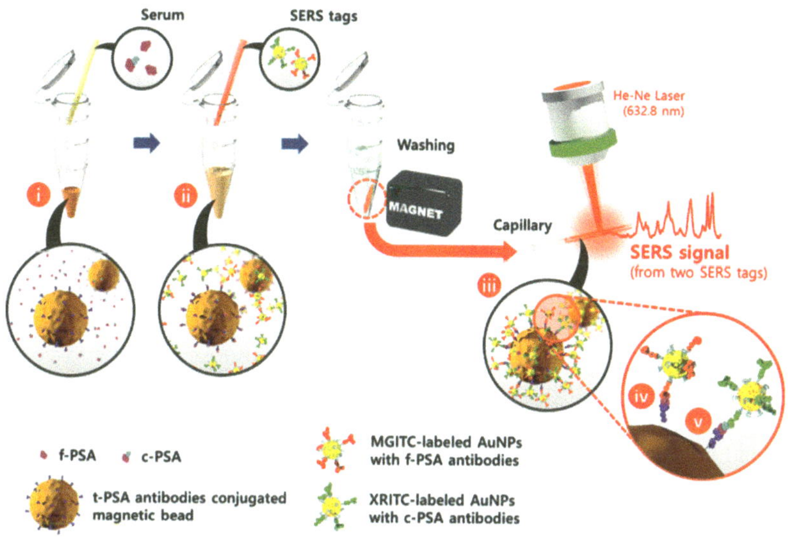

FIGURE 10.22 Scheme illustration of a SERS immunoassay based on magnetic beads and two types of AuNPs tags. *Reprinted from Cheng, Z., Choi, N., Wang, R., Lee, S., Moon, K.C., Yoon, S.Y., et al., 2017. Simultaneous detection of dual prostate specific antigens using surface-enhanced raman scattering-based immunoassay for accurate diagnosis of prostate cancer. ACS Nano 11, 4926–4933.*

complex was formed in the presence of targets. Thus the simultaneous detection of dual prostate-specific antigens was achieved with adorable sensitivity and a wide linear range (Cheng et al., 2017).

In another study, magnate nanoparticles (MNPs) were used for construction of MNPs core–AuNPs satellite assemblies to detect PSA. MNPs were modified with PSA aptamer for specific recognition of PSA. AuNPs were immobilized with Raman reporter molecules and DNA sequences, which were complementary to PSA aptamer. MNPs and AuNPs were cross-linked to inhibit SERS. However, in the presence of PSA, the strong interaction between aptamer and PSA led to the dissolution of the core–satellite assemblies. After separation by an external magnetic field, the SERS signals from the supernatant was determined, which was corresponding to the amount of PSA. A low detection limit of 5.0 pg mL^{-1} was obtained with a wide linear range (Yang et al., 2017).

10.7 DYNAMIC LIGHT SCATTERING SIGNAL-READOUT

10.7.1 Introduction

Dynamic light scattering (DLS), also known as photon correlation spectroscopy (PCS), is a method of measuring fluctuations in light intensity over time. With the advantages of accuracy, rapidity, good repeatability, etc., DLS has become a commonly used method in nanotechnology. With the development of instrument and data processing techniques, current DLS instruments not only have the function of measuring particle size, but also the abilities to measure zeta potential and the molecular weight of macromolecules.

The basic principles of DLS can be described as follows. Brownian motion of particles causes fluctuations in light intensity. Small particles suspended in the liquid keep irregular movement. The speed of Brownian motion depends on the size of the particles and the viscosity of the medium in which the particles are present (e.g., water, organic solvents, etc.). Research indicates that the smaller the particles and the smaller the viscosity of the medium are, the faster the Brownian motion is. When the light passes through the colloid, the particles scatter the light, and the light signal can be detected at a certain angle. The detected signal is the result of the superposition of multiple scattered photons, which has statistical significance. Instantaneous light intensity is not a fixed value, fluctuating around an average value, but the fluctuation amplitude is associated with particle size. The light intensity at a certain time is considered to be the same in a very short period of time compared with the light intensity at another time. As mentioned earlier, the particle velocity of the Brownian motion is related to the particle size (Stokes–Einstein equation).

If large particles are measured, the intensity of the scattered spot will also fluctuate slowly due to their slow motion. Similarly, if small particles are measured, the density of the scattered spots will fluctuate rapidly as they move quickly. The correlation between large particles and small particles can be expressed by a function. It can be seen that the rate of decay of the correlation function is related to particle size, and the decay rate of small particles is much faster than that of large particles. Finally, the particle size and distribution were calculated by the variation of light intensity and the correlation function of light intensity.

The distribution coefficient (particle dispersion index, PDI) is needed when analyzing data obtained from experiments. It reflects the uniformity of particle size and is an important indicator of particle size characterization. If the value is less than 0.05, the systems are monodispersed, such as the standard for some emulsions. When less than 0.08, the system is near monodispersed, but DLS can only be analyzed by a single exponential decay method, which could not provide higher resolution. If the value is located in the range of 0.08–0.7, this is a moderate dispersion system. It is the most optimal application of the algorithm. Finally, if the value is larger than 0.7, this is a very wide size distribution system. It is probably not suitable for light scattering method of analysis.

10.7.2 Nanoparticles-Enabled Dynamic Light Scattering Assay

Gold nanoparticles (AuNPs) show strong light scattering characteristics in the surface plasmon wavelength region. Compared with the polymer beads, AuNPs can scatter light 100 to 1000 times more strongly (Yguerabide and Yguerabide, 1998; Jain et al., 2006). AuNP is an excellent optical probe used for biological imaging and biological molecular detection based on light scattering (El-Sayed et al., 2005; Kang et al., 2014). Through strong light scattering characteristics of AuNPs, together with DLS technology, Zheng and coworkers developed an assay of nanoparticles-activated DLS (NanoDLSay) for the analysis of chemical and biological targets (Liu et al., 2008; Dai et al., 2008; Zheng et al., 2015). The complete program of the two-steps NanoDLSay is shown in Fig. 10.23. The first step was to mix a small amount of serum sample directly with the citrate-coated AuNPs solution. After a certain period of incubation (5–20 min), the average particle size (D1) of the mixed solution was measured. In the second step, the polyclonal rabbit antihuman immunoglobulin G (IgG) was added to the solution to detect the relative amount of human IgG in the protein crown. Because the human IgG antibody existed in the protein crown, the addition of antihuman IgG antibodies caused the formation of large aggregates. After 5–20 min of temperature breeding, DLS was used to measure the average particle size of the solution (D2) again. The more the human IgG existed in the protein crown, the larger the average particle size was. The ratio between the second step (D2) and the average particle size measured in the first step (D1) was calculated and expressed as the test score to evaluate the relative amount of human IgG in the nanoparticle protein crown. The new test could distinguish prostate cancer patients from noncancer patients, the specificity

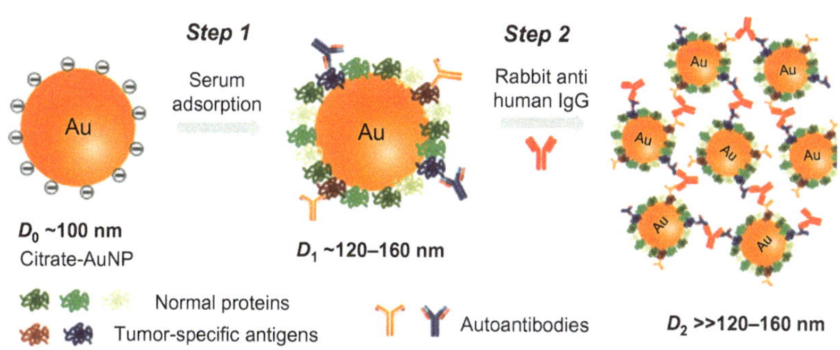

FIGURE 10.23 Illustration of a two-step NanoDLSay to analyze the relative amount of human IgG adsorbed to citrate-capped AuNPs for early stage prostate cancer detection. *Reprinted from Zheng, T., Pierre-Pierre, N., Yan, X., Huo, Q., Almodovar, A.J.O., Valerio, F., Rivera-Ramirez, I., Griffith, E., Decker, D.D., Chen, S., et al. 2015. Gold nanoparticle-enabled blood test for early stage cancer detection and risk assessment. ACS Appl. Mat. Interfaces 7, 6819–6827.*

was 90%–95%, and the sensitivity was 50%, which was a significant improvement compared with prostate-specific antigen (PSA) detection in routine prostate cancer screening.

10.7.3 Dynamic Light Scattering Coupled With Immunoassay

Although the DLS-based approach shows better performance, there are two major disadvantages that may limit their early diagnostic capabilities. One limitation is that the current methods are based on size changes of a single GNP, therefore they are susceptible to nonspecific adsorption of external disruptors. Another limitation is that the currently developed methods rely on analyte binding-induced nanoparticle aggregation, so they are difficult to control, which is difficult for the determination in practical application (Liu et al., 2008). Li et al. combined ELISA platform and DLS technology to detect trace tumor marker protein, named DLS-linked immunosorbent assay (DLS-LISA), and proved its practicability in clinical samples (Li et al., 2016a). The principle of this method can be seen in Fig. 10.24. In order to achieve this method, the modified GNP of manganese dioxide (MnO_2) nanosheet was used. In the presence of the target biomarker, the antibody-MnO_2-GNP conjugates were pulled down to the reactants. When the MnO_2 nanosheets were decomposed, a number of GNPs wrapped in the nanocomplex were released into the solution and could be measured directly by the

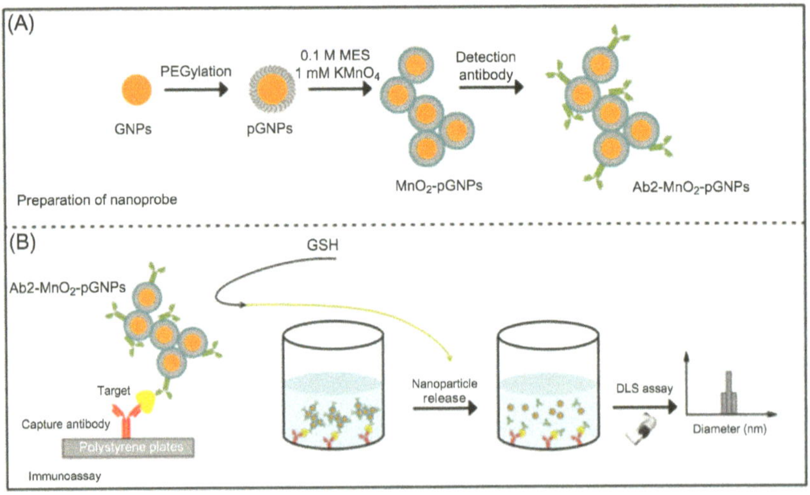

FIGURE 10.24 Representation of the DLS-linked immunosorbent assay (DLS-LISA) for protein detection performed in one hole of the 96-well polystyrene (PS) plates. (A) The preparation of activatable nanoprobe. (B) The detection process of DLS-LISA. *Reprinted from Li, C., Ma, J., Fan, Q., Tao, Y., Li, G., 2016. Dynamic light scattering (DLS)-based immunoassay for ultra-sensitive detection of tumor marker protein. Chem. Commun. 52, 7850–7853.*

DLS. Compared with the previous DLS-based approach, DLS-ELISA has several prominent features. First, the activated and GNP-rich nanoprobes can provide direct and amplified DLS signal readings, and make them completely different from the previous concept, that relied on nanoparticle aggregation induced by analytes, which may greatly increase the sensitivity of the proposed immunoassay. Second, the prepared probe is almost unaffected by the complex sample, which basically eliminates the external interference and lowers the detection limit. Third, it can be used almost for any biochemical test that depends on the immune response at the solid/liquid interface.

10.8 CONCLUSION

Compared with some other kinds of biosensors, spectrometric biosensors are a promising and exciting diagnostic sensing platform because these methods do not require complicated instruments and can be detected in a short time, even visible to the naked eye. At the same time, the emergence of nanotechnology and nanomaterials has provided a powerful impetus for the development of advanced spectrometric biosensing systems. They either participate in signal amplification systems or directly act as signal conversion components to develop diverse biosensors that meet different needs. Different application scenarios in clinical diagnosis have different requirements for the timeliness, convenience, and sensitivity of the sensing system. The diversity of nanomaterials and the diversity of spectrometric sensing technologies provide unlimited possibilities for this purpose.

REFERENCES

Aldewachi, H., Chalati, T., Woodroofe, M.N., Bricklebank, N., Sharrack, B., Gardiner, P., 2018. Gold nanoparticle-based colorimetric biosensors. Nanoscale 10, 18–33.

Bartelmess, J., Quinn, S.J., Giordani, S., 2015. Carbon nanomaterials: multi-functional agents for biomedical fluorescence and Raman imaging. Chem. Soc. Rev. 44, 4672–4698.

Chen, X., Yang, L., Ma, Q., 2018. Recent advances in quantum dot-based electrochemiluminescence sensor. J. Mater. Chem. C 6, 942–959.

Cheng, Z., Choi, N., Wang, R., Lee, S., Moon, K.C., Yoon, S.Y., et al., 2017. Simultaneous detection of dual prostate specific antigens using surface-enhanced raman scattering-based immunoassay for accurate diagnosis of prostate cancer. ACS Nano 11, 4926–4933.

Chinen, A.B., Guan, C.M., Ferrer, J.R., Barnaby, S.N., Merkel, T.J., Mirkin, C.A., 2015. Nanoparticle probes for the detection of cancer biomarkers, cells, and tissues by fluorescence. Chem. Rev. 115, 10530–10574.

Cialla-May, D., Zheng, X.S., Weber, K., Popp, J., 2017. Recent progress in surface-enhanced Raman spectroscopy for biological and biomedical applications: from cells to clinics. Chem. Soc. Rev. 46, 3945–3961.

Dai, Q., Liu, X., Coutts, J., Austin, L., Huo, Q., 2008. A one-step highly sensitive method for DNA detection using dynamic light scattering. J. Am. Chem. Soc. 130, 8138–8139.

Deng, W., Xie, F., Baltar, H.T., Goldys, E.M., 2013. Metal-enhanced fluorescence in the life sciences: here, now and beyond. Phys. Chem. Chem. Phys. 15, 15695–15708.

Ehsani, M., Chaichi, M.J., Hosseini, S.N., 2017. Comparison of CuO nanoparticle and CuO/MWCNT nanocomposite for amplification of chemiluminescence immunoassay for detection of the hepatitis B surface antigen in biological samples. Sens. Actuat. B 247, 319–328.

El-Sayed, I.H., Huang, X., El-Sayed, M.A., 2005. Surface plasmon resonance scattering and absorption of anti-EGFR antibody conjugated gold nanoparticles in cancer diagnostics: applications in oral cancer. Nano. Lett. 5, 829–834.

Grebenik, E.A., Kostyuk, A.B., Deyev, S.M., 2016. Upconversion nanoparticles and their hybrid assemblies for biomedical applications. Russian Chem. Rev. 85, 1277–1296.

He, M., Wang, K., Wang, W.J., Yu, Y.L., Wang, J.H., 2017. A smart DNA machine for carcinoembryonic antigen detection by exonuclease III-assisted target recycling and DNA walker cascade amplification. Anal. Chem. 89, 9292–9298.

Hiep, N.H., Jeho, P., Kang, S., Moonil, K., 2015. Surface plasmon resonance: a versatile technique for biosensor applications. Sensors 15, 10481–10510.

Hildebrandt, N., Spillmann, C.M., Algar, W.R., Pons, T., Stewart, M.H., Oh, E., et al., 2017. Energy transfer with semiconductor quantum dot bioconjugates: a versatile platform for biosensing, energy harvesting, and other developing applications. Chem. Rev. 117, 536–711.

Jain, P.K., Lee, K.S., El-Sayed, I.H., El-Sayed, M.A., 2006. Calculated absorption and scattering properties of gold nanoparticles of different size, shape, and composition: applications in biological imaging and biomedicine. J. Phys. Chem. B 110, 7238–7248.

Kang, B., Austin, L.A., El-Sayed, M.A., 2014. Observing real-time molecular event dynamics of apoptosis in living cancer cells using nuclear-targeted plasmonically enhanced raman nanoprobes. ACS Nano 8, 4883-3892.

Li, J., Ma, H., Wu, D., Li, X., Zhao, Y., Zhang, Y., et al., 2015. A label-free electrochemiluminescence immunosensor based on $KNbO_3$-Au nanoparticles@Bi_2S_3 for the detection of prostate specific antigen. Biosens. Bioelectron. 74, 104–112.

Li, C., Ma, J., Fan, Q., Tao, Y., Li, G., 2016a. Dynamic light scattering (DLS)-based immunoassay for ultra-sensitive detection of tumor marker protein. Chem. Commun. 52, 7850–7853.

Li, H., Shi, L., Sun, D.E., Li, P., Liu, Z., 2016c. Fluorescence resonance energy transfer biosensor between upconverting nanoparticles and palladium nanoparticles for ultrasensitive CEA detection. Biosens. Bioelectron. 86, 791–798.

Li, J., Cao, Y., Hinman, S.S., Mckeating, K.S., Guan, Y., Hu, X., et al., 2017a. Efficient label-free chemiluminescent immunosensor based on dual functional cupric oxide nanorods as peroxidase mimics. Biosens. Bioelectron. 100, 304–311.

Li, K., Liu, W., Ni, Y., Li, D., Lin, D., Su, Z., et al., 2017b. Technical synthesis and biomedical applications of graphene quantum dots. J. Mater. Chem. B 5, 4811–4826.

Li, N.L., Jia, L.P., Ma, R.N., Jia, W.L., Lu, Y.Y., Shi, S.S., et al., 2017c. A novel sandwiched electrochemiluminescence immunosensor for the detection of carcinoembryonic antigen based on carbon quantum dots and signal amplification. Biosens. Bioelectron. 89, 453–460.

Lisi, S., Scarano, S., Fedeli, S., Pascale, E., Cicchi, S., Ravelet, C., et al., 2016. Toward sensitive immuno-based detection of tau protein by surface plasmon resonance coupled to carbon nanostructures as signal amplifiers. Biosens. Bioelectron. 93, 289–292.

Liu, X., Dai, Q., Austin, L., Coutts, J., Knowles, G., Zou, J., et al., 2008. A one-step homogeneous immunoassay for cancer biomarker detection using gold nanoparticle probes coupled with dynamic light scattering. J. Am. Chem. Soc. 130, 2780–2782.

Liu, M., Wu, J., Yang, K., Zong, C., Lei, J., Ju, H., 2016. Proximity hybridization-regulated chemiluminescence resonance energy transfer for homogeneous immunoassay. Talanta 154, 455–460.

Lou, Z., Han, H., Zhou, M., Wan, J., Sun, Q., Zhou, X., et al., 2017. Fabrication of magnetic conjugation clusters via intermolecular assembling for ultrasensitive surface plasmon resonance (SPR) detection in wide-range concentration. Anal. Chem. 89, 13472–13479.

Miao, H., Wang, L., Zhuo, Y., Zhou, Z., Yang, X., 2016. Label-free fluorimetric detection of CEA using carbon dots derived from tomato juice. Biosens. Bioelectron. 86, 83–89.

Nie, G., Wang, Y., Tang, Y., Zhao, D., Guo, Q., 2018. A graphene quantum dots based electrochemiluminescence immunosensor for carcinoembryonic antigen detection using poly(5-formylindole)/reduced graphene oxide nanocomposite. Biosens. Bioelectron. 101, 123–128.

Nguyen, P.D., Cong, V.T., Baek, C., Min, J., 2017. Fabrication of peptide stabilized fluorescent gold nanocluster/graphene oxide nanocomplex and its application in turn-on detection of metalloproteinase-9. Biosens. Bioelectron. 89, 666–672.

Piriya, V.S.A., Joseph, P., Daniel, S.C.G.K., Lakshmanan, S., Kinoshita, T., Muthusamy, S., 2017. Colorimetric sensors for rapid detection of various analytes. Mater. Sci. Eng. C. Mater. Biol. Appl. 78, 1231–1245.

Rizwan, M., Mohdnaim, N.F., Ahmed, M.U., 2018. Trends and advances in electrochemiluminescence nanobiosensors. Sensors 18, 166.

Sapsford, K.E., Berti, L., Medintz, I.L., 2006. Materials for fluorescence resonance energy transfer analysis: beyond traditional donor-acceptor combinations. Angew. Chem. 45 (28), 4562–4589.

Shi, J., Tian, F., Jing, L., Yang, M., 2015. Nanoparticle based fluorescence resonanceenergy transfer (FRET) for biosensing applications. J. Mater. Chem. 6989–7005.

Song, E., Han, W., Li, J., Jiang, Y., Cheng, D., Song, Y., et al., 2014. Magnetic encoded fluorescent multifunctional nanospheres for simultaneous multicomponent analysis. Anal. Chem. 86, 9434–9442.

Sun, X., Lei, Y., 2017. Fluorescent carbon dots and their sensing applications. Trac Trends in Anal. Chem. 89, 163–180.

Tang, L., Li, J., 2017. Plasmon-based colorimetric nanosensors for ultrasensitive molecular diagnostics. ACS Sens. 2, 857.

Tian, F., Lyu, J., Shi, J., Yang, M., 2016. Graphene and graphene-like two-denominational materials based fluorescence resonance energy transfer (FRET) assays for biological applications. Biosens. Bioelectron. 89, 123–135.

Tiwari, A., Dhoble, S.J., 2017. Recent advances and developments on integrating nanotechnology with chemiluminescence assays. Talanta 180, 1–11.

Wang, C., Hou, F., Ma, Y., 2015. Simultaneous quantitative detection of multiple tumor markers with a rapid and sensitive multicolor quantum dots based immunochromatographic test strip. Biosens. Bioelectron. 68, 156–162.

Wang, F., Cao, S., Yan, R., Wang, Z., Wang, D., Yang, H., 2017. Selectivity/specificity improvement strategies in surface-enhanced raman spectroscopy analysis. Sensors 17, 2689.

Wegner, K.D., Hildebrandt, N., 2015. Quantum dots: bright and versatile in vitro and in vivo fluorescence imaging biosensors. Chem. Soc. Rev. 44, 4792–4834.

Wei, L., Wang, X., Li, C., Li, X., Yin, Y., Li, G., 2015. Colorimetric assay for protein detection based on "nano-pumpkin" induced aggregation of peptide-decorated gold nanoparticles. Biosens. Bioelectron. 71, 348–352.

Wu, Q., Sun, Y., Zhang, D., Li, S., Zhang, Y., Ma, P., et al., 2017. Ultrasensitive magnetic field-assisted surface plasmon resonance immunoassay for human cardiac troponin I. Biosens. Bioelectron. 96, 288–293.

Xu, H., Liu, B., Meng, Y., 2017. Ultrasensitive chemiluminescence assay for the lung cancer biomarker cytokeratin 21-1 via a dual amplification scheme based on the use of encoded gold nanoparticles and a toehold-mediated strand displacement reaction. Microchim. Acta. 184, 3953–3959.

Yang, K., Hu, Y., Dong, N., Zhu, G., Zhu, T., Jiang, N., 2017. A novel SERS-based magnetic aptasensor for prostate specific antigen assay with high sensitivity. Biosens. Bioelectron. 94, 286–291.

Yang, X., Zhuo, Y., Zhu, S., Luo, Y., Feng, Y., Xu, Y., 2015. Selectively assaying CEA based on a creative strategy of gold nanoparticles enhancing silver nanoclusters' fluorescence. Biosens. Bioelectron. 64, 345–351.

Yguerabide, J., Yguerabide, E.E., 1998. Light-scattering submicroscopic particles as highly fluorescent analogs and their use as tracer labels in clinical and biological applications. Anal. Biochem. 262, 137–156.

Zeng, Y., Hu, R., Wang, L., Gu, D., He, J., Wu, S.Y., et al., 2017. Recent advances in surface plasmon resonance imaging: detection speed, sensitivity, and portability. Nanophotonics 6 (5), 1017–1030.

Zhang, A., Huang, C., Shi, H., Guo, W., Zhang, X., Xiang, H., et al., 2017. Electrochemiluminescence immunosensor for sensitive determination of tumor biomarker CEA based on multifunctionalized Flower-like Au@BSA nanoparticles. Sens. Actuat. B 238, 24–31.

Zhang, D., Huang, L., Liu, B., Ni, H., Sun, L., Su, E., et al., 2018. Quantitative and ultrasensitive detection of multiplex cardiac biomarkers in lateral flow assay with core-shell SERS nanotags. Biosens. Bioelectron. 106, 204–211.

Zhang, J., Wang, S., Gao, N., Feng, D., Wang, L., Chen, H., 2015. Luminescence energy transfer detection of PSA in red region based on Mn(2 +)-enhanced NaYF4:Yb, Er upconversion nanorods. Biosens. Bioelectron. 72, 282–287.

Zhao, W., Brook, M.A., Li, Y.F., 2008. Design of gold nanoparticle-based colorimetric biosensing assays. Chembiochem 9, 2363–2371.

Zheng, T., Pierre-Pierre, N., Yan, X., Huo, Q., Almodovar, A.J.O., Valerio, F., et al., 2015. Gold nanoparticle-enabled blood test for early stage cancer detection and risk assessment. ACS Appl. Mat. Interfaces. 7, 6819–6827.

Zheng, Y., Zhang, D., Sna, S., Li, H., Lin, J.M., 2017. Ultra-weak chemiluminescence enhanced by facilely synthesized nitrogen-rich quantum dots through chemiluminescence resonance energy transfer and electron hole injection. Chem. Commun. 53, 5657–5660.

Zhou, Z.M., Feng, Z., Zhou, J., Fang, B.Y., Qi, X.X., Ma, Z.Y., et al., 2015. Capillary electrophoresis-chemiluminescence detection for carcino-embryonic antigen based on aptamer/graphene oxide structure. Biosens. Bioelectron. 64, 493–498.

FURTHER READING

Li, C., Yang, Y., Wu, D., Li, T., Yin, Y., Li, G., 2016b. Improvement of enzyme-linked immunosorbent assay for multicolor detection of biomarkers. Chem. Sci. 7, 3011–3016.

Chapter 11

Other Signal-Readout Technologies

Xiaoli Zhu[1] and Chao Li[2]
[1]*Center for Molecular Recognition and Biosensing, School of Life Sciences, Shanghai University, Shanghai, P. R. China,* [2]*State Key Laboratory of Pharmaceutical Biotechnology, Department of Biochemistry, Nanjing University, Nanjing, P. R. China*

Chapter Outline

11.1 DISTANCE SIGNAL-READOUT

The method uses geometrical distance as the signal, which builds relationship between the recorded length signals with the concentration of target protein. The distance-based signal-readout method is of low cost and equipment-independent, and can be integrated into a portable analytical device. Moreover, this method allows quantitative detection of various proteins by the naked eye. In this signal output mode, the addition of target starts a specific reaction, producing a color band proportional to the target concentration. Matching the length of the color stripe to the concentration of the target in the sample allows for quantitative analysis of the target by the naked eye. In addition, the ruler on the path helps analysis, similar to temperature signal-readout. The distance output does not require other electronic equipment and is less susceptible than the traditional intensity measurement to the experimenter and is therefore less likely to be misread.

Based on the abovementioned advantages, distance-based signal-readout has received a lot of attention over the years. In 1985, the first distance reading device was developed to detect theophylline based on the competitive immunoassay by Zuk et al. (1985) (Fig. 11.1A). During the test, the sample was first mixed with a horseradish peroxidase-labeled target. The capture antibody was immobilized on the strip of paper, which was inserted into the

Nano-inspired Biosensors for Protein Assay with Clinical Applications.
DOI: https://doi.org/10.1016/B978-0-12-815053-5.00011-8

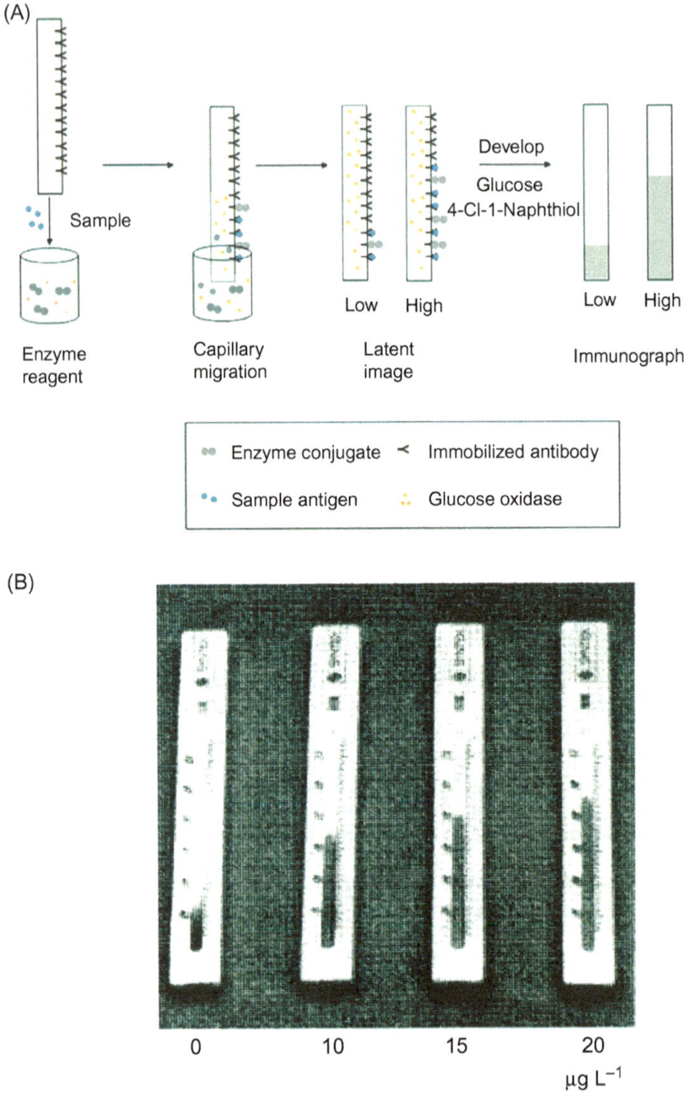

FIGURE 11.1 The first distance signal-readout device. (A) Competitive immunoassay to detect theophylline. The strategy of enzymatic reaction to form a colored band of the mixture of sample and enzyme binds to the immobilized antibody. (B) Distance-based theophylline quantitative analysis evaluation. *(A) Reprinted from Tian, T., Li, J., Song, Y., Zhou, L., Zhu, Z., Yang, C. J. 2016. Distance-based microfluidic quantitative detection methods for point-of-care testing. Lab. Chip., 16(7), 1139–1151. (B) Reprinted from Vaughan, L., Milavetz, G., Ellis, E., Szefler, S., Conboy, K., Weinberger, M., et al. 1986. Multicentre evaluation of disposable visual measuring device to assay theophylline from capillary blood sample. The Lancet, 327(8474), 184–186.*

mixture of target and enzyme-labeled target, allowing the liquid to flow along the strip by capillary action. Because the density of the capture antibody was constant on the band, when the target was absent, all enzyme-labeled targets can bind to the capture antibody at the bottom of the band. And in the presence of the target, the capture antibody would be occupied by the target, leading to the forward movement of the enzyme-labeled target. Hence, the distance on the band was related to the target concentration. The strip was then immersed in a solution containing glucose and an enzyme substrate, where hydrogen peroxide was triggered by the oxidation of glucose under glucose oxidase. Next, the substrate was converted to a blue-gray precipitate which can be considered as a ribbon. The higher theophylline concentration is, the longer the color band will be. By using this method, Zuk et al. successfully quantified theophylline with a 15 min two-step incubation in whole-blood. This method also decreased the effect of enzyme stability by measuring the distribution of the enzyme instead of its activity. Then, Vaughan et al. (1986) evaluated the accuracy of the device (Fig. 11.1B). The consequence was the same as the routine therapy monitoring with a coefficient of variation between 5% and 10%. Afterwards, Litman's group simplified the device to one-step while maintaining good performance (Chen et al., 1987). In this optimized assay, the antibody along with glucose oxidase was immobilized on the band. The enzyme and substrate were mixed with the ascorbate salt as a color inhibitor to delay color formation until the mixture reaches the band.

In the signal-readout of the distance, the key step is to convert the molecular signal into a visualized distance signal (Fig. 11.2). A traditional method is capillary action which employs capillary substrates such as cellulose and fiber for reading. Under the capillary force, the target flows along the

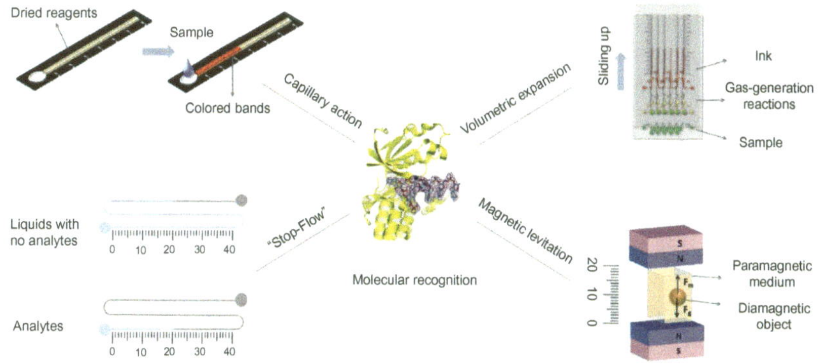

FIGURE 11.2 Volume expansion through capillary action "stops flow" and magnetism, converting the signal to distance-based magnetic levitation readings. *Reprinted from Tian, T., Li, J., Song, Y., Zhou, L., Zhu, Z., Yang, C. J. 2016. Distance-based microfluidic quantitative detection methods for point-of-care testing. Lab. Chip., 16(7), 1139–1151.*

channel when it is introduced and relys on the enzymatic reaction to react with the dry reagent. In 1990, Allen et al. developed an enzyme-based method for measuring cholesterol. This instrument-free technology has been adapted for the quantitative measurement of high-density lipoprotein (HDL). With the development of in situ lipoprotein separation method which is incorporated into the test for HDL determination, the sensitivity of the assay system has been adjusted to a significant range of 250–1000 mg L^{-1} rather than 1000–4000 mg L^{-1} in the earlier reported assay for the total cholesterol (Liu et al., 1993). However, in addition to relying on enzymatic reactions, some analyses also depend on precipitation (Cate et al., 2013), aggregation (Dungchai et al., 2013; Song et al., 2013), or complexes (Yamada et al., 2015). Therefore, the higher the target concentration, the stronger the signal produced. As a consequence, the ribbon's travel distance is related to the amount of target. Equipment manufactured on this principle is of low cost and are easy to manufacture because most of them are made of thread (Nilghaz et al., 2014) and paper (Allen et al., 1990; Chen et al., 1987; Liu et al., 1993; Vaughan et al., 1986b; Zuk et al., 1985).

Another way to get the signal output is to use a visual strip of ink from a gas reaction. The main principle of gas production is the decomposition of H_2O_2 into nontoxic O_2, where the reactants and products are environmentally friendly (Sonnleitner, 2012; Zhu et al., 2015). In a sealed device, the gas moves the ink strip along the strip by a distance consistent with the target concentration. Also, catalysts such as catalase (Song et al., 2012) and PtNPs (Li et al., 2015a; Song et al., 2014; Zhu et al., 2014) are introduced to achieve high-sensitivity signal amplification. The "stop-flow" of the control fluid also shows the reading of the distance. In the presence of target, the elastic channel shrinks due to molecular recognition and a decrease in surface tension of the target (Zhang et al., 2015) or corresponding receptor (Chatterjee et al., 2012), causing the liquid to stop. Therefore, the high target concentration may result in a short traveled distance. In the case of maglev, the signal output of the distance determines the equilibrium position of the sample between the magnets by the balance of gravity and magnetism, where the magnetism is affected by the analyte overall density or beads labeled (Mirica et al., 2010; Subramaniam et al., 2015). Therefore, the detection of the target can be transformed into the analysis of suspended length. Thus, this method has been widely used in various fields, and multiple species of analysts are able to be analyzed with low sample consumption.

In the detection of proteins, enzyme-linked immunosorbent assay (ELISA) has been widely used as a clinical gold standard. However, the traditional ELISA method is a time-consuming process and requires a large number of instruments. In addition, conventional ELISA also meets the challenge for detecting ultralow concentrations of species. Unquestionably, the distance signal-readout is an excellent candidate to provide a good platform for the development of novel sensors. The group of Qin proposed a

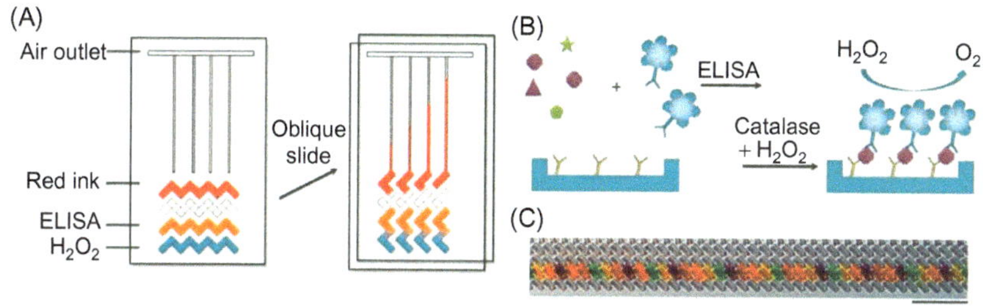

FIGURE 11.3 (A) Operation of the V-Chip. On the left is a view of a preassembled V-Chip where ink and H_2O_2 can be placed in advance and the flow channel is in a horizontal position. ELISA assays can be performed in designated lanes. The slanted slide interrupts the flow path and forms a reactive structure on the right side so that the catalase reacts with H_2O_2 and pushes the ink strip to a certain distance. (B) Schematic diagram of the V-Chip reaction principle. (C) 50 sample wells are loaded with images of different color food dyes based on the swab tip. Source: *Reprinted from Song, Y., Zhang, Y., Bernard, P. E., Reuben, J. M., Ueno, N. T., Arlinghaus, R. B., Qin, L. 2012. Multiplexed volumetric bar-chart chip for point-of-care diagnostics. Nat. Commun. 3, 1283.*

multiplexed volumetric bar-chart chip (V-Chip) without the need for optical instruments or plotting steps or any data processing (Song et al., 2012; Fig. 11.3). The V-Chip presented an on-chip visualized bar chart as a distance signal, based on the volumetric measurement of oxygen generation. The method employed catalase as the ELISA probe, and used the antibody-functionalized silica nanoparticles as the probe. Then, hydrogen peroxide was added to release oxygen with the reaction of catalase. The generated oxygen gathered in the microfluidics channels and pushed preloaded inked bars. The walking distance was related with the concentration of the corresponding ELISA target and each individual inked bar independently indicates the amount of catalase (George, 1947). This is a quantitative and multiplexed approach to detect target protein biomarkers. In order to optimize this technique, they used platinum nanoparticles (PtNPs) to further the experimental design (Song et al., 2014). As we know, PtNPs can efficiently catalyze H_2O_2 transformation into oxygen gas, which can overcome the drawbacks of high cost and low stability during catalase preparation. They used the PtNP-based V-Chip to detect the lung cancer biomarker CYFRA 21-1 in a buffer and serum based on a standard sandwich ELISA in a range of 0.5-50 ng mL^{-1}. Hence, PtNP-based V-Chip has the potential for clinical diagnosis of non-small cell lung cancer (NSCLC). Based on the previous reports, Qin et al. have come up with more inspiring methods to achieve better protein detection using volume-competitive bar-graph chip (CV chip) (Li et al., 2015b). CV chip also shows great potential in clinical diagnosis as it can distinguish small variations in human chorionic gonadotropin (hCG) less than 1.3-fold, which is close to the clinical pregnancy index (20 mIU mL^{-1} – 1.4 ng mL^{-1}). Despite the above advantages, the CV chip is limited to qualitative or semiquantitative analysis, Qin et al. further developed a digital volumetric bar-graph chip (DV chip) based on the prior art, allowing quantification by the naked eye without a calibration test target (Li et al., 2015a). The DV chip system displayed a digital ink bar chart to directly identify the target, where the competition results for H_2O_2 producing O_2 in the target and control samples are shown by 0 and 1. The upward bar representing "1" showed that the target sample produces more O_2, When the concentration of target was lower than that of the control sample, the more produced O_2 in the control side resulted in a downward bar (representing a "0"). This was flexible and versatile because the control concentration and dynamic range could be modified according to the resolution and criticality requirements of a biomarker. Using a DV chip, a CEA sample with a minimal variation between 1 ng mL^{-1} (13 pM) and 1.5 ng mL^{-1} (19.5 pM) was distinguished. In addition, type B natriuretic peptide (BNP) as a biomarker of heart failure (HF) is analyzed as a model and has been used in clinical practice to assess disease states with two cutoffs (29 and 116 pM). The 20 patients' plasma detected by the DV chip had the same clinical results with a LOD of less than 5 pM, indicating that this versatile DV chip system with versatility, visible readings, and very low LOD holds the potential for personalized diagnosis.

The output of the distance has many advantages for instrumentless and visual analysis. The distance-based determination simplifies the assay by recording the length of the color band, which does not require instrumentation and is not susceptible to interpretations when compared to most intensity-based measurements. In addition, the flexibility of distance reading devices can be further enhanced by changing the dynamic range by changing the design of specific reagents (Dungchai et al., 2010; Rattanarat et al., 2013; Wang et al., 2010). Since most distance-based measurements rely on reading the length of a color stripe, the chip's channel is designed to be as narrow as possible for greater sensitivity. Due to the longer colored bands appearing in a given time, a greater change in length between different target concentrations is achieved, which increases the detection sensitivity and resolution. The size of the sample area can also be optimized upon request.

However, there are some challenges that need to be addressed before distance-based output can be widely used and commercialized. First, better reproducibility is required, especially for assays where the stability of the catalyst is a key factor in gas production. Second, environmental conditions can also affect the signal output, thus hindering the application in harsh conditions. Third, how to massively produce device panels with high quality remains a challenge.

11.2 PRESSURE SIGNAL-READOUT

Pressure as an output signal has the advantages of convenient and quick output, including pressure gauge, electronic balance, and cantilever arm and so on.

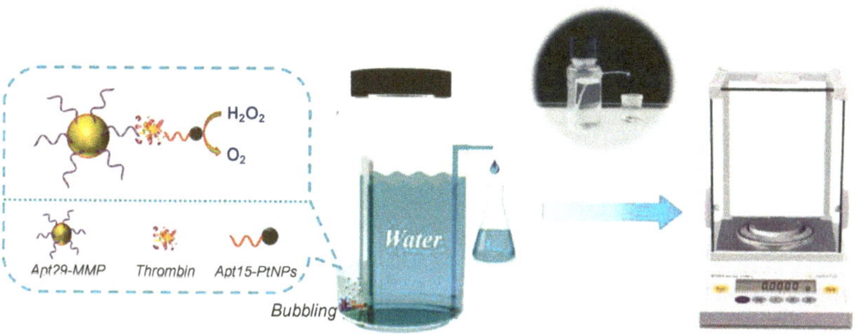

FIGURE 11.4 The principle of using the electronic balance as a thrombin-sensitive protein sensor for signal readings. *Reprinted from Wang, A., Ma, X., Ye, Y., Luo, F., Guo, L., Qiu, B., et al. 2018. A Simple and Convenient Aptasensor for Protein Using Electronic Balance as Readout. Anal. Chem., 90 (2), 1087–1091*

The pressure signal as the output reading can be quickly identified and used as a small, home-made device.

Lin et al. used the most common electronic balance in the lab as an output device to develop a new aptamer sensor for protein quantification (Fig. 11.4; Wang et al., 2018). They chose thrombin as the target, using two aptamers at different sites (one aptamer immobilized on the surface of magnetic particles and the other aptamer labeled onto platinum nanoparticles (PtNPs)) to recognize the protein. Both aptamers and thrombin specifically bound to form a sandwich structure. Based on which, the platinum particles were connected with the magnetic particles and were easily separated by the magnet. Thus, the captured platinum nanoparticles effectively catalyzed the decomposition of H_2O_2, producing a large amount of O_2 that expelled a certain amount of water in the discharge device because the pressure in the vial was higher than the pressure outside the vial. By using pressure as signal output, accurate measurement of water weight by electronic balance could be undertaken. Since the concentration of PtNPs was proportional to the concentration of thrombin, a close relationship between water and target could be established. Water weight increased with increasing thrombin concentration in the range of 0–100 nM with a limit of detection of 2.8 nM.

In addition, because PtNPs were not contaminated with H_2O_2, the sensitivity can be adjusted to meet specific requirements by changing reaction time. When the electronic balance is used as a signal-readout, the pressure difference caused by the continuous O_2 generation can be pushed to water in the drain; thus, the pressure signal can be converted to an amplified weight signal.

The manometer is a portable device for sensitive and quantitative detection of pressure and is widely used for common measurements. With its portability and sensitivity, the gauge is suitable for a wide range of protein assays. Gas generation can lead to a significant increase in the pressure inside a closed container, which can easily be detected using a pressure

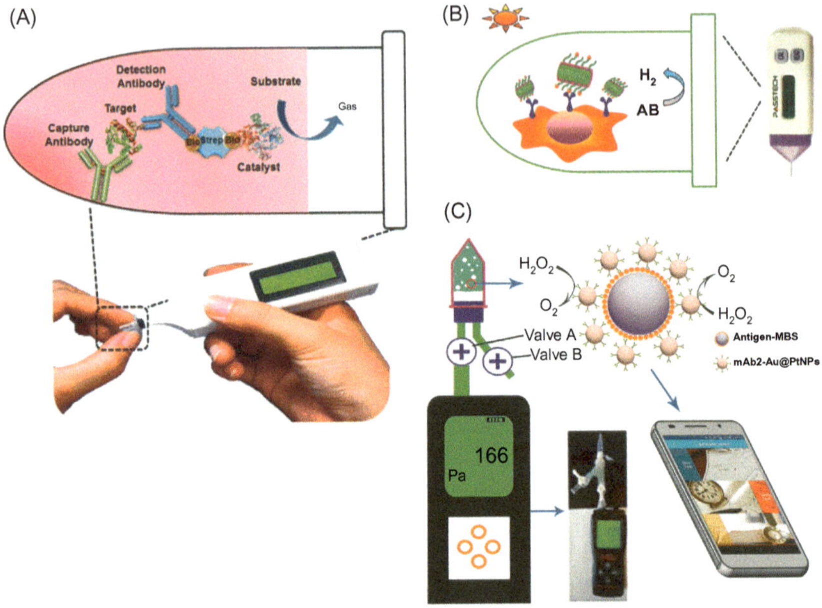

FIGURE 11.5 (A) Working principle of the bioanalysis where pressure is used as a signal output. High-sensitivity of the assay can be achieved by converting molecular recognition signals into measurable pressure signals. (B) Rapid and sensitive detection of NFs in cancer cells using a hand-held manometer. (C) Schematic of a barometer-based biosensor. *(A) Reprinted from Zhu, Z., Guan, Z., Liu, D., Jia, S., Li, J., Lei, Z., et al. 2015. Translating molecular recognition into a pressure signal to enable rapid, sensitive, and portable biomedical analysis. Angew. Chem. Int. Edit., 54(36), 10448–10453. (B) Reprinted from Ding, E., Hai, J., Li, T., Wu, J., Chen, F., Wen, Y., et al. 2017. Efficient Hydrogen-Generation CuO/Co3O4 Heterojunction Nanofibers for Sensitive Detection of Cancer Cells by Portable Pressure Meter. Anal. Chem., 89(15), 8140–8147. (C) Reprinted from Fu, Q., Wu, Z., Du, D., Zhu, C., Lin, Y., & Tang, Y. 2017. Versatile Barometer Biosensor Based on Au@ Pt Core/Shell Nanoparticle Probe. ACS sensors, 2(6), 789–795.*

gauge. Pressure-based biosensors can be applied by combining biological components and pressure signal-readouts with gas signal amplification.

By catalytically decomposing hydrogen peroxide in the seal into oxygen, Yang et al. developed a pressure-based method for high-sensitivity bioanalysis (Zhu et al., 2015). As shown in Fig. 11.5A, a sandwich bioassay consists of a capture antibody bound to a solid support, a target antigen from the sample and a detection antibody labeled with a catalyst, such as an enzyme or catalytic nanoparticle. Upon introduction of the substrate, a rapidly catalyzed gas-generating reaction takes place, with consequent gas production resulting in significant pressure increase of the sealing device. The increase in pressure can be sensed with its own pressure gauge, which is made up of a digital barometric pressure sensor, a lithium battery, and an LCD monitor for displaying air pressure. This pressure-based approach is then applied to detect the disease biomarker C-reactive protein (CRP) by converting CRP/antibody recognition into a pressure signal, enabling the rapid and ultrasensitive quantification of CRP (Ji et al., 2016).

Based on the same principle, Lin et al. developed a platform for the highly sensitive quantification of thrombin (Yang et al., 2015). Catalytic nanoparticles were loaded onto silica nanospheres, and then the resulting complex was modified with antibodies before carrying out the immunoassay-based assay (Wang et al., 2017). After disruption, the released catalytic nanoparticles subsequently triggered a quantitative measurement of the gas reaction by the differential pressure gauge and an increase in the target-related pressure.

Recently, Lu et al. reported a hand-held manometer method that used rapid and sensitive detection of cancer cells (Fig. 11.5B; Ding et al., 2017). In this approach, the folate-conjugated CuO/Co_3O_4 heterojunction nanofibers (NFs) have two functions: one is to provide a specific ligand for folate receptors overexpressed in cancer cells, the other is the efficient photocatalytic activity that can decompose the borane–ammonia complex to produce nontoxic H_2. Tang et al. used core–shell Au@Pt nanoparticles as the catalyst for O_2 generation, and developed pneumatic biosensors for the detection of various targets (Fig. 11.5C; Fu et al., 2017). Smartphone software was further developed to calculate and transmit results so that stress technology is more conducive to real-time detection at home.

Bioanalysis that uses pressure as the output signal has two main advantages. Firstly, ultrasensitive detection can be achieved through two powerful amplification processes. When the gas is generated, the volume of the reaction system will expand by 2–3 orders of magnitude, resulting in a significant increase in the pressure in the closed system. Moreover, the active catalyst (e.g., PtNP) can catalyze 10^6 substrate molecules per second. Both of them cause significant signal amplification to exceed 10^{10} times within a few minutes of response. Secondly, the pressure-based measurement does not contain electromagnetic radiation noise and magnetic field noise. This unique feature simplifies the design of the device and further facilitates highly sensitive and accurate detection.

11.3 PIEZOELECTRIC BIOSENSORS

Piezoelectric sensors make use of the piezoelectric effect produced by some dielectrics under stress. Piezoelectric effect refers to the fact that when some dielectrics undergo deformation (including bending and telescopic deformation) due to an external force in a certain direction, they will generate charges on the surface due to the polarization of internal charges. The piezoelectric material as a dielectric can be classified into piezoelectric single-crystal, piezoelectric polycrystal, and organic piezoelectric material. Piezoelectric sensors most commonly use piezoelectric ceramics and piezoelectric crystals in various types. Pizeozelectric biosensor (PEBS) is a novel biosensor that uses piezoelectric crystals as signal transducers and bioactive substances as recognition molecules (e.g., DNA, antibodies/antigens, enzymes, microbial cells, etc.). It combines electronics, mechanics, molecular biology, and other disciplines, with the characteristics of sensitive and rapid response, high specificity, simplicity, label-free, easy automation, and integration. It uses piezoelectric media to activate sound waves and immobilize biologically active materials on its surface. Through the interaction of sound waves with the surrounding environment, the sensor can convert the changes in the surface and system properties of piezoelectric media such as quartz crystals during the reaction into frequency signals that can be quantitatively detected. The piezoelectric medium has both elasticity and electrical characteristics, so the sensor can simultaneously sense the mechanical and electrical characteristics of the external environment.

In 1876, Jacques and Pierre Curie, the Curie brothers, first discovered the piezoelectric phenomena of some crystals such as quartz. Later in 1959 Sauerbrey deduced the Sauerbrey equation about the mass and resonant frequency shifts carried on the crystal surface in the gas phase: $\Delta F = -KF^2 \Delta M/A$, where ΔF is the change of the vibration frequency (Hz) after the crystal absorbing foreign substances, K is a constant, A is the area covered by the adsorbate, F is the fundamental frequency of the piezoelectric crystal (MHz), and ΔM is the mass of the adsorbed material. From the equations, it can be seen that when the quartz crystal is oscillating in the gas phase, ΔF is in a simple linear relationship with ΔM. If the crystal is made to selectively adsorb exogenous bioactive substances, a piezoelectric biosensor is able to be produced.

PEBS consists of piezoelectric crystals coated with biometric substances, oscillators, and frequency meters. The most commonly used piezoelectric crystals are AT-cut quartz crystals with a certain resonant frequency (9 MHz, 27 MHz). Piezoeletric quartz crystal (PQC), also known as quartz crystal microbalance (QCM), is typically 10–16 mm in diameter, and approximately 0.16 mm thick. It contains gold films with a diameter of 3–8 mm on both sides, forming a gold electrode to facilitate modification and fixation of the identification element. Piezoelectric detection systems generally have two oscillator circuits. One is a crystal detection oscillation circuit. The other is a crystal reference oscillation circuit. The reference oscillator is not coated with biometric substances in order to correct the influence of interference factors and eliminate the influence of some errors. The piezoelectric crystal's resonant frequency and frequency change are accomplished by a frequency counter, which is then subjected to computer analysis and results reporting.

Nowadays PEBS using QCM for protein assay is increasingly applied in clinical assays. Examples can be coagulation factor VIII (Yao et al., 2013), IgG and albumin levels in human serum (Jaruwongrungsee et al., 2015), breast cancer biomarker CM15.3 (Wang et al., 2014), and the like. Piezoelectric immunosensor is an important kind of PEBS used for protein assay. Fig. 11.6 shows three formats of protein assay using this method.

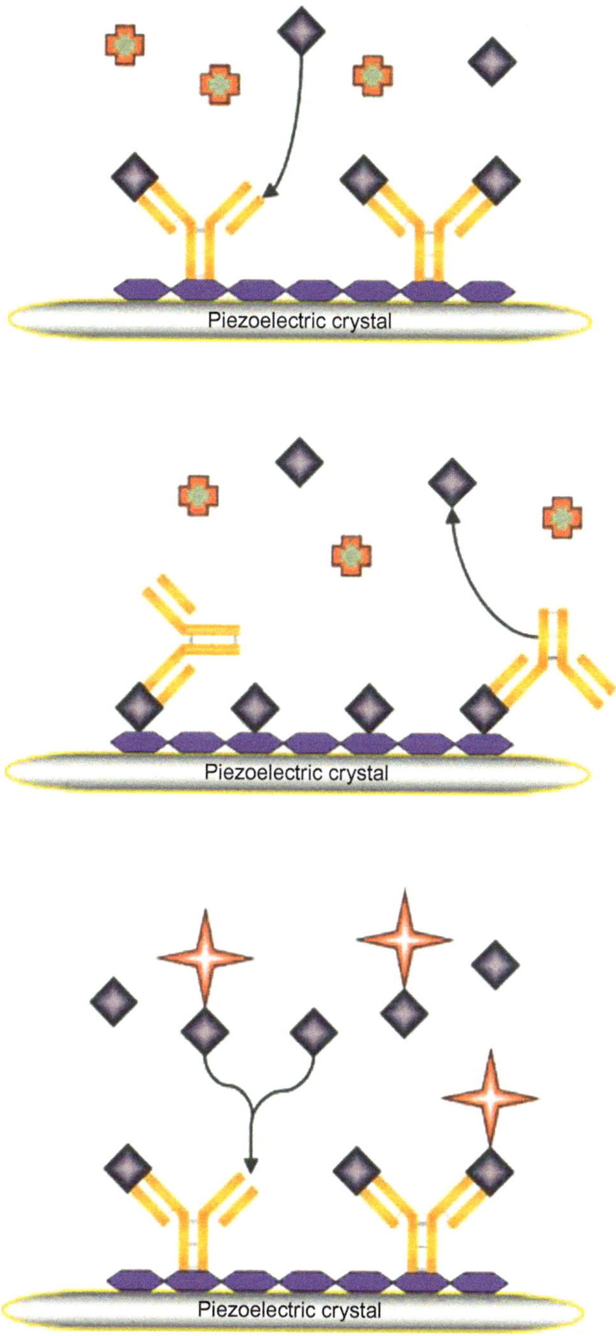

FIGURE 11.6 The antibody immobilized on the crystal surface captures analytes of interest, leading to a frequency reduction (top). The immobilized antibody is substituted by free antigen and leaves the surface, which results in an increase in frequency (center). Labeled antigen competes with free antigen for the accessible binding site on the antibodies (Vaughan and Guilbault, 2007). *Reprinted from Vaughan, R. D., & Guilbault, G. G. Piezoelectric immunosensors. In Piezoelectric Sensors, Springer, Berlin, Heidelberg, 2007, 237–280.*

Detection of protein using QCM has a large number of advantages such as being label-free, low cost, highly sensitive, and having facile operation. Hence, QCM has been widely used as a standard noninvasive tool. Particularly in the liquid phase, the QCM can detect mass loading as well as changes in solution viscosity and density near the electrode sensitively. In 1972, Shons et al. first reported the detection of cow serum IgG antibody using QCM, which initiated the development of the QCM immunoassay (Shons et al., 1972). Subsequently, Muratsugu et al. innovated a piezoelectric immunosensor for the detection of human serum albumin (HSA) (Muratsugu et al., 1993). The traditional method of flow cell could detect HSA in the range of 0.1 to 100 $\mu g\ mL^{-1}$, and the immunosensor was sensitive enough to monitor low levels of albuminuria. In addition, the immunosensor only responded to HSA and did not reply to bovine serum albumin, indicating that the specificity for HSA was high. They utilized the relationship between the frequency change (ΔF) and adsorption of piezoelectrically active quartz crystal (ΔM) per unit area to signal-readout. During the test, ΔM was estimated with radioisotope-labeled anti-HSA or HSA. When HSA was bound to anti-HSA supported by the crystal or anti-HSA was added onto the surface, values of $|\Delta F/\Delta M|$ were larger than reported before.

The QCM can provide multidimensional piezoelectric impedance analysis based on the resonant frequency (f_0), motional resistance (R_1), etc. which enable it to be a powerful quantitative tool to explore various modified substances on an electrode and investigate the electrode-modification processes in situ. Nevertheless, the piezoelectric signal-readout using QCM immunosensors has still faced challenges of limited detection sensitivity. Hence, some strategies have been developed to improve the detection sensitivity of QCM immunosensors. Traditional QCM biosensors use 5–10 MHz crystals with limited sensitivity. Subsequently, in order to improve the sensitivity of QCM biosensors, QCMs at 15-MHz (Albyn, 2001), 27-MHz (Furusawa et al., 2009; Nishino et al., 2004; Takahashi et al., 2009), 50-MHz (Lederer et al., 2010), 62-MHz (Ping et al., 2008), 96-MHz (Williams et al., 2007), and higher fundamental frequencies (Ogi et al., 2009) were reported. However, QCM at fundamental frequencies higher than ca. 40 MHz need special fabrication and engineering and have very thin crystal wafer which is limited for wide applications. Xie et al. used a 35-MHz QCM to monitor the binding process of solution LIS to ACE in situ adsorbed at a 1-dodecanethiol (C12SH)-modified Au electrode and the binding and separation rate constants (k_1 and k_{-1}) (Su et al., 2011). Since the adsorption amount of ACE is in the submicrogram scale, their method only requires a very low consumption of the expensive target. The QCM results are quantitatively supported by surface plasmon resonance (SPR) experiments (Fig. 11.7). This method could have presented an easily popularized experimental platform to investigate for biomedical and biosensor applications.

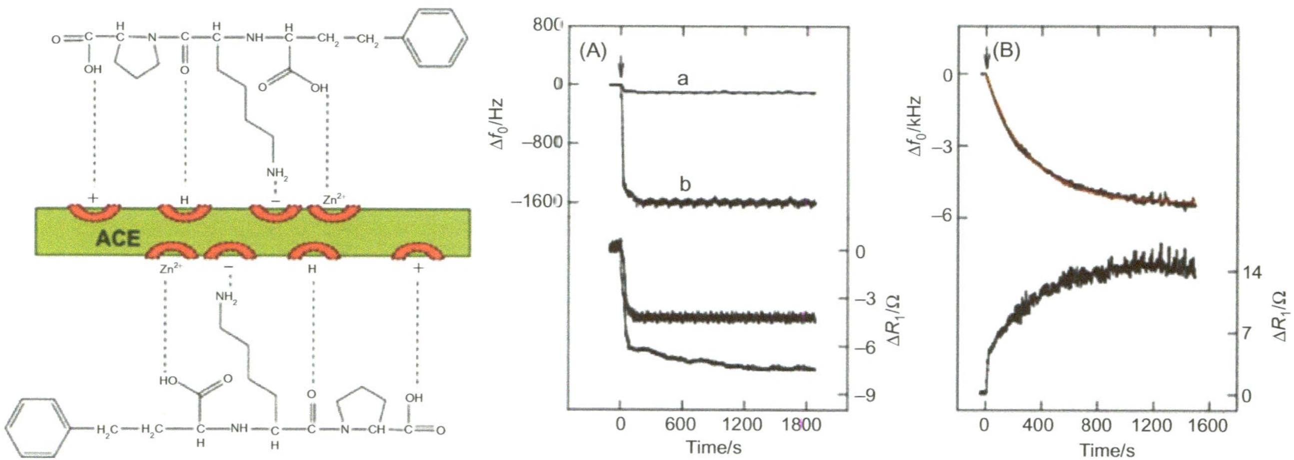

FIGURE 11.7 The left diagram shows the mechanism of the binding of ACE with LIS. The right diagram indicates the change of frequency. (A) QCM responses of adsorption of 10 mmol L^{-1} C_{12}-SH onto bare 9-MHz (a) and 35-MHz (b) QCM Au electrodes in absolute ethanol. The arrow indicates the moment of addition of C_{12}-SH. (B) 35-MHz QCM responses at C_{12}-SH/Au to the addition of 2.15 μmol L^{-1} ACE into PBS. The arrow indicates the moment of ACE addition. *Reprinted from Su, Z., Chen, L., Liu, Y., He, X., Zhou, Y., Xie, Q., Yao, S. 2011. 35 MHz quartz crystal microbalance and surface plasmon resonance studies on the binding of angiotensin converting enzyme with lisinopril. Biosens. Bioelectron., 26(7), 3240–3245.*

In the liquid phase, ultrasensitive QCM measurements could be achieved using high-frequency crystal (Uttenthaler et al., 2001). The effective area of the electrode surface could also be increased with molecular imprinted polymers, porous films, and multilayers (Ávila et al., 2008; Dunér et al., 2008). In addition, the introduction of nanoplatforms such as nanoparticles (Jin et al., 2009; Lyu et al., 2008; Shan et al., 2014) and vesicles (Hildebrand et al., 2002; Mahon et al., 2010) can realize the amplification on the electrode surface.

11.4 MAGNETIC SIGNAL-READOUT

Magnetic nanoparticles (MNPs) have been widely used in the field of biomedicine, including drug delivery, magnetic resonance imaging, hyperthermia, and cell isolation (Colombo et al., 2012; Xu et al., 2015). In addition, MNPs are involved in the development of various models using electrochemical, optical, or piezoelectric measurements. In recent years, magnetic biosensors using MNPs as signal output tags have drawn wide attention in analytical chemistry. Compared with other signal output methods, the use of MNPs as biosensing tags has the following advantages. MNPs are easy to be manipulated under the applied magnetic field. In addition, unlike electrical or optical sensing techniques, the tiny magnetic background signal of a biological sample results in a high signal-to-noise ratio in magnetic induction, which can greatly increase the sensitivity of the assay. These features make magnetic biosensors a good platform for the detection of proteins (Lee et al., 2015). Of these magnetic sensors, magnetic sensing methods are receiving more attention nowadays because of their outstanding advantages over heterogeneous sensors. Currently, magnetic biosensors rely on one of two sensing mechanisms: one based on directly reading magnetic signals from MNP and the other on optical signals measuring MNP under magnetic stirring (Stefan et al., 2016). For the first method, magnetic signals often detected in magnetic biosensors mainly include permeability, magnetic susceptibility, and magnetic relaxation (Alcantara et al., 2016). For the second method, photomagnetic detection utilizes the magnetic and intrinsic optical anisotropy of the MNP or MNPs cluster linkage, such that the scattering and absorption of light by the MNP and MNP clusters is correspondingly changed to magnetically-induced MNPs with the application of an oscillating magnetic field (Donolato et al., 2015; Fock et al., 2016). This magneto-optical approach shows a larger detection signal, simplifying setup and eliminating the need for polarizers.

For protein detection, Kim et al. reported an MRS-based biosensing platform that formed HCG in solution by antigen–antibody interaction through target-induced MNP aggregation (Stefan et al., 2016). In the meantime, Ling and colleagues proposed a multifunctional MRSw-based sensing platform for the simultaneous detection of different protein biomarkers. In addition, MRSw-based biosensing strategies were also developed for monitoring tumor-associated enzyme activity. There are two strategies for measuring enzyme activity using MRSw. The first approach is the target-induced MNP aggregation, similar to the protein assay, it has been successfully used to measure enzyme activities of telomerase (Grimm et al., 2004; Perez et al., 2008), caspase 3 (Yuan et al., 2016), and peroxidase activities (Perez et al., 2004). Another approach is crucial for the reverse switch, in which substrate enzymatic cleavage can cause the breakdown of preformed MNP clusters and further convert the enzymatic activity to an enhanced T_2 signal. Lee et al. reported a diagnostic MRSw-based sensor that detects and analyzes cancer cells by detecting changes in T_2 in biological samples following MNPs targeting (Lee et al., 2009). Later, Haun and colleagues further improved the sensor with bio-orthogonal nanoparticles (Haun et al., 2010). This method uses 1,2,4,5-tetrazine (Tz) and transcyclooctene (TCO) as coupling agents to mediate cell-to-MNP binding, where the TCO-antibody conjugate can promote Tz-coated MNPs. Recently, Bamrungsap et al. used aptamers as surrogates for antibodies to achieve pattern-specific aptamer-modified MNPs and targeted cancer cells (Bamrungsap et al., 2012). By this strategy, the corresponding change of T_2 relaxation time was recorded to build disposable magnetic nanosensor.

Magnetic signal-readout has also been widely used to design magnetic biosensors. Park et al. (2012) prepared magneto-optic nanoparticle probes for the analysis of MMPs in which specific peptide substrate sequences with proteolytic ligands for MMPs and activatable fluorescent labels are simultaneously conjugated to the surface of magnetic nanocrystals, as a targeting moiety and fluorescence signal converter (Fig. 11.8A). In the presence of MMPs, it can restore the quenching fluorescence of the photomagnetic nanoprobes and additional MR imaging of cancer cells as contrast agent. Subsequently, Ranzoni et al. reported that nanoclusters formed from MNPs based on interparticle binding can measure the PSA's magneto-optic immunosensor by optical scattering of external magnetic rotation frequencies (Fig. 11.8B). Subsequently, a series of photomagnetic washable biosensors were developed for the determination of C-reactive protein (Daynès et al., 2015) and thrombin (Uddin et al., 2016). Recently, Chen et al. proposed a magnetic immunosensor with naked eye detection (Fig. 11.8C; Chen et al., 2015). Aggregation of immunomagnetic MNPs is induced by the target to produce MNP nanoclusters on the sidewalls of a sample tube with an external magnetic field, resulting in a noticeable color change from yellow to brown which can be easily identified with the naked eye. The sensor shows excellent detection capability directly in spiked urine and patient serum samples and can be used to directly measure AFP and CEA.

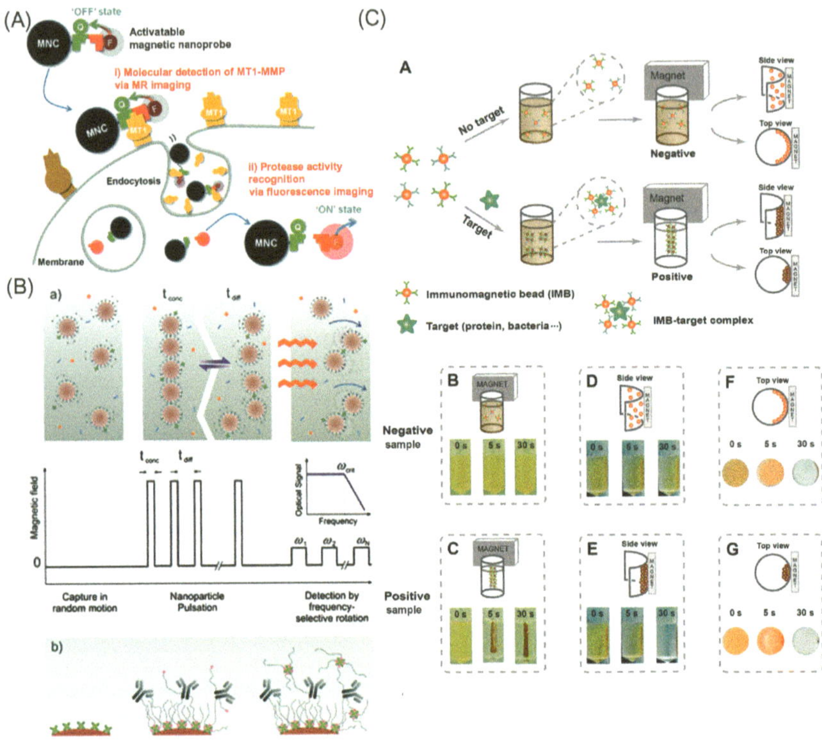

FIGURE 11.8 (A) Photomagnetic nanoprobes detect MT1-MMP anchored on invasive cancer cells by MR imaging and sensitively recognize the proteolytic activity of MT1-MMP by fluorescence imaging. (B) Disposable magnetic immunosensor for PSA. (C) Immunomagnetic aggregation sensor for naked eye detection of protein biomarkers. *(A) Reprinted from Park, J., Yang, J., Lim, E. K., Kim, E., Choi, J., Ryu, J. K., et al. 2012. Anchored Proteinase-Targetable Optomagnetic Nanoprobes for Molecular Imaging of Invasive Cancer Cells. Angew. Chem. Int. Edit., 51(4), 945–948. (B) Reprinted from Ranzoni, A., Sabatte, G., van IJzendoorn, L. J., & Prins, M. W. 2012. One-step homogeneous magnetic nanoparticle immunoassay for biomarker detection directly in blood plasma. ACS Nano, 6(4), 3134–3141. (C) Reprinted from Chen, Y., Xianyu, Y., Sun, J., Niu, Y., Wang, Y., & Jiang, X. 2016. One-step detection of pathogens and cancer biomarkers by the naked eye based on aggregation of immunomagnetic beads. Nanoscale, 8(2), 1100-1107.*

11.5 SMELL SIGNAL-READOUT

The signal outputs of many existing methods depend on complex operation processes and expensive instruments and devices. Visual recognition has been widely used in various substances detection without the need for costly instrument. However, visual detection relies on light sources, which limits its applications in many aspects. Now a new way of relying on olfactory output has appeared, which may make up for the drawback of visual recognition.

The combination of multiple olfactory signals in the olfactory bulb can be used to identify odor. These signals are collected through various olfactory receptor neurons (OSN) expressing different olfactory receptors (Malnic et al., 1999; Zou and Buck, 2006). The human nose can express a series of olfactory receptors (about 390 species), and can identify and distinguish specific olfactory receptors, and its resolution is equivalent to that of a single carbon atom (Malnic et al., 2004). The normal receptor/odorant interaction in the human olfactory system can detect extremely low concentration target odorants with high selectivity and sensitivity. The sensitivity is several orders of magnitude higher than the ability of visual stimuli. The sensitivity and versatility make olfactory to be a promising method of signal output. However, few examples of biosensing using olfactory output are available (Kuhnt et al., 2014; Kuhnt et al., 2015; Mohapatra and Phillips, 2012; Xu et al., 2014).

Duncan and coworkers introduced a highly efficient sensor that provided olfactory output, which meant that there was no need for instrumental help for detection and it was not dependent on visual recognition (Duncan et al., 2017). The principle can be seen in Fig. 11.9. The sensor used nanoparticles for reversible binding and inhibition of lipase. The introduction of bacteria resulted in the competitive binding with nanoparticles toward lipase, and the activity of the enzyme was restored, which could catalyze aromatic substrate molecules to produce aroma smell. The system used the human sense of smell as the output, and the detection speed was fast (15 min), and the sensitivity was very high (10^2 CFU mL^{-1}).

In addition, Kwon and coworkers reported the creation of the prototype of the human nose with hOR-conjugate graphene micropattern (GM) geometry, which was called the multiplexed superbio electronic nose (MSB-nose)

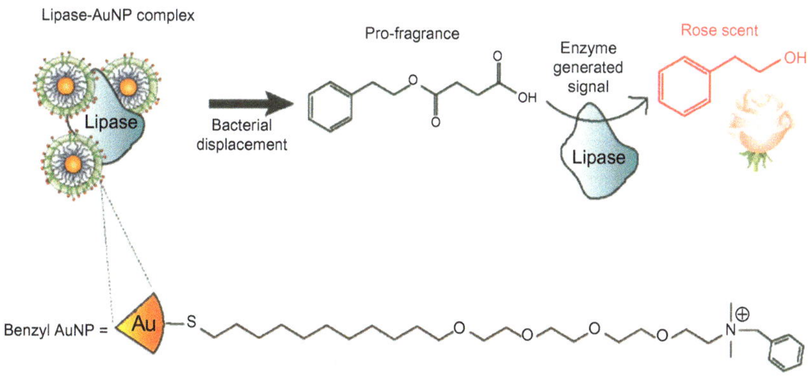

FIGURE 11.9 Schematic representation of sensor elements used in this study. Cationic AuNPs bind with the anionic enzyme inhibiting the catalysis of the pro-fragrance into scent. Bacteria presenting in solution compete for the AuNP surface and displace the enzyme, inducing the production of the rose fragrance. *Reprinted from Duncan, B., Le, N. D., Alexander, C., Gupta, A., Yesilbag Tonga, G., Yazdani, M., et al. 2017. Sensing by Smell: Nanoparticle–Enzyme Sensors for Rapid and Sensitive Detection of Bacteria with Olfactory Output. ACS Nano, 11(6), 5339–5343.*

(Jang et al., 2015). MSB-nose can operate steadily as a liquid ion gated field-effect transistor (FET) system, and has excellent mixed odor discrimination ability.

Although there are not many sensors for olfactory output, and few are used to detect proteins, its high sensitivity and simple operation will make olfactory output a common way for signal output in the future.

REFERENCES

Albyn, K.C., 2001. Extension of the enthalpy of sublimation for adipic acid to temperatures below 80°C. J. Chem. Eng. Data 46 (6), 1415–1416.

Alcantara, D., Lopez, S., Garcíamartin, M.L., Pozo, D., 2016. Iron oxide nanoparticles as magnetic relaxation switching (MRSw) sensors: current applications in nanomedicine. Nanomedicine. 12 (5), 1253–1262.

Allen, M.P., Delizza, A., Ramel, U., Jeong, H., Singh, P., 1990. A noninstrumented quantitative test system and its application for determining cholesterol concentration in whole-blood. Clin. Chem. 36 (9), 1591–1597.

Ávila, M., Zougagh, M., Ríos, Á., Escarpa, A., 2008. Molecularly imprinted polymers for selective piezoelectric sensing of small molecules. Trac Trends Anal. Chem. 27 (1), 54–65.

Bamrungsap, S., Chen, T., Shukoor, M.I., Chen, Z., Sefah, K., Chen, Y., et al., 2012. Pattern recognition of cancer cells using aptamer-conjugated magnetic nanoparticles. ACS Nano 6 (5), 3974–3981.

Cate, D.M., Dungchai, W., Cunningham, J.C., Volckens, J., Henry, C.S., 2013. Simple, distance-based measurement for paper analytical devices. Lab. Chip. 13 (12), 2397–2404.

Chatterjee, D., Mansfield, D.S., Anderson, N.G., Subedi, S., Woolley, A.T., 2012. "Flow valve" microfluidic devices for simple, detectorless and label-free analyte quantitation. Anal. Chem. 84 (16), 7057–7063.

Chen, R., Li, T.M., Merrick, H., Parrish, R.F., Bruno, V., Kwong, A., et al., 1987. An internal clock reaction used in a one-step enzyme immunochromatographic assay of theophylline in whole-blood. Clin. Chem. 33 (9), 1521–1525.

Chen, Y., Xianyu, Y., Sun, J., Niu, Y., Wang, Y., Jiang, X., 2015. One-step detection of pathogens and cancer biomarkers by the naked eye based on aggregation of immunomagnetic beads. Nanoscale 8 (2), 1100–1107.

Colombo, M., Carregal-Romero, S., Casula, M.F., Gutierrez, L., Morales, M.P., Boehm, I.B., et al., 2012. ChemInform abstract: biological applications of magnetic nanoparticles. ChemInform 43 (35), no-no.

Daynès, A., Temurok, N., Gineys, J.P., Cauet, G., Nerin, P., Baudry, J., et al., 2015. Fast magnetic field-enhanced linear colloidal agglutination immunoassay. Anal. Chem. 87 (15), 7583.

Ding, E., Hai, J., Li, T., Wu, J., Chen, F., Wen, Y., et al., 2017. An efficient hydrogen-generation CuO/Co_3O_4 heterojunction nanofibers for sensitive detection of cancer cells by portable pressure meter. Anal. Chem. 89 (15).

Donolato, M., Antunes, P., Bejhed, R.S., Zardán, G.D.L.T.T., Fw, Ø., Strömberg, M., et al., 2015. Novel readout method for molecular diagnostic assays based on optical measurements of magnetic nanobead dynamics. Anal. Chem. 87 (3), 1622–1629.

Duncan, B., Le, N.D.B., Alexander, C., Gupta, A., Tonga, G.Y., Yazdani, M., et al., 2017. Sensing by smell. Nanoparticle-enzyme sensors for rapid and sensitive detection of bacteria with olfactory output. ACS Nano 11 (6), 5339.

Dunér, G., Anderson, H., Myrskog, A., Hedlund, M., Aastrup, T., Ramström, O., 2008. Surface-confined photopolymerization of pH-responsive acrylamide/acrylate brushes on polymer thin films. Langmuir ACS J. Surf. Colloids 24 (14), 7559–7564.

Dungchai, W., Chailapakul, O., Henry, C.S., 2010. Use of multiple colorimetric indicators for paper-based microfluidic devices. Anal. Chim. Acta 674 (2), 227–233.

Dungchai, W., Sameenoi, Y., Chailapakul, O., Volckens, J., Henry, C.S., 2013. Determination of aerosol oxidative activity using silver nanoparticle aggregation on paper-based analytical devices. Analyst 138 (22), 6766.

Fock, J., Parmvi, M., Strömberg, M., Svedlindh, P., Donolato, M., Hansen, M.F., 2016. Comparison of optomagnetic and AC susceptibility readouts in a magnetic nanoparticle agglutination assay for detection of C-reactive protein. Biosensors Bioelectron. 88, 94–100.

Fu, Q., Wu, Z., Du, D., Zhu, C., Lin, Y., Tang, Y., 2017. Versatile barometer biosensor based on Au@Pt core/shell nanoparticle probe. ACS Sensors 2 (6), 789.

Furusawa, H., Komatsu, M., Okahata, Y., 2009. In situ monitoring of conformational changes of and peptide bindings to calmodulin on a 27 MHz quartz-crystal microbalance. Anal. Chem. 81 (5), 1841.

George, P., 1947. Reaction between catalase and hydrogen peroxide. Nature 159 (4054), 41.

Grimm, J., Perez, J.M., Josephson, L., Weissleder, R., 2004. Novel nanosensors for rapid analysis of telomerase activity. Cancer Res. 64 (2), 639–643.

Haun, J.B., Devaraj, N.K., Hilderbrand, S.A., Lee, H., Weissleder, R., 2010. Bioorthogonal chemistry amplifies nanoparticle binding and enhances the sensitivity of cell detection. Nat. Nanotechnol. 5 (9), 660–665.

Hildebrand, A., Schaedlich, A., Rothe, U., Neubert, R.H., 2002. Sensing specific adhesion of liposomal and micellar systems with attached carbohydrate recognition structures at lectin surfaces. J. Colloid Interface Sci. 249 (2), 274.

Jang, J., Kwon, O.S., Song, H.S., Park, S.J., Lee, S.H., An, J.H., et al., 2015. An ultrasensitive, selective, multiplexed super-bioelectronic nose that mimics the human sense of smell. Nano. Lett. 15 (10), 6559–6567.

Jaruwongrungsee, K., Waiwijit, U., Wisitsoraat, A., Sangworasil, M., Pintavirooj, C., Tuantranont, A., 2015. Real-time multianalyte biosensors based on interference-free multichannel monolithic quartz crystal microbalance. Biosensors Bioelectron. 67, 576–581.

Ji, T., Liu, D., Liu, F., Li, J., Ruan, Q., Song, Y., et al., 2016. A pressure-based bioassay for the rapid, portable and quantitative detection of C-reactive protein. Chem. Commun. 52 (54), 8452–8454.

Jin, X., Jin, X., Chen, L., Jiang, J., Shen, G., Yu, R., 2009. Piezoelectric immunosensor with gold nanoparticles enhanced competitive immunoreaction technique for quantification of aflatoxin B1. Biosensors Bioelectron. 24 (8), 2580–2585.

Kuhnt, T., Herrmann, A., Benczedi, D., Weder, C., Foster, E.J., 2014. Controlled fragrance release from galactose-based pro-fragrances. RSc Adv. 4 (92), 50882–50890.

Kuhnt, T., Herrmann, A., Benczedi, D., Foster, E.J., Weder, C., 2015. Functionalized cellulose nanocrystals as nanocarriers for sustained fragrance release. Polym. Chem. 6 (36), 6553–6562.

Lederer, T., Stehrer, B.P., Bauer, S., Jakoby, B., Hilber, W., 2010. An electrowetting on dielectrics based lab-on-a-chip utilizing an integrated high fundamental frequency quartz crystal resonator as a biosensor. Proc. Eng. 5, 959–964.

Lee, H., Yoon, T.J., Figueiredo, J.L., Swirski, F.K., Weissleder, R., 2009. From the cover: rapid detection and profiling of cancer cells in fine-needle aspirates. Proc. Natl. Acad. Sci. USA 106 (30), 12459–12464.

Lee, H., Shin, T.H., Cheon, J., Weissleder, R., 2015. Recent developments in magnetic diagnostic systems. Chem. Rev. 115 (19), 10690.

Li, Y., Xuan, J., Song, Y., Wang, P., Qin, L., 2015a. A microfluidic platform with digital readout and ultra-Low detection limit for quantitative point-of-care diagnostics. Lab. Chip. 15 (16), 3300–3306.

Li, Y., Xuan, J., Xia, T., Han, X., Song, Y., Cao, Z., et al., 2015b. Competitive volumetric bar-chart chip with real-time internal control for point-of-care diagnostics. Anal. Chem. 87 (7), 3771.

Liu, V.Y., Lin, T.Y., Schrier, W., Allen, M., Singh, P., 1993. AccuMeter noninstrumented quantitative assay of high-density lipoprotein in whole blood. Clin. Chem. 39 (9), 1948–1952.

Lyu, Y.K., Lim, K.R., Lee, B.Y., Kim, K.S., Lee, W.Y., 2008. Microgravimetric lectin biosensor based on signal amplification using carbohydrate-stabilized gold nanoparticles. Chem. Commun. 39 (39), 4771–4773.

Mahon, E., Aastrup, T., Barboiu, M., 2010. Dynamic glycovesicle systems for amplified QCM detection of carbohydrate-lectin multivalent biorecognition. Chem. Commun. 46 (14), 2441.

Malnic, B., Hirono, J., Sato, T., Buck, L.B., 1999. Combinatorial receptor codes for odors. Cell 96 (5), 713–723.

Malnic, B., Godfrey, P.A., Buck, L.B., 2004. The human olfactory receptor gene family. Proc. Natl. Acad. Sci. USA 101 (8), 2584.

Mirica, K.A., Phillips, S.T., Mace, C.R., Whitesides, G.M., 2010. Magnetic levitation in the analysis of foods and water. J. Agric. Food. Chem. 58 (11), 6565–6569.

Mohapatra, H., Phillips, S.T., 2012. Using smell to triage samples in point-of-care assays. Angew. Chem. Int. Ed. Engl. 51 (44), 11145–11148.

Muratsugu, M., Ohta, F., Miya, Y., Hosokawa, T., Kurosawa, S., Kamo, N., et al., 1993. Quartz crystal microbalance for the detection of microgram quantities of human serum albumin: relationship between the frequency change and the mass of protein adsorbed. Anal. Chem. 65 (20), 2933–2937.

Nilghaz, A., Ballerini, D.R., Fang, X.Y., Shen, W., 2014. Semiquantitative analysis on microfluidic thread-based analytical devices by ruler. Sensors Actuat. B Chem. 191 (2), 586–594.

Nishino, H., Murakawa, A., Toshiaki Mori, A., Okahata, Y., 2004. Kinetic studies of AMP-dependent phosphorolysis of amylopectin catalyzed by Phosphorylase B on a 27 MHz quartz-crystal microbalance. J. Am. Chem. Soc. 126 (45), 14752.

Ogi, H., Nagai, H., Naga, H., Fukunishi, Y., Hirao, M., Nishiyama, M., 2009. 170-MHz electrodeless quartz crystal microbalance biosensor: capability and limitation of higher frequency measurement. Anal. Chem. 81 (19), 8068–8073.

Park, J., Yang, J., Lim, E.K., Kim, E., Choi, J., Ryu, J.K., et al., 2012. Anchored proteinase-targetable optomagnetic nanoprobes for molecular imaging of invasive cancer cells. Angew. Chem. Int. Ed. Engl. 124 (4), 945–948.

Perez, J.M., Simeone, F.J., Tsourkas, A., Lee Josephson, A., Weissleder, R., 2004. Peroxidase substrate nanosensors for MR imaging. Nano. Lett. 4 (1), 119–122.

Perez, J.M., Grimm, J., Josephson, L., Weissleder, R., 2008. Integrated nanosensors to determine levels and functional activity of human telomerase. Neoplasia 10 (10), 1066.

Ping, K., Patwardhan, A., Allara, D., Tadigadapa, S., 2008. Human serum albumin adsorption study on 62-MHz miniaturized quartz gravimetric sensors. Anal. Chem. 80 (15), 5930–5936.

Rattanarat, P., Dungchai, W., Cate, D.M., Siangproh, W., Volckens, J., Chailapakul, O., et al., 2013. A microfluidic paper-based analytical device for rapid quantification of particulate chromium. Anal. Chim. Acta 800 (19), 50.

Ranzoni, A., Sabatte, G., van IJzendoorn, L.J., Prins, M.W., 2012. One-step homogeneous magnetic nanoparticle immunoassay for biomarker detection directly in blood plasma. ACS Nano. 6 (4), 3134–3141.

Shan, W., Pan, Y., Fang, H., Guo, M., Nie, Z., Huang, Y., et al., 2014. An aptamer-based quartz crystal microbalance biosensor for sensitive and selective detection of leukemia cells using silver-enhanced gold nanoparticle label. Talanta 126 (126), 130–135.

Shons, A., Dorman, F., Najarian, J., 1972. An immunospecific microbalance. J. Biomed. Mater. Res. 6 (6), 565–570.

Song, S.H., Lim, C.S., Shin, S., 2013. Migration distance-based platelet function analysis in a microfluidic system. Biomicrofluidics 7 (6), 64101.

Song, Y., Zhang, Y., Bernard, P.E., Reuben, J.M., Ueno, N.T., Arlinghaus, R.B., et al., 2012. Multiplexed volumetric bar-chart chip for point-of-care diagnostics. Nat. Commun. 3 (4), 1283.

Song, Y., Xia, X., Wu, X., Wang, P., Qin, L., 2014. Integration of platinum nanoparticles with a volumetric bar-chart chip for biomarker assays. Angew. Chem. 53 (46), 12451.

Sonnleitner, B., 2012. Automated Measurement and Monitoring of Bioprocesses: Key Elements of the M 3 C Strategy. Springer Berlin Heidelberg.

Stefan, S., Beatriz, P., Parak, W.J., Sergio, L.M., Katerina, S., Jan, D., et al., 2016. homogeneous biosensing based on magnetic particle labels. Sensors 16 (6), 828.

Su, Z., Chen, L., Liu, Y., He, X., Zhou, Y., Xie, Q., et al., 2011. 35 MHz quartz crystal microbalance and surface plasmon resonance studies on the binding of angiotensin converting enzyme with lisinopril. Biosensors Bioelectron. 26 (7), 3240–3245.

Subramaniam, A.B., Gonidec, M., Shapiro, N.D., Kresse, K.M., Whitesides, G.M., 2015. Metal-amplified density assays, (MADAs), including a density-linked immunosorbent assay (DeLISA). Lab. Chip. 15 (4), 1009.

Takahashi, S., Iida, M., Furusawa, H., Shimizu, Y., Ueda, T., Okahata, Y., 2009. Real-time monitoring of cell-free translation on a quartz-crystal microbalance. J. Am. Chem. Soc. 131 (26), 9326.

Tian, T., Li, J., Song, Y., Zhou, L., Zhu, Z., Yang, C.J., 2016. Distance-based microfluidic quantitative detection methods for point-of-care testing. Lab. Chip. 16 (7), 1139–1151.

Uddin, R., Burger, R., Donolato, M., Fock, J., Creagh, M., Hansen, M.F., et al., 2016. Lab-on-a-disc agglutination assay for protein detection by optomagnetic readout and optical imaging using nano- and micro-sized magnetic beads. Biosensors Bioelectron. 85, 351–357.

Uttenthaler, E., Schräml, M., Mandel, J., Drost, S., 2001. Ultrasensitive quartz crystal microbalance sensors for detection of M13-phages in liquids. Biosensors Bioelectron. 16 (9-12), 735.

Vaughan, R.D., Guilbault, G.G., 2007. Piezoelectric immunosensors. Piezoelectric Sensors. Springer, Berlin, Heidelberg, pp. 237–280.

Vaughan, L., Milavetz, G., Ellis, E., Szefler, S., Conboy, K., Weinberger, M., et al., 1986. Multicentre evaluation of disposable visual measuring device to assay theophylline from capillary blood sample. The Lancet 327 (8474), 184–186.

Wang, Q., Li, R., Shao, K., Lin, Y., Yang, W., Guo, L., et al., 2017. A portable immunosensor with differential pressure gauges readout for alpha fetoprotein detection. Sci. Rep. 7, 45343.

Wang, A., Ma, X., Ye, Y., Luo, F., Guo, L., Qiu, B., et al., 2018. A simple and convenient aptasensor for protein using an electronic balance as a readout. Anal. Chem. 90 (2), 1087–1091.

Wang, W., Wu, W.Y., Wang, W., Zhu, J.J., 2010. Tree-shaped paper strip for semiquantitative colorimetric detection of protein with self-calibration. J. Chromatogr. A. 1217 (24), 3896–3899.

Wang, X., Yu, H., Lu, D., Zhang, J., Deng, W., 2014. Label free detection of the breast cancer biomarker CA15.3 using ZnO nanorods coated quartz crystal microbalance. Sens. Actuat. B Chem. 195, 630–634.

Williams, R.D., Upadhyayula, A.K., Bhethanabotla, V.R., 2007. High frequency thickness shear mode devices for organic vapor sensing. Sens. Actuat. B Chem. 122 (2), 635–643.

Xu, L., Yan, W., Ma, W., Kuang, H., Wu, X., Liu, L., et al., 2015. SERS encoded silver pyramids for attomolar detection of multiplexed disease biomarkers. Adv. Mater. 27 (10), 1706–1711.

Xu, Y., Zhang, Z., Ali, M.M., Sauder, J., Deng, X., Giang, K., et al., 2014. Turning tryptophanase into odor-generating biosensors. Angew. Chem. Int. Ed. Engl. 53 (10), 2620–2622.

Yamada, K., Henares, T.G., Suzuki, K., Citterio, D., 2015. Distance-based tear lactoferrin assay on microfluidic paper device using interfacial interactions on surface-modified cellulose. ACS Appl. Mater. Interfaces 7 (44), 24864.

Yang, W., Li, R., Wang, Q., Wei, Q., Fu, C., Lin, Z., et al., 2015. A micro-pressure sensor-based analytic platform and its application in thrombin quantification. Anal. Methods 7 (19), 7985–7988.

Yao, C., Ling, Q., Fu, W., 2013. Detection of fibrinogen and coagulation factor VIII in plasma by a quartz crystal microbalance biosensor. Sensors 13 (6), 6946–6956.

Yuan, Y., Ding, Z., Qian, J., Zhang, J., Xu, J., Dong, X., et al., 2016. Casp3/7-instructed intracellular aggregation of Fe3O4 nanoparticles enhances T2 MR imaging of tumor apoptosis. Nano. Lett. 16 (4).

Zhang, S., Ying, L., Yu, P., Tong, C., Zhou, W., Zhang, W., et al., 2015. In vitro release of cupric ion from intrauterine devices: influence of frame, shape, copper surface area and indomethacin. Biomed. Microdevices. 17 (1), 1–7.

Zhu, Z., Guan, Z., Jia, S., Lei, Z., Lin, S., Zhang, H., et al., 2014. Au@Pt nanoparticle encapsulated target-responsive hydrogel with volumetric bar-chart chip readout for quantitative point-of-care testing. Angew. Chem. 53 (46), 12503.

Zhu, Z., Guan, Z., Liu, D., Jia, S., Li, J., Lei, Z., et al., 2015. Translating molecular recognition into a pressure signal to enable rapid, sensitive, and portable biomedical analysis. Angew. Chem. 127 (36), 10594–10599.

Zou, Z., Buck, L.B., 2006. Combinatorial effects of odorant mixes in olfactory cortex. Science 311 (5766), 1477–1481.

Zuk, R.F., Ginsberg, V.K., Houts, T., Rabbie, J., Merrick, H., Ullman, E.F., et al., 1985. Enzyme immunochromatography--a quantitative immunoassay requiring no instrumentation. Clin. Chem. 31 (7), 1144–1150.

Chapter 12

Signal Amplification

Chang Feng[1] and Xiaoli Zhu[2]

[1]State Key Laboratory of Pharmaceutical Biotechnology, Department of Biochemistry, Nanjing University, Nanjing, P. R. China, [2]Center for Molecular Recognition and Biosensing, School of Life Sciences, Shanghai University, Shanghai, P. R. China

Chapter Outline

12.1 INTRODUCTION

Ultrasensitive detection of proteins is highly required in various application areas including clinical diagnostics, food safety, and environmental protection. The detection limit of these protein biosensors is continuously being pushed down by the improvement of existing analytical technologies. However, although the sensitive detection can now be achieved by a combined use of mass spectrometric techniques, various microscopies (TEM, SEM, AFM), surface plasmon resonance (SPR), and surface enhanced Raman spectroscopy (SERS), etc. (Chiang et al., 2011; Giessibl, 2003; Nie and Emory, 1997; Taniguchi et al., 2009), the ultimate challenge in sensing is to develop methodologies that permit detection of low abundance proteins in a complex mixture.

In general, the sensitivity of an analytical protocol is determined by the correlation between the analyte concentration and the intensity of output

Nano-inspired Biosensors for Protein Assay with Clinical Applications.
DOI: https://doi.org/10.1016/B978-0-12-815053-5.00012-X

signal (Scrimin and Prins, 2011). The importance of this correlation has been perfectly illustrated by the emergence of enzyme-linked immunosorbent assays (ELISAs) (Berg et al., 2015; Mugoni et al., 2013). In ELISA, molecular recognition is correlated with the activity of immobilized enzymes, which generate detectable signals (Haab, 2006). The conceptual novelty lies in an enzyme-based signal amplification strategy: the event of protein binding recruits reporter enzymes that are able to generate a multitude of reporter molecules. The increased concentration of reporter molecules renders detection using fluorescence or UV/Vis-spectroscopy feasible. However, the lowest detection limit by using ELISAs is typically in the picomole range, which is still far away from the requirements.

Biological assays need signal amplification to achieve high sensitivity, in order to detect low abundant bioanalytes. In this chapter, we intend to highlight the innovative methods to amplify weak input signals for ultrasensitive detection. Some of these methods depend on nucleic acid amplification, others are hybrid methods relying on the use of amplification components, such as enzymes and synthetic materials. As boundary conditions we have chosen to focus on molecular signal amplification, which implies we will discuss how a single analyte molecule is able to affect the properties of a multitude of reporter molecules. Moreover, highly sensitive analytical techniques relying on physical changes will be discussed. Nonetheless, the molecule-based approaches are so versatile that only key examples from the literature are discussed, chosen by the criterion of permitting an illustration of the underlying concept and a discussion of the advantages and critical issues. This chapter is divided into three parts (nucleic acid amplification-based signal amplification, enzyme-aided signal amplification, "1-to-N" binding-based signal amplification) relying on the signal amplification. These sections are preceded by a section in which signal amplification in the most frequently applied protocols is analyzed quantitatively.

12.2 NUCLEIC ACID-BASED SIGNAL AMPLIFICATION

Amplification techniques based on nucleic acids have enabled us to use a small amount of biological samples for molecular diagnosis, which are widely used in the field of recombinant DNA technologies and molecular biology (e.g., polymerase chain reaction, PCR) (Gill and Ghaemi, 2008). Therefore, the combined use of DNA amplification for protein detection could achieve high sensitivity when analyzing cancer-related biomarkers, infectious disease, and biochemical processes.

12.2.1 PCR-Based Signal Amplification

PCR is one of the widely used amplification techniques due to its high sensitivity and good reproducibility (Mullis and Faloona, 1987). The efficacy of

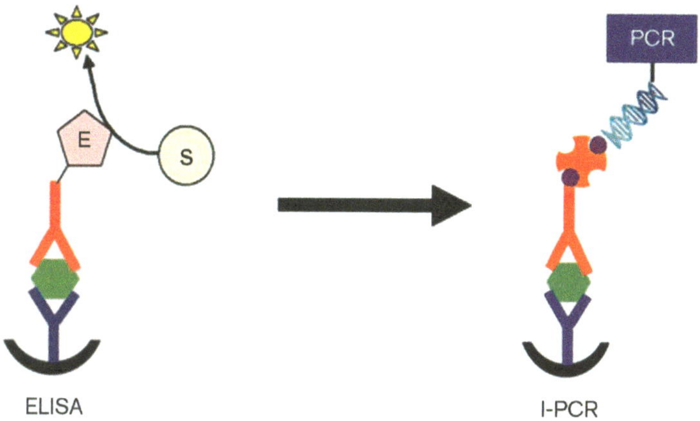

FIGURE 12.1 Schematic representation of the immune-PCR for protein detection. *Reprinted from Mehta, P.K., Raj, A., Singh, N.P., Khuller, G.K. 2014. Detection of potential microbial antigens by immuno-PCR (PCR-amplified immunoassay). J Med Microbiol. 63, 627–641.*

PCR is based on its ability to amplify a specific DNA segment through a pair of primers. The amplification capability of PCR allows the production of large amounts of specific DNA products, which can be detected by various methods (Erlich et al., 1991; Gibbs, 1991; Saiki et al., 1985). In 1992, Sano et al. described a new technique using PCR to detect specific proteins, which they called immuno-PCR. Immuno-PCR is basically similar to ELISA, which detects an antigen–antibody reaction, but instead of using an enzyme-conjugated antibody, the antibody is labeled with a DNA fragment, which can be amplified by using PCR (Fig. 12.1; Sano et al., 1992). Immuno-PCR combines the versatility of ELISA with the exponential amplification power and sensitivity of PCR, which is independent of surface, thus leading to an increase in sensitivity compared with traditional ELISA.

12.2.2 Isothermal Nucleic Acid Amplification

Although immuno-PCR was shown to be significantly more sensitive than ELISA, the requirements for thermal cycling and product separation by gel electrophoresis have restricted the widespread adoption of immuno-PCR as an alternative to ELISA and have precluded its utility in immunohistochemical or array formats. On the other hand, several isothermal amplification techniques have been developed in the two past decades without using a thermocycler machine, including helicase-dependent amplification (HDA), rolling circle amplification (RCA), and strand displacement amplification (SDA), such as enzyme-free SDA (e.g., toehold-mediated strand displacement (TMSD) amplification) and enzyme-mediated SDA (e.g., polymerase-mediated SDA) (Ali et al., 2014; Walker et al., 1992; Vincent et al., 2004; Zhang and Seelig, 2011). These non-PCR-based methods have been developed according to some new findings in molecular biology of DNA/RNA synthesis and some accessory proteins and their mimicking in vitro for nucleic acid amplification.

RCA is a enzymatic process that can be used to generate extremely long single-stranded DNA (ssDNA) with repeating sequences, which provide a large number of sites to link DNA detection probes. Aiming at the detection need of low abundant proteins and the problems existing in the amplification techniques by PCR and nanoparticle probes, Cheng et al. designed a cascade signal amplification strategy for ultrasensitive detection of proteins combining an advanced amplification technique, rolling circle amplification (RCA), with quantum dot tags as well as two general signal amplification methods including the multiplex binding of the biotin–strepavidin system and anodic stripping voltammetric detection (ASV) (Fig. 12.2; Cheng et al., 2010).

Additionally, different immunoassay-based methods have been devised to detect protein targets. Immuno-loop-mediated isothermal amplification (ILAMP) was proposed by Joo et al., and comprises of two main steps, including immune assay and loop-mediated isothermal amplification, followed by ultrasensitive detection of amplified signal (Fig. 12.3; Bourke et al., 2015; Hsieh et al., 2012; Notomi et al., 2000; Pourhassan-Moghaddam et al., 2013). Due to the rapidness, simplicity, and affordability with no need

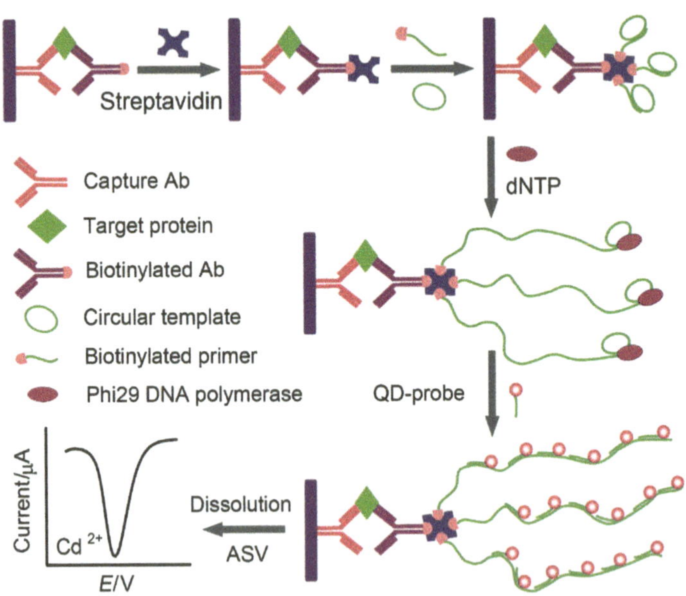

FIGURE 12.2 Schematic representation of the cascade signal amplification strategy for protein detection. *Reprinted from Cheng, W., Yan, F., Ding, L., Ju, H., Yin, Y., 2010. Cascade signal amplification strategy for subattomolar protein detection by rolling circle amplification and quantum dots tagging. Anal. Chem. 82, 3337–3342.*

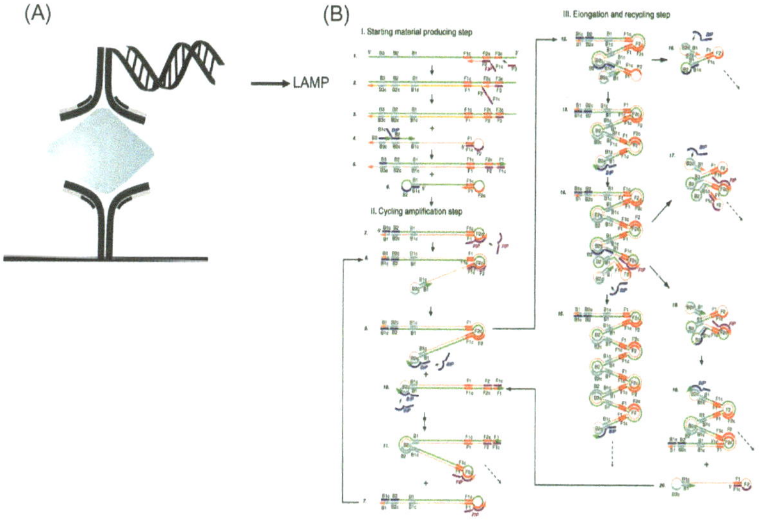

FIGURE 12.3 The principles of (A) iLAMP and (B) LAMP reaction. *Reprinted from (A) Pourhassan-Moghaddam, M., Rahmati-Yamchi, M., Akbarzadeh, A., Daraee, H., Nejati-Koshki, K., Hanifehpour, Y., et al., 2013. Protein detection through different platforms of immuno-loop-mediated isothermal amplification. Nanoscale Res. Lett. 8, 485. (B) Notomi, T., Okayama, H., Masubuchi, H., Yonekawa, T., Watanabe, K., Amino, N., et al., 2000. Loop-mediated isothermal amplification of DNA. Nucleic Acids Res. 28, E63.*

for expert personnel and thermal cycler instruments, the iLAMP method can be an important alternative for point-of-care diagnostic techniques, particularly in low-resource laboratories and poverty-stricken areas.

12.2.3 Enzyme-Free Nucleic Acid Amplification

Moreover, enzyme-free nucleic acid isothermal amplification strategies have been developed for better clinical performance. Li et al. reported a new signal amplification strategy for highly sensitive and enzyme-free methods to assay proteins based on the target-driven self-assembly of stacking deoxyribonucleic acids (DNA) on an electrode surface (Fig. 12.4; Cao et al., 2015). In the sensing process, the integration of target protein with the aptamer probe is used as a triggering point for a scheduled cycle of DNA hairpin assembly consisting of hybridization, displacement, and target regeneration. After repeated assembly, a large number of DNA duplexes can accordingly be formed on the electrode surface, and then switch on a succeeding propagation of self-assembled DNA concatemers that exhibit further signal enhancement. Catalytic hairpin assembly (CHA), a procedure of the accelerated hybridization between two DNA hairpins catalyzed by a DNA input, is one of the most widely studied and used enzyme-free catalytic DNA circuits for amplified detection of nucleic acids and small molecules (Liao et al., 2014). Therefore, Li et al. have

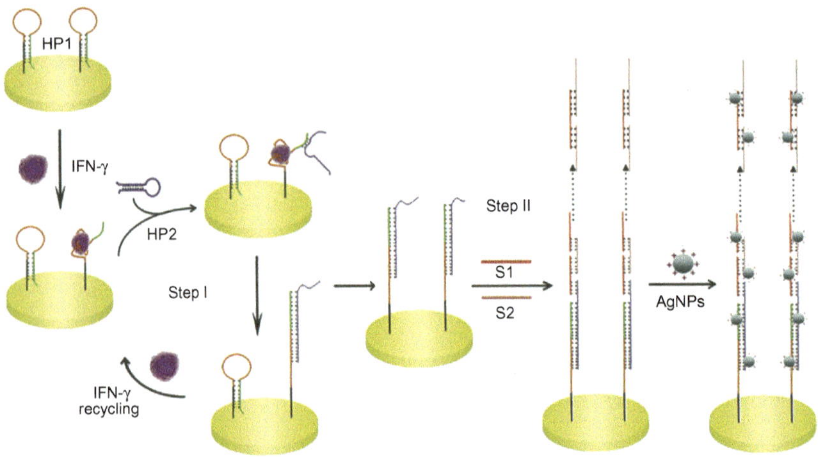

FIGURE 12.4 Schematic illustration of the mechanism to detect IFN-γ based on target-driven self-assembly of stacking DNA. *Reprinted from Cao, Y., Chen, W., Han, P., Wang, Z., Li, G., 2015. Target-driven self-assembly of stacking deoxyribonucleic acids for highly sensitive assay of proteins. Anal. Chim. Acta. 890, 1–6.*

successfully developed a universal strategy to construct protein-responsive CHA systems, which indicated that a rationally designed protein – DNA binding complex can be used as an effective catalyst to induce CHA reactions (Tang et al., 2015). The strategy is not only highly versatile and robust, but also compatible with a wide range of affinity interactions, and can easily be tailored for many practical applications, such as determination of serum and cellular proteins (Fig. 12.5).

12.3 ENZYME-BASED SIGNAL AMPLIFICATION

12.3.1 Natural Protein Enzyme

Catalysis using protein enzymes is the most widely used strategy for signal amplification. A few enzymes, e.g., horseradish peroxidase (HRP) and alkaline phosphatase (ALP), have been investigated. HRP (horseradish peroxidase) is a 44,173.9-Dalton glycoprotein with six lysine residues which can be conjugated to a labeled molecule. It produces a colored, fluorimetric, or luminescent derivative of the labeled molecule when incubated with a proper substrate, allowing it to be detected and quantified. It has been widely used in ELISA which is often used in conjugates (molecules that have been joined genetically or chemically) to determine the presence of a molecular target. In addition, various strategies based on the catalysis of HRP have been developed. For example, Gao et al. present a novel signal amplification technology (EASE) combining horseradish peroxidase (HRP) via ultrafast deposition of polydopamine that can be plugged into virtually all common

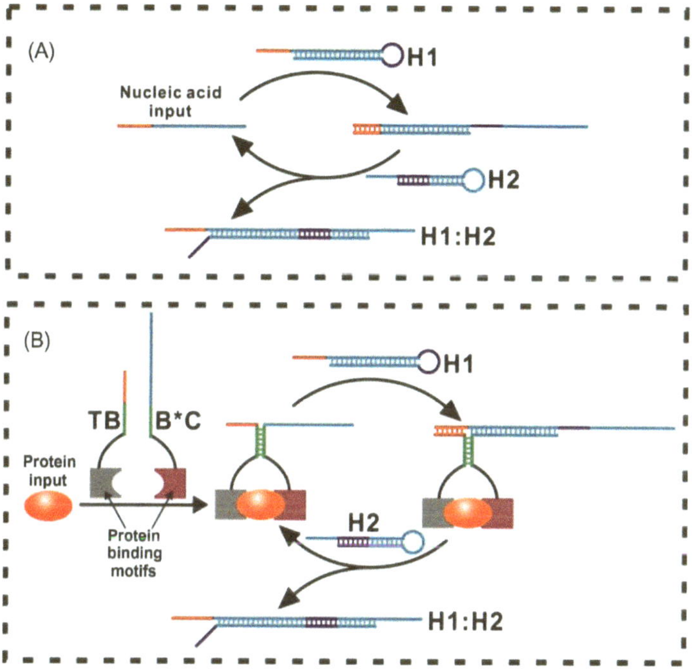

FIGURE 12.5 Schematic illustrating the principles of catalytic hairpin assembly (A) and protein-responsive catalytic hairpin assembly (B). *Reprinted from Tang, Y., Lin, Y., Yang, X., Wang, Z., Le, X.C., Li, F., 2015. Universal strategy to engineer catalytic DNA hairpin assemblies for protein analysis. Anal Chem. 87, 8063–8066.*

biodetection and bioimaging techniques, and can enhance their sensitivities by approximately three orders of magnitude (Li et al., 2017). In the first round of amplification, HRP molecules bound to the target can catalyze localized deposition of PDA. The PDA layer can in turn capture a large number of HRP molecules that are capable of catalyzing the conversion of chromogenic substrates which significantly improved the limit of detection by 1266-fold compared with conventional ELISA (Fig. 12.6). Additionally, through integrating with, for example, nucleic acid amplification, HRP may produce enhanced amplification efficiency which achieves the dual-amplification for the ultrasensitive detection of analytes (Cheng et al., 2014; Sheng et al., 2015). For instance, Ding et al. developed a simple, sensitive, and specific chemiluminescence immunoassay (CLIA) strategy for protein detection by integrating rolling circle amplification (RCA), multiplex binding of the biotin-streptavidin (B-SA) system, and enzymatic amplification. Thousands of repeated sequences are generated upon RCA for hybridizing with biotinylated detection probes (Chen et al., 2015). Then the enhanced chemiluminescence (CL) signal has been generated through binding of streptavidin horseradish peroxidases (ST-HRPs) and biotinylated detection probes,

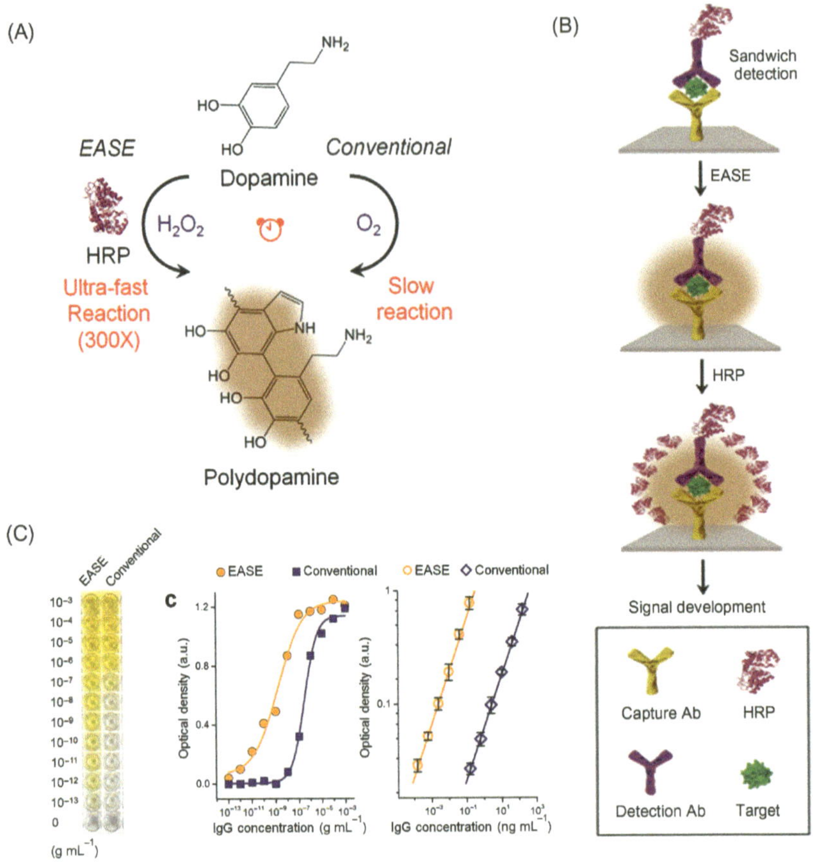

FIGURE 12.6 (A) Schematic illustration of the EASE technology. (B) Schematic illustration of the signal enhancement process. (C) Visual assessment of the detection sensitivity of ELISA-EASE using mouse IgG as a model target in comparison with conventional ELISA. *Reprinted from Li, J., Baird, M.A., Davis, M.A., Tai, W., Zweifel, L.S., Adams Waldorf, K.M., et al., 2017. Dramatic enhancement of the detection limits of bioassays via ultrafast deposition of polydopamine. Nat. Biomed. Eng. 1, 0082.*

which subsequently catalyze the oxidation of luminol by H_2O_2 and yield (Fig. 12.7). A simple, robust, cost-efficient, highly sensitive, and specific platform has been provided for the detection of human PRL, which may become a powerful tool for the detection of other protein targets by changing antibodies in the future.

Alkaline phosphatase (ALP) is a homodimeric protein enzyme that has the physiological role of dephosphorylating compounds under alkaline pH environments and is found across a multitude of organisms, prokaryotes and eukaryotes alike, with the same general function but in different structural forms suitable to the environment they function in. Alkaline phosphatase has

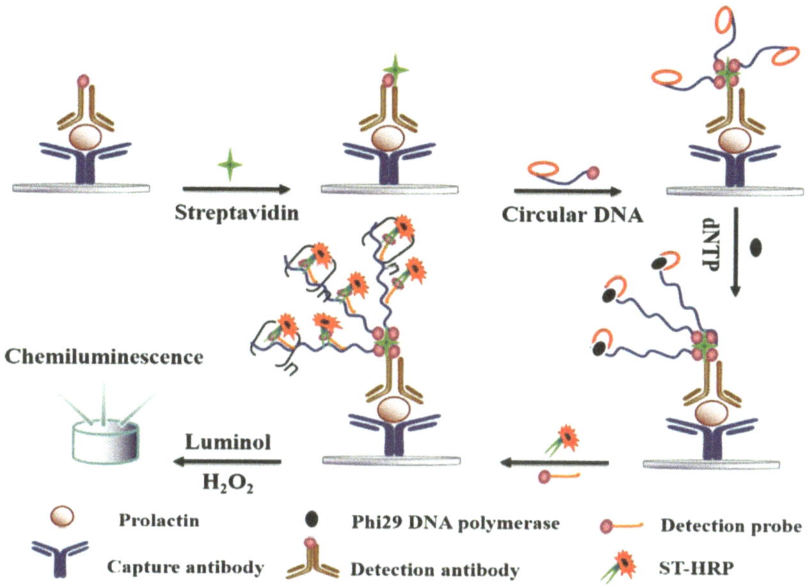

FIGURE 12.7 Schematic representation of chemiluminescence immunoassay for human prolactin based on rolling circle amplification and biotin–streptavidin system with enzymatic amplification. *Reprinted from Chen, H., Wu, S., Dong, F., Cheng, W., Li, Q., Ding, S., et al., 2015. A novel chemiluminescence immunoassay for highly sensitive and specific detection of protein using rolling circle amplification and the multiplex binding system. Sens. Actuat. B 221, 328–333.*

become a useful tool in molecular biology laboratories, since DNA normally possesses phosphate groups at the 5'-end. Another important use of alkaline phosphatase is as a label for enzyme immunoassays such as HRP, which is another popular enzyme used in bioanalysis, because of the inherent advantages of ALP including its high stability and activity and small size. For example, Lai et al. proposed an ultrasensitive multiplexed immunosensor by integrating ALP-labeled antibody functionalized gold nanoparticles (AuNPs) with enzyme AuNP-catalyzed deposition of silver nanoparticles at a disposable immunosensor array, achieving the dual detection of immunoglobulin G (Lai et al., 2011). Mori et al. developed an ALP-based technique to amplify the fluorescence signal through designing a fluorescent substrate that acquires membrane permeability upon dephosphorylation by ALP (Fig. 12.8; Nobori et al., 2018). This method could successfully amplify fluorescence signal to give a much stronger signal than the cells labeled using a conventional fluorophore-modified antibody. Moreover, other natural protein enzymes, such as glucose oxidase (GOx), with high catalytical efficiency have also been widely employed in signal amplification for the detection of protein by means of the enzymatic oxidation mechanism using optical and electrical methods (Liu et al., 2014).

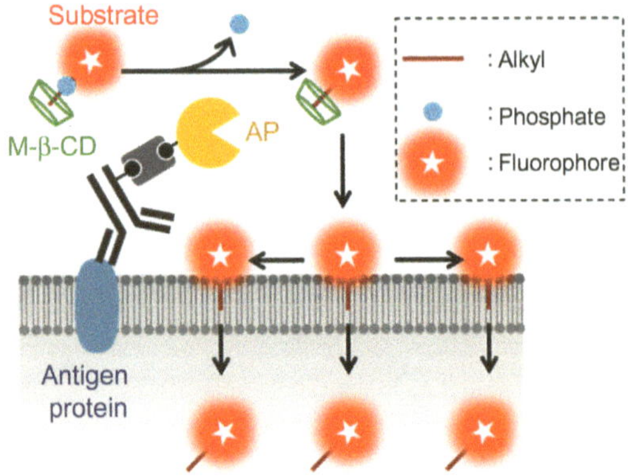

FIGURE 12.8 Mechanism of fluorescence signal amplification by catalyzed reporter deposition (CARP) method. *Reprinted from Nobori, T., Tosaka, K., Kawamura, A., Joichi, T., Kamino, K., Kishimura, A., et al., 2018. Alkaline phosphatase-catalyzed amplification of a fluorescence signal for flow cytometry. Anal. Chem. 90, 1059–1062.*

12.3.2 DNAzyme

Deoxyribozymes, also called DNA enzymes, DNAzymes, or catalytic DNA, are DNA oligonucleotides that are capable of performing a specific chemical reaction via catalytic function which can be use as amplifying labels for the development of optical or electronic sensors. Compared with protein enzymes and RNA ribozymes, DNAzyme has the advantages of better thermal stability, cost-efficiency, and constructability. The most commonly used DNAzyme in biosensors is hemin/G-quadruplex DNAzyme which was first proposed by Sen et al. in 2001 (Travascio et al., 2001). In this molecule, the complexation of hemin with a guanine-rich single-stranded nucleic acid possesses a G-quadruplex structure that catalyzes the oxidation of 2,2'-azino-bis (3-ethylbenzothiozoline)-6-sulfonic acid ($ABTS^{2-}$) by H_2O_2 to form the respective colored radical product, $ABTS^{-}$. In addition, the hemin/G-quadruplex structure can also catalyze the oxidation of luminol by H_2O_2 and the generation of chemiluminescence (Fig. 12.9; Willner et al., 2008). Therefore, this DNAzyme was used for the colorimetric or chemiluminescent detection of nucleic acids, or to follow the activity of telomerase, a versatile marker for cancer cells. For instance, a new binding-induced and label-free colorimetric method for protein detection has been developed on the basis of an autonomous assembly of hemin/G-quadruplex DNAzyme amplification strategy by Zou et al. (Wu et al., 2015). In the presence of target protein, a stable DNA–protein complex is formed when two proximity probes bind to the protein simultaneously. Then the complex triggers an autonomous cross-opening of the two functional hairpin structures, resulting in the formation of numerous hemin/G-quadruplex DNAzymes. The generated DNAzymes

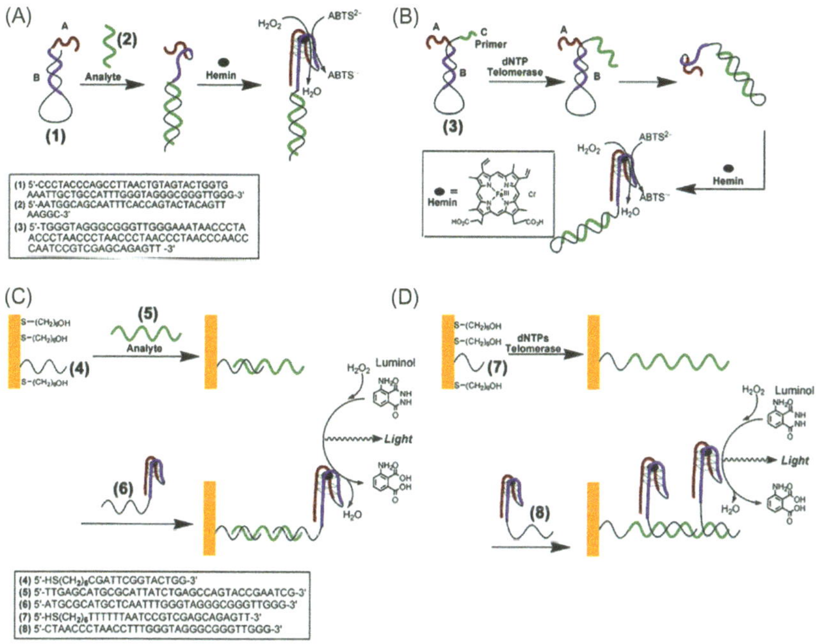

FIGURE 12.9 (A) Analysis of a DNA by a hairpin that generates the horseradish peroxidase-mimicking DNAzyme. (B) Detection of telomerase activity by a hairpin structure that yields the horseradish peroxidase-mimicking DNAzyme. (C) and (D) The chemiluminescence analysis of DNA or telomerase activity on surfaces by the use of nucleic acid-functionalized DNAzyme units as labels. *Reprinted from Willner, I., Shlyahovsky, B., Zayats, M., Willner, B. 2008. DNAzymes for sensing, nanobiotechnology and logic gate applications. Chem. Soc. Rev., 37(6), 1153–1165.*

catalyze the oxidation of colorless 2,2'-azino-bis (3-ethylbenzothiazoline-6-sulfonic acid) ($ABTS^{2-}$) to the green-colored $ABTS^{2-}$ with the presence of H_2O_2, thus producing the amplified colorimetric detection of target (Fig. 12.10). This method does not require any chemical modification of DNA, which makes it technically label-free and cost-effective.

Furthermore, DNAzymes exhibited nucleic acid cleavage activities in the presence of added cofactors, which have been used for the specific scission of DNA sequences. For example, nucleic acid sequences that specifically bind Pb^{2+}, Mg^{2+}, or Cu^{2+} ions, UO^{2+} or histidine could yield supramolecular coiled structures that cleave specific DNA sequences (Fig. 12.11). Utilizing these properties, predesigned nucleic acid sequences were tethered to the DNAzyme structures to yield functional units for the amplified detection of the cofactors and for target DNA/RNA, or alternatively protein. For example, by adopting a well-designed DNAzyme, Li et al. proposed an integrated strategy to comprehensively analyze tumor-associated membrane

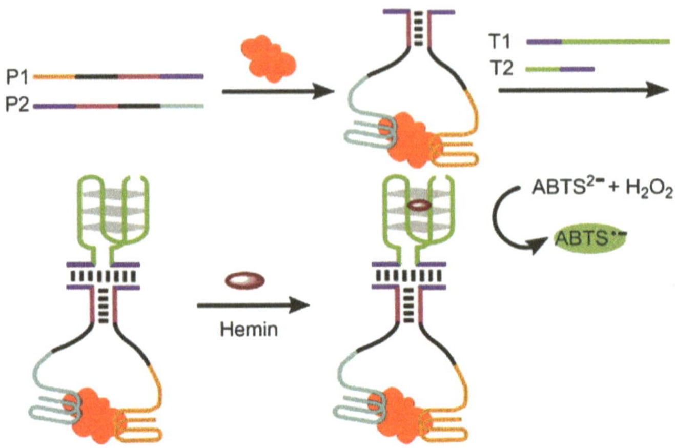

FIGURE 12.10 Schematic illustration of binding-induced colorimetric assay for thrombin detection without amplification. *Reprinted from Wu, H., Zhang, K., Liu, Y., Wang, H., Wu, J., Zhu, F., et al., 2015. Binding-induced and label-free colorimetric method for protein detection based on autonomous assembly of hemin/G-quadruplex DNAzyme amplification strategy. Biosens. Bioelectron. 64, 572–578.*

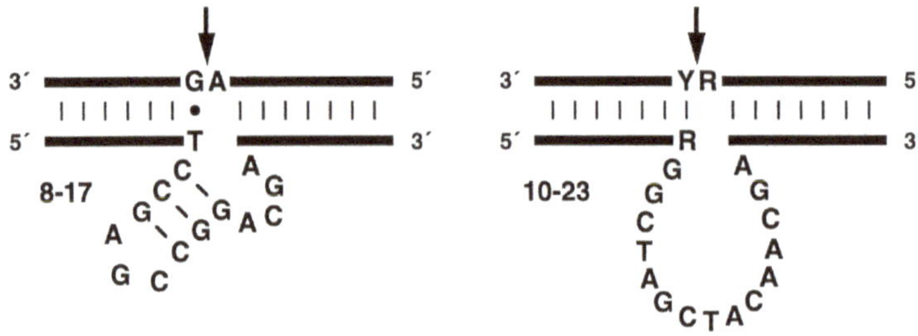

FIGURE 12.11 Composition of the DNAzyme catalytic motifs. The DNA enzyme (bottom strand) binds the RNA substrate (top strand) through Watson–Crick pairing. Cleavage occurs at the position indicated by the arrow. R 5 A or G; Y 5 U or C. *Reprinted from Santoro, S.W., Joyce, G.F., 1997. A general purpose RNA-cleaving DNA enzyme. Proc. Natl. Acad. Sci. U. S. A. 94, 4262–4266.*

proteins (TMPs), which achieved in situ imaging and amplified detection of TMPs in a single system and was capable of providing both the locational and the quantitative information of TMPs (Fig. 12.12; Chen et al., 2018). In comparison with some conventional methods, this DNAzyme-based method possesses the advantages of two functions in one integrated system (two-in-one): high sensitivity and nondestructivity. Therefore, the concept of dual-labeling in this work also expands its application and provides a reference for the research on DNAzyme.

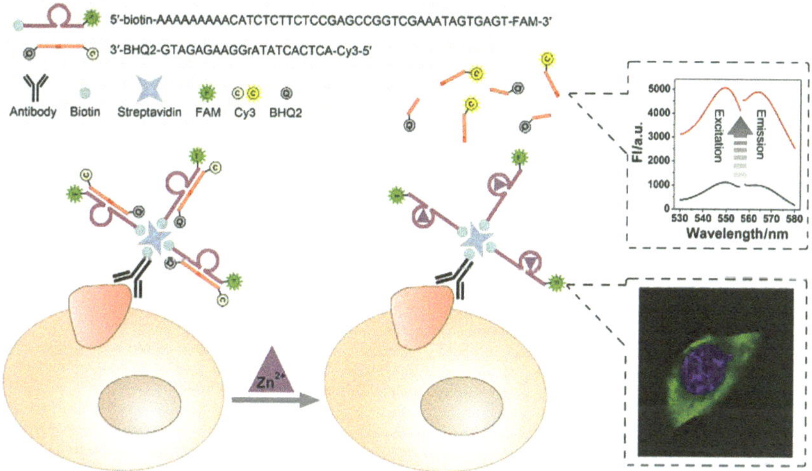

FIGURE 12.12 Scheme of the DNAzyme-based nondestructive analysis of tumor-associated membrane protein integrating imaging and amplified detection. *Reprinted from Chen, X., Zhao, J., Chen, T., Gao, T., Zhu, X., Li, G., 2018. Nondestructive analysis of tumor-associated membrane protein integrating imaging and amplified detection in situ based on dual-labeled DNAzyme. Theranostics 8, 1075–1083.*

12.3.3 Nucleic Acid Tool Enzyme

In the above section "Nucleic Acid-Based Signal Amplification," for both PCR and isothermal nucleic acid amplification, DNA polymerase is the key element. In fact, other kinds of nucleic acid tool enzymes also can be involved in signal amplification. Nucleic acid tool enzymes can be divided briefly into two categories: exonuclease and endonuclease. For the former, exonucleases are enzymes that work by cleaving nucleotides one at a time from the end (exo) of a polynucleotide chain. A hydrolyzing reaction that breaks phosphodiester bonds at either the 3'- or the 5'-end occurs in double-strand DNA (dsDNA), or ssDNA. In addition, duplex specific nuclease (DSN), obtained from the hepatopancreas of the Kamchatka crab (*Paralithodes camtschaticus*), can hydrolyze dsDNA or DNA in DNA–RNA heteroduplexes, and is practically inactive toward ssDNA, or single- or double-stranded RNA (Shagin et al., 2002). Therefore, DSN was utilized to recycle the target-assisted cleavage of Taqman probes, leading to significant fluorescence signal amplification. For example, Zhang et al. proposed a new method for sensitive detection of microphthalmia-associated transcription factor (MITF) based on "OFF-state" and "ON-state" equilibrium of a well-designed probe and duplex-specific nuclease signal amplification (Fig. 12.13; Zhang et al., 2017). By taking advantage of the high amplification efficiency of DSN-aided recycling, high sensitivity of MITF is realized with a detection limit as low as 1.1 pM.

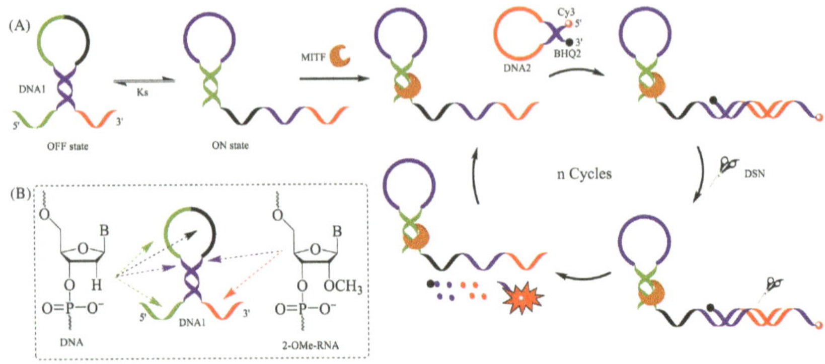

(C) DNA1: 5'-CACGTGGAGGATCCCACGTGTCTTCAGGATCCTCAGCAGCATG-3'

DNA2: 5'- Cy3-GGATCCCATGCTGCTGAGGATCC-BHQ2-3'

FIGURE 12.13 (A) Schematic diagram showing the principle of DSN-based amplification fluorescence assay for the detection of MITF. (B) The structure of the DNA1: the sequences of DNA or 2-OMe-RNA were marked. (C) The oligonucleotides sequence used in this strategy. The colors of the sequences are the same as given in part A and part B and the underlined parts mark the hybridization part of DNA1 and DNA2. *Reprinted from Zhang, K., Wang, K., Zhu, X., Xie, M., Zhang, X., 2017. A new method for sensitive detection of microphthalmia-associated transcription factor based on "OFF-state" and "ON-state" equilibrium of a well-designed probe and duplex-specific nuclease signal amplification. Biosens. Bioelectron. 87, 299–304.*

Endonucleases are enzymes that cleave the phosphodiester bond within a polynucleotide chain such as Deoxyribonuclease I which cuts DNA relatively nonspecifically (without regard to sequence), while many, typically called restriction endonucleases or restriction enzymes, cleave only at very specific nucleotide sequences (Meselson and Yuan, 1968). Endonucleases differ from exonucleases, which cleave the ends of recognition sequences instead of the middle (endo) portion. Nicking endonucleases, known as type II restriction enzymes, are a special family of restriction endonucleases. Like restriction endonucleases, they recognize short specific DNA sequences and cleave DNA at a fixed position relative to the recognition sequence (Xu et al., 2007). However, unlike other restriction endonucleases, nicking endonucleases cleave only one specific strand instead of both strand. In recent years, nicking enzymes have attracted considerable research efforts due to their excellent capability to mediate signal amplification. Some novel methods called nicking endonuclease signal amplification (NESA) have been developed and successfully applied for the highly sensitive detection of nucleic acids, potassium, proteins, and small molecules (Li et al., 2008; Li et al., 2012; Zhu et al., 2011; Xu et al., 2011). For example, a new protein assay is developed on the basis of hairpin probe and nicking enzyme-assisted signal amplification strategy (Fig. 12.14; Xue et al., 2012). The metastable state hairpin probe with short loop and long stem is designed to contain a protein

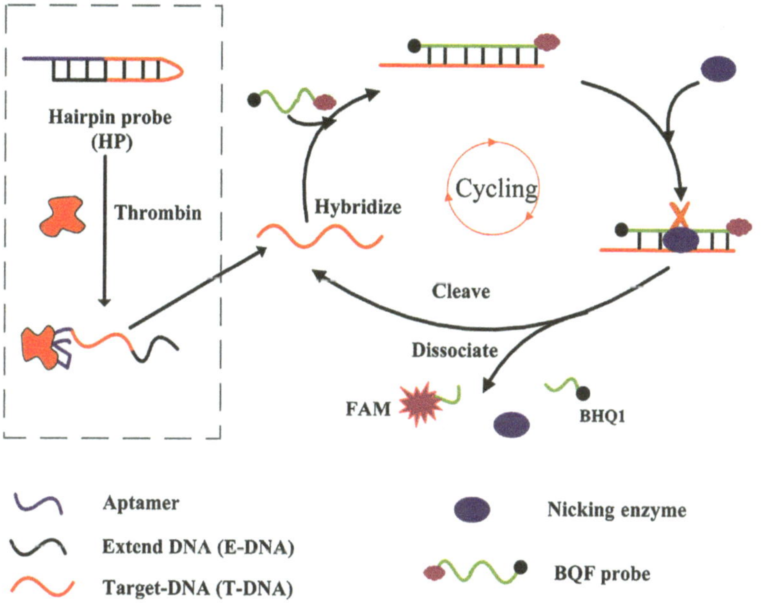

FIGURE 12.14 Schematic representation of homogeneous aptamer and nicking enzyme-assisted fluorescence signal amplification (NEFSA) assay for protein. *Reprinted from Xue, L., Zhou, X., Xing, D., 2012. Sensitive and homogeneous protein detection based on target- triggered aptamer hairpin switch and nicking enzyme assisted fluorescence signal amplification. Anal. Chem. 84, 3507–3513.*

aptamer for target protein–thrombin recognition. Introduction of target protein into the assay leads to the formation change of the hairpin probe from hairpin shape to open form, thus facilitating the hybridization between the hairpin probe and BQF probe. The fluorescence signal is amplified through continuous enzyme cleavage which can detect thrombin specifically with a detection limit as low as 100 pM.

With demand for the detection of ultratrace-level protein growing, higher amplification is greatly and even urgently required. By combining the enzyme with other enzymes such as polymerase through delicate design, it is expected to achieve more efficient signal amplification. On the basis of our previously developed net-like rolling circle amplification technique (NRCA) (Zhu et al., 2015), Li et al. proposed an immuno-NRCA strategy for platelet-derived growth factor (PDGF) detection (Feng et al., 2017). In this strategy, the events of antigen–antibody reactions on the interface can be transferred to highly efficient isothermal nucleic acid amplification in solution. Therefore, this system possesses the advantages of both solution- and interface-based assays (Fig. 12.15). That is, this system can be applied for complex samples without pretreatment, and has an ultrahigh sensitivity in the meantime. Benefiting from the ultrahigh sensitivity, only one drop of

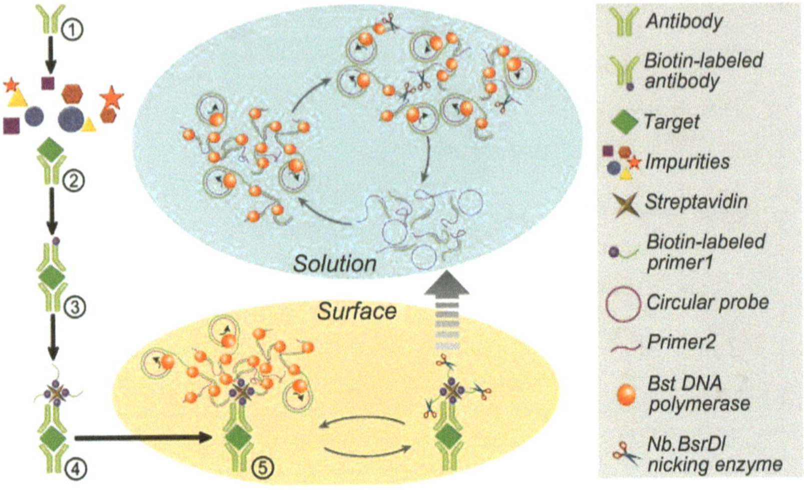

FIGURE 12.15 Schematic presentation of the principle of immuno-NRCA assay. *Reprinted from Feng, C., Bo, B., Mao, X., Shi, H., Zhu, X., Li, G., 2017. From interface to solution: integrating immunoassay with netlike rolling circle amplification for ultrasensitive detection of tumor biomarker. Theranostics 7, 31–39.*

blood is required in the assay. Also, owing to the isothermal amplification, this system has no requirement for a thermal cycler, making it cost-efficient and easy to operate.

12.3.4 Nanozyme

Over the past few decades, researchers have established artificial enzymes as highly stable and low-cost alternatives to natural enzymes in a wide range of applications. However, some nanomaterials, such as fullerene derivatives, gold nanoparticles, rare earth nanoparticles, and ferromagnetic nanoparticles, have been found to exhibit unexpected enzyme-like activity (Chen et al., 2006; Natalio et al., 2012; Wang et al., 2012; Wei and Wang, 2008). The term "nanozymes" was initially coined by Scrimin, Pasquato, and coworkers to describe their thiol monolayer protected gold clusters with outstanding ribonuclease-like activity (Manea et al., 2004). Since then, considerable advances have been made in this area due to the tremendous progress in nanoresearch and the unique characteristics of nanomaterials (Fig. 12.16; Wei and Wang, 2013). These nanomaterial-based artificial enzymes (nanozymes) have already found wide applications in numerous fields, including biosensing, immunoassays, cancer diagnostics and therapy, neuroprotection, stem cell growth, and pollutant removal.

Among these, magnetic nanoparticles possess intrinsic peroxidase mimicking activity which has allowed their application in detecting proteins due

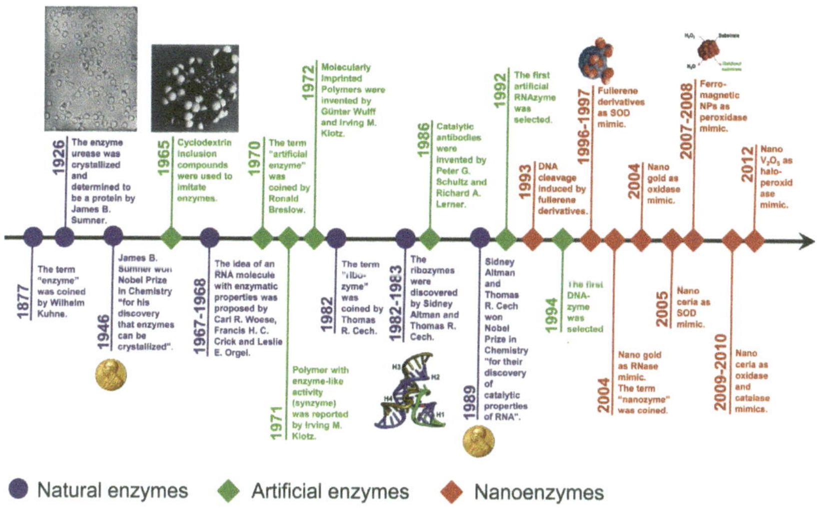

FIGURE 12.16 A brief timeline for the development of enzymes. *Reprinted from Wei, H., Wang, E., 2013. Nanomaterials with enzyme-like characteristics (nanozymes): next-generation artificial enzymes. Chem. Soc. Rev. 42, 6060–6093.*

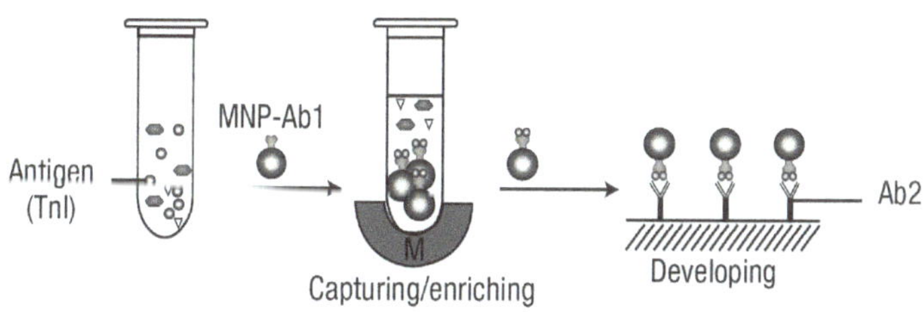

FIGURE 12.17 Schematic representation for Fe_3O_4 MNPs-based immunoassay. *Reprinted from Gao, L.Z., Zhuang, J., Nie, L., Zhang, J.B., Zhang, Y., Gu, N., et al., 2007. Intrinsic peroxidase-like activity of ferromagnetic nanoparticles. Nat. Nanotechnol. 2, 577–583.*

to their low-cost and robust activity (Xie et al., 2012). For example, Fe_3O_4 MNPs are highly effective as a catalyst due to a higher binding affinity for the substrate 3,3,5,5-tetramethylbenzidine than horseradish peroxidase (HRP). At the same molar concentration, the Fe_3O_4 MNPs exhibited an activity level 40 times higher than HRP. Therefore, a novel immunoassay has been developed based on this finding, in which antibody-modified MNPs provide three functions: capture, separation, and detection (Fig. 12.17; Gao et al., 2007). Additionally, AuNPs-catalyzed oxidation of glucose is combined with a HRP-catalyzed reaction as well as an electrocatalytic reaction to result in cascade reactions in the electrolyte. Thus, the intensity of the electrocatalytic signals has a quantitative relation with the concentration of glucose, and favors the sensitive detection of glucose. Li et al. have

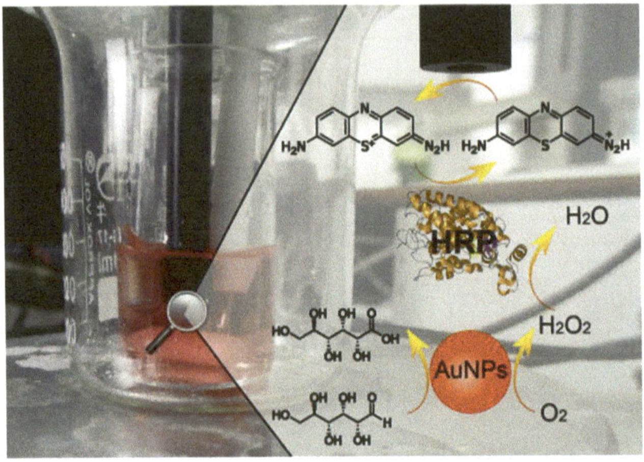

FIGURE 12.18 Schematic illustration of the colloidal AuNPs-based electrochemical detection system. *Reprinted from Chen, G., Tong, H., Gao, T., Chen, Y., Li, G., 2014. Direct application of gold nanoparticles to one-pot electrochemical biosensors. Anal. Chim. Acta. 849, 1–6.*

successfully integrated the electrochemical technique with colloidal AuNPs to develop a sensitive and reusable glucose biosensor (Fig. 12.18; Chen et al., 2014). Citrate-stabilized colloidal AuNPs are employed as the electrolyte directly for the first time, while the electrode is unmodified. Making use of cascade reactions initiated by the catalysis of AuNPs and ended by electrocatalysis, sensitivity detection of glucose is achieved. Furthermore, cerium oxide nanoparticles (nanoceria) possess intrinsic oxidase-like activity at acidic pH values, and can quickly oxidize a series of organic substrates without any oxidizing agent (e.g., hydrogen peroxide) (Asati et al., 2011; Tarnuzzer et al., 2005). The unique pH-dependent oxidase-like activity of cerium oxide nanoparticles in aqueous media makes them a powerful tool for a wide range of potential applications in biotechnology and environmental chemistry. For instance, Perez et al. proposed a nanoceria-based detection approach that performs an immunoassay and can identify the presence and concentration of a target faster and cheaper than using traditional ELISA (Fig. 12.19; Asati et al., 2009). In nanoceria-based ELISA, the oxidase-like activity of nanoceria facilitates the direct oxidation of TMB without the need for HRP or hydrogen peroxide which makes nanoceria an effective nanocatalyst and detection tool in immunoassays.

12.4 "1-TO-N" BINDING-BASED SIGNAL AMPLIFICATION

12.4.1 Biotin–Streptavidin System

Biotin–streptavidin system (BAS) has been introduced and developed for application purposes and to try to understand the special nature of this

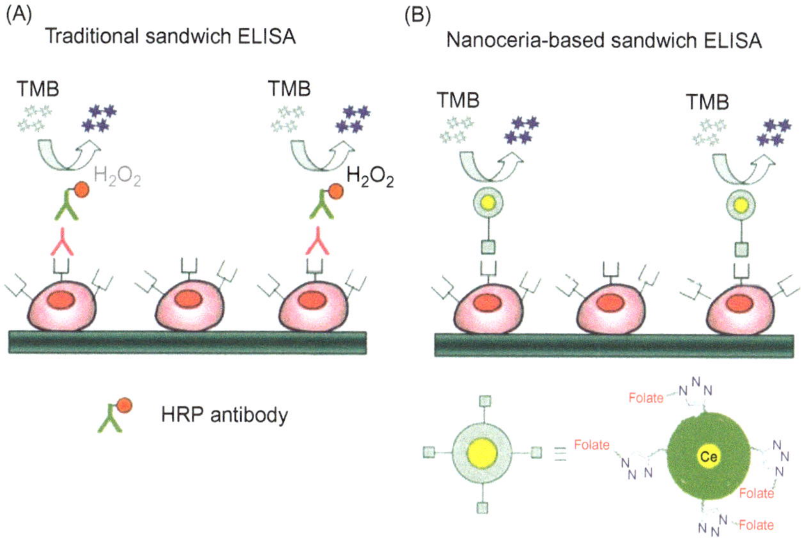

FIGURE 12.19 Comparison of traditional ELISA (A) and nanoceria-based ELISA (B). *Reprinted from Asati, A., Santra, S., Kaittanis, C., Nath, S., Perez, J.M., 2009. Oxidase-like activity of polymer- coated cerium oxide nanoparticles. Angew Chem., 121(13), 2344–2348.*

ultrahigh-affinity interaction (Fig. 12.20). Streptavidin is composed of four identical subunits and each subunit binds a single biotin molecule exhibiting the epitome of high-affinity protein–ligand interactions, which is widely used in western blotting and immunoassays conjugated to some reporter molecule, such as ELISA and makes "1-to-3" signal amplification achieved. For example, Yuan et al. developed a signal amplificatory electrochemical immunoassay with biotin–streptavidin conjunction and multienzymatic-based substrate recycling. (Fig. 12.21; Yuan et al., 2010). Owing to four identical binding sites of streptavidin to biotin, amounts of biotinylated alkaline phosphatase (bio-AP) were attached, and this "1:3" signal amplification strategy improved the catalytic performance and detection limit of the proposed immunosensor.

12.4.2 Nanocarrier

With the achievements of nanotechnology and nanoscience, nanomaterial-based signal amplifications hold great promise in realizing high sensitivity and selectivity for in situ or online detection of biomolecules due to the rapid analysis procedure and easy miniaturization (Lei and Ju, 2012). A lot of nanomaterials, especially carbon and metal-based NPs are excellent candidates as carriers to enhance the probe area due to their unique properties, such as high specific surface area, which can be employed as supports at the

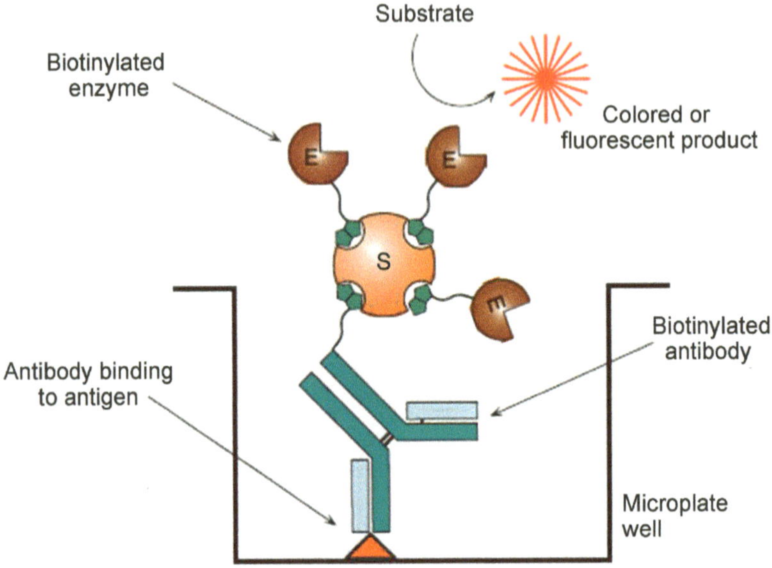

FIGURE 12.20 Biotin–streptavidin system.

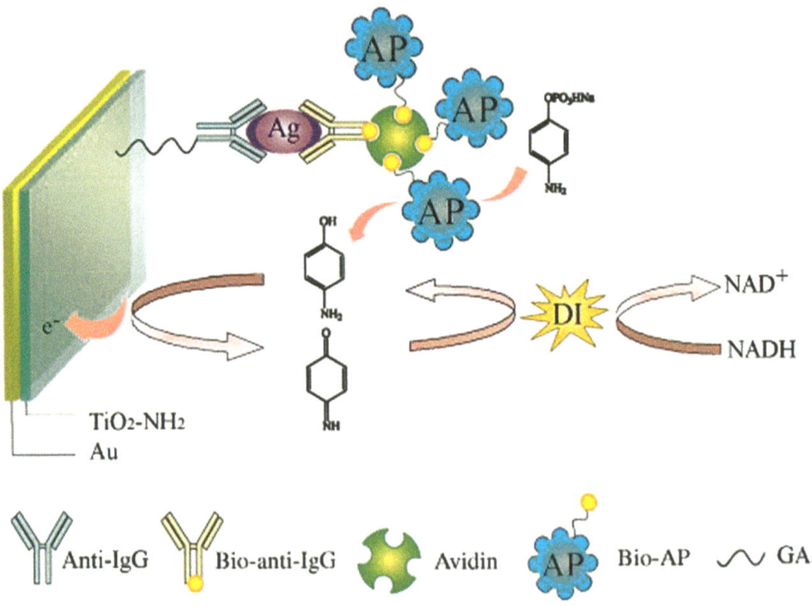

FIGURE 12.21 Schematic illustration of assembly process with the enzyme bioaffinity immunosensor. *Reprinted from Yuan, Y., Yuan, R., Chai, Y., Zhuo, Y., Bai, L., Liao, Y., 2010. An electrochemical enzyme bioaffinity electrode based on biotin-streptavidin conjunction and bienzyme substrate recycling for amplification. Anal. Biochem. 405, 121–126.*

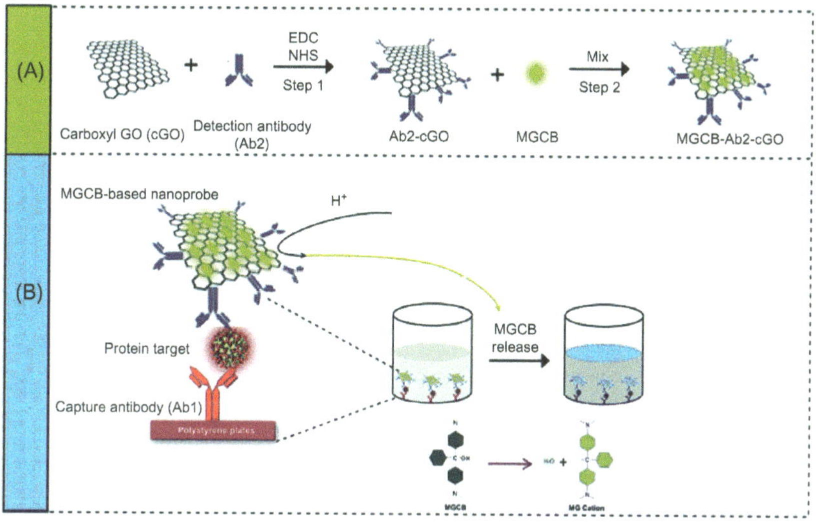

FIGURE 12.22 Schematic representation of the allochroic-cGO linked immunosorbent assay (ALISA) for protein detection performed in 96-well polystyrene (PS) plates. *Reprinted from Li, C., Yang, Y., Wu, D., Li, T., Yin, Y., Li, G., 2016. Improvement of enzyme-linked immunosorbent assay for the multicolor detection of biomarkers. Chem. Sci. 7, 3011–3016.*

bottom for concentrating the capture molecules and/or largely loading signal molecules. For example, Li et al. report a new signal amplification scheme based on allochroic molecule-modified carboxyl graphene oxide (cGO), which can be used to develop a multicolor immunoassay named as allochroic-cGO linked immunosorbent assay (ALISA) (Li et al., 2016). They took advantage of high adsorption levels and a wide selection of allochroic molecules through π-π stacking interactions with allochroic-cGO that can be released into the solution and lead to a strong color change, thus the simultaneous colorimetric detection based on the amplification ratio of target to signal being 1-to-N could be successfully achieved and the detection limit reached a picogram level for the first time (Fig. 12.22). In addition, due to the small size, high surface-to-volume ratio, and good biocompatibility, AuNPs have been also used as carriers of the signaling molecules for amplification detection of DNA and protein targets. For example, using fluorescent (Foligo) on AuNPs as a reporter, a new signal amplification strategy has realized sensitive detection of target proteins by designing a novel DNA nanomachine (Fig. 12.23; Zhang et al., 2015). This nanomachine achieves high density, three-dimensional DNA tracks on AuNPs and the operation of the nanomachine, powered by enzymatic cleavage of conjugated oligonucleotides, cleaves hundreds of oligonucleotides in response to a single protein binding event, enhancing the sensitivity.

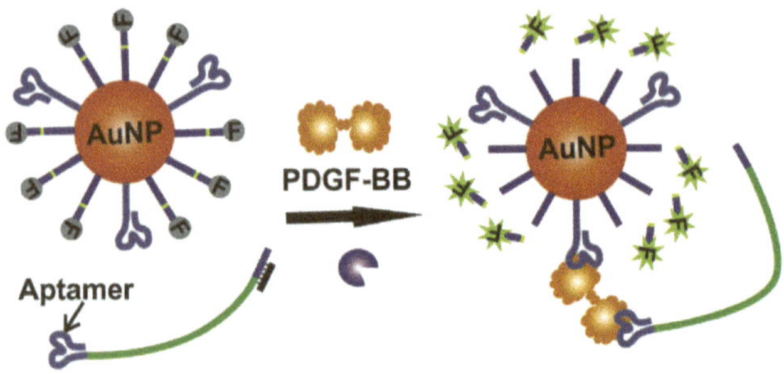

FIGURE 12.23 Schematic of the nanomachine responsive to PDGF-BB. *Reprinted from Zhang, H., Lai, M., Zuehlke, A., Peng, H., Li, X.F., Le, X.C., 2015. Binding-induced DNA nanomachines triggered by proteins and nucleic acids. Angew. Chem. Int. Ed. 54, 14326–14330.*

12.5 CONCLUSION

In this chapter, we have highlighted several examples to summarize signal amplification strategies for protein assays. Among these, various isothermal nucleic acid amplification techniques and signal amplification strategies have been successively developed for the simple and rapid detection of nucleic acids with remarkably high sensitivity. Benefiting from rapid advances in biotechnology, chemistry, and nanotechnology, these signal amplification methods have been expanded to detect targets ranging from DNA and RNA to cells, proteins, small molecules, and even ions. In addition, signal amplification can also be accomplished to generate an amplified signal via natural enzymes, DNAzymes, nucleases, and nanoenzymes, which depend on the high catalytical efficiency of the enzymes. Therefore, the enzymatically catalyzed amplification strategy has enabled the signal amplification for the ultrasensitive detection of protein. Moreover, the biofunctionalization of nanomaterials have become one of the hottest fields of research due to the need for ultrasensitive bioassays through signal amplification. The biofunctional nanoparticles can be used as carriers or tracers, catalysts, electronic conductors, and optical emitters, and can produce a synergic effect among catalytic activity, conductivity, and biocompatibility to perform significant signal amplification. With the demand in life sciences and clinical diagnosis, the ultimate goal of this field is the utilization of nanomaterials and biomaterials which not only enhance the biosensing capabilities compared with conventional platforms, but also bring out new approaches, such as miniaturization, reagent-less biosensing, and single-molecule detection. In conclusion, with the development of physics, chemistry, biology, and nanoscience, new avenues for more intelligent signal amplification techniques for multiplexed protein detection are expected, and simple strategies with tremendous amplification will be developed and applied to track biomarkers in vivo.

REFERENCES

Ali, M.M., Li, F., Zhang, Z., Zhang, K., Kang, D.K., Ankrum, J.A., et al., 2014. Rolling circle amplification: a versatile tool for chemical biology, materials science and medicine. Chem. Soc. Rev. 43, 3324–3341.

Asati, A., Santra, S., Kaittanis, C., Nath, S., Perez, J.M., 2009. Oxidase-like activity of polymer-coated cerium oxide nanoparticles. Angew Chem. 121 (13), 2344–2348.

Asati, A., Kaittanis, C., Santra, S., Perez, J.M., 2011. pH-tunable oxidase-like activity of cerium oxide nanoparticles achieving sensitive fluorigenic detection of cancer biomarkers at neutral pH. Anal. Chem. 83, 2547–2553.

Berg, B., Cortazar, B., Tseng, D., Ozkan, H., Feng, S., Wei, Q., et al., 2015. Cellphone-based hand-held microplate reader for point-of-care testing of enzyme-linked immunosorbent assays. ACS Nano 9, 7857–7866.

Bourke, T.W., McKenna, J.P., Coyle, P.V., Shields, M.D., Fairley, D.J., 2015. Diagnostic accuracy of loop-mediated isothermal amplification as a near-patient test for meningococcal disease in children: an observational cohort study. Lancet Infect. Dis. 15, 552–558.

Cao, Y., Chen, W., Han, P., Wang, Z., Li, G., 2015. Target-driven self-assembly of stacking deoxyribonucleic acids for highly sensitive assay of proteins. Anal. Chim. Acta. 890, 1–6.

Chen, G., Tong, H., Gao, T., Chen, Y., Li, G., 2014. Direct application of gold nanoparticles to one-pot electrochemical biosensors. Anal. Chim. Acta. 849, 1–6.

Chen, H., Wu, S., Dong, F., Cheng, W., Li, Q., Ding, S., et al., 2015. A novel chemiluminescence immunoassay for highly sensitive and specific detection of protein using rolling circle amplification and the multiplex binding system. Sens. Actuat. B 221, 328–333.

Chen, J., Patil, S., Seal, S., McGinnis, J.F., 2006. Rare earth nanoparticles prevent retinal degeneration induced by intracellular peroxides. Nat. Nanotechnol. 1, 142–150.

Chen, X., Zhao, J., Chen, T., Gao, T., Zhu, X., Li, G., 2018. Nondestructive analysis of tumor-associated membrane protein integrating imaging and amplified detection in situ based on dual-labeled DNAzyme. Theranostics 8, 1075–1083.

Cheng, W., Yan, F., Ding, L., Ju, H., Yin, Y., 2010. Cascade signal amplification strategy for subattomolar protein detection by rolling circle amplification and quantum dots tagging. Anal. Chem. 82, 3337–3342.

Cheng, W., Zhang, W., Yan, Y., Shen, B., Zhu, D., Lei, P., et al., 2014. A novel electrochemical biosensor for ultrasensitive and specific detection of DNA based on molecular beacon mediated circular strand displacement and rolling circle amplification. Biosens. Bioelectron. 62, 274–279.

Chiang, C.K., Chen, W.T., Chang, H.T., 2011. Nanoparticle-based mass spectrometry for the analysis of biomolecules. Chem. Soc. Rev. 40, 1269–1281.

Erlich, H.A., Gelfand, D., Sninsky, J.J., 1991. Recent advances in the polymerase chain reaction. Science 252, 1643–1651.

Feng, C., Bo, B., Mao, X., Shi, H., Zhu, X., Li, G., 2017. From interface to solution: integrating immunoassay with netlike rolling circle amplification for ultrasensitive detection of tumor biomarker. Theranostics 7, 31–39.

Gao, L.Z., Zhuang, J., Nie, L., Zhang, J.B., Zhang, Y., Gu, N., et al., 2007. Intrinsic peroxidase-like activity of ferromagnetic nanoparticles. Nat. Nanotechnol. 2, 577–583.

Gibbs, R.A., 1991. Polymerase chain reaction techniques. Curr. Opin. Biotechnol. 2, 69–75.

Giessibl, F.J., 2003. Advances in atomic force microscopy. Rev. Mod. Phys. 75, 949–983.

Gill, P., Ghaemi, A., 2008. Nucleic acid isothermal amplification technologies: a review. Nucleos. Nucleot. Nucl. 27, 224–243.

Haab, B.B., 2006. Applications of antibody array platforms. Curr. Opin. Biotechnol. 17, 415–421.

Hsieh, K., Patterson, A.S., Ferguson, B.S., Plaxco, K.W., Soh, H.T., 2012. Rapid, sensitive, and quantitative detection of pathogenic DNA at the point of care through microfluidic electrochemical quantitative loop-mediated isothermal amplification. Angew. Chem. Int. Ed. 51, 4896–4900.

Lai, G., Yan, F., Wu, J., Leng, C., Ju, H., 2011. Ultrasensitive multiplexed immunoassay with electrochemical stripping analysis of silver nanoparticles catalytically deposited by gold nanoparticles and enzymatic reaction. Anal. Chem. 83, 2726–2732.

Lei, J., Ju, H., 2012. Signal amplification using functional nanomaterials for biosensing. Chem. Soc. Rev. 41, 2122–2134.

Li, C., Yang, Y., Wu, D., Li, T., Yin, Y., Li, G., 2016. Improvement of enzyme-linked immunosorbent assay for the multicolor detection of biomarkers. Chem. Sci. 7, 3011–3016.

Li, J., Fu, H.E., Wu, L.J., Zheng, A.X., Chen, G., Yang, H.H., 2012. General colorimetric detection of proteins and small molecules based on cyclic enzymatic signal amplification and hairpin aptamer probe. Anal. Chem. 84, 5309–5315.

Li, J., Baird, M.A., Davis, M.A., Tai, W., Zweifel, L.S., Adams Waldorf, K.M., et al., 2017. Dramatic enhancement of the detection limits of bioassays via ultrafast deposition of polydopamine. Nat. Biomed. Eng. 1, 0082.

Li, J.J., Chu, Y., Lee, B.Y.H., Xie, X.S., 2008. Enzymatic signal amplification of molecular beacons for sensitive DNA detection. Nucleic Acids Res. 36, e36.

Liao, Y., Huang, R., Ma, Z., Wu, Y., Zhou, X., Xing, D., 2014. Target-triggered enzyme-free amplification strategy for sensitive detection of microRNA in tumor cells and tissues. Anal. Chem. 86, 4596–4604.

Liu, D., Yang, J., Wang, H.F., Wang, Z., Huang, X., Wang, Z., et al., 2014. Glucose oxidase-catalyzed growth of gold nanoparticles enables quantitative detection of attomolar cancer biomarkers. Anal. Chem. 86, 5800–5806.

Manea, F., Houillon, F.B., Pasquato, L., Scrimin, P., 2004. Nanozymes: gold-nanoparticle-based transphosphorylation catalysts. Angew Chem. Int. Ed. 43, 6165–6169.

Meselson, M., Yuan, R., 1968. DNA restriction enzyme from *E. coli*. Nature 217, 1110–1114.

Mugoni, V., Medana, C., Santoro, M.M., 2013. 13C-isotope-based protocol for prenyl lipid metabolic analysis in zebrafish embryos. Nat. Protoc. 8, 2337–2347.

Mullis, K.B., Faloona, F.A., 1987. Specific synthesis of DNA in vitro via a polymerase-catalyzed chain reaction. Methods Enzymol. 155, 335–350.

Natalio, F., André, R., Hartog, A.F., Stoll, B., Jochum, K.P., Wever, R., et al., 2012. Vanadium pentoxide nanoparticles mimic vanadium haloperoxidases and thwart biofilm formation. Nat. Nanotechnol. 7, 530–535.

Nie, S., Emory, S.R., 1997. Probing single molecules and single nanoparticles by surface-enhanced Raman scattering. Science 275, 1102–1106.

Nobori, T., Tosaka, K., Kawamura, A., Joichi, T., Kamino, K., Kishimura, A., et al., 2018. Alkaline phosphatase-catalyzed amplification of a fluorescence signal for flow cytometry. Anal. Chem. 90, 1059–1062.

Notomi, T., Okayama, H., Masubuchi, H., Yonekawa, T., Watanabe, K., Amino, N., et al., 2000. Loop-mediated isothermal amplification of DNA. Nucleic Acids Res. 28, E63.

Pourhassan-Moghaddam, M., Rahmati-Yamchi, M., Akbarzadeh, A., Daraee, H., Nejati-Koshki, K., Hanifehpour, Y., et al., 2013. Protein detection through different platforms of immuno-loop-mediated isothermal amplification. Nanoscale Res. Lett. 8, 485.

Saiki, R.K., Scharf, S., Faloona, F., Mullis, K.B., Horn, G.T., Erlich, H.A., et al., 1985. Enzymatic amplification of beta-globin genomic sequences and restriction site analysis for diagnosis of sickle cell anemia. Science 230, 1350–1354.

Sano, T., Smith, C.L., Cantor, C.R., 1992. Immuno-PCR: very sensitive antigen detection by means of specific antibody-DNA conjugates. Science 258, 120–122.

Scrimin, P., Prins, L.J., 2011. Sensing through signal amplification. Chem. Soc. Rev. 40, 4488–4505.

Shagin, D.A., Rebrikov, D.V., Kozhemyako, V.B., Altshuler, I.M., Shcheglov, A.S., 2002. A novel method for SNP detection using a new duplex-specific nuclease from crab hepatopancreas. Genome Res. 12, 1935–1942.

Sheng, Q., Cheng, N., Bai, W., Zheng, J., 2015. Ultrasensitive electrochemical detection of breast cancer cells based on DNA-rolling-circle-amplification-directed enzyme-catalyzed polymerization. Chem. Commun. 51, 2114–2117.

Tang, Y., Lin, Y., Yang, X., Wang, Z., Le, X.C., Li, F., 2015. Universal strategy to engineer catalytic DNA hairpin assemblies for protein analysis. Anal Chem. 87, 8063–8066.

Taniguchi, K., Kajiyama, T., Kambara, H., 2009. Quantitative analysis of gene expression in a single cell by qPCR. Nat. Methods 6, 503–506.

Tarnuzzer, R.W., Colon, J., Patil, S., Seal, S., 2005. Vacancy engineered ceria nanostructures for protection from radiation-induced cellular damage. Nano Lett. 5, 2573–2577.

Travascio, P., Witting, P.K., Mauk, A.G., Sen, D., 2001. The peroxidase activity of a hemin-DNA oligonucleotide complex: free radical damage to specific guanine bases of the DNA. J. Am. Chem. Soc. 123, 1337–1348.

Vincent, M., Xu, Y., Kong, H., 2004. Helicase-dependent isothermal DNA amplification. EMBO Rep. 5, 795–800.

Walker, G.T., Little, M.C., Nadeau, J.G., Shank, D.D., 1992. Isothermal in vitro amplification of DNA by a restriction enzyme/DNA polymerase system. Proc. Natl. Acad. Sci. USA. 89, 392–396.

Wang, Z., Liu, H., Yang, S.H., Wang, T., Liu, C., Cao, Y.C., 2012. Nanoparticle based artificial RNA silencing machinery for antiviral therapy. Proc. Natl. Acad. Sci. USA. 109, 12387–12392.

Wei, H., Wang, E., 2008. Fe_3O_4 magnetic nanoparticles as peroxidase mimetics and their applications in H_2O_2 and glucose detection. Anal. Chem. 80, 2250–2254.

Wei, H., Wang, E., 2013. Nanomaterials with enzyme-like characteristics (nanozymes): next-generation artificial enzymes. Chem. Soc. Rev. 42, 6060–6093.

Willner, I., Shlyahovsky, B., Zayats, M., Willner, B., 2008. DNAzymes for sensing, nanobiotechnology and logic gate applications. Chem. Soc. Rev. 37 (6), 1153–1165.

Wu, H., Zhang, K., Liu, Y., Wang, H., Wu, J., Zhu, F., et al., 2015. Binding-induced and label-free colorimetric method for protein detection based on autonomous assembly of hemin/G-quadruplex DNAzyme amplification strategy. Biosens. Bioelectron. 64, 572–578.

Xie, J., Zhang, X., Wang, H., Zheng, H., Huang, Y., 2012. Analytical and environmental applications of nanoparticles as enzyme mimetics. TrAC-Trends Anal. Chem. 39, 114–129.

Xu, L., Zhu, Y., Ma, W., Kuang, H., Liu, L., Wang, L., et al., 2011. Sensitive and specific DNA detection based on nicking endonuclease-assisted fluorescence resonance energy transfer amplification. J. Phys. Chem. C 115, 16315–16321.

Xu, S.Y., Zhu, Z., Zhang, P., Chan, S.H., Samuelson, J.C., Xiao, J., et al., 2007. Discovery of natural nicking endonucleases Nb.BsrDI and Nb.BtsI and engineering of top-strand nicking variants from BsrDI and BtsI. Nucleic Acids Res. 35, 4608–4618.

Xue, L., Zhou, X., Xing, D., 2012. Sensitive and homogeneous protein detection based on target-triggered aptamer hairpin switch and nicking enzyme assisted fluorescence signal amplification. Anal. Chem. 84, 3507–3513.

Yuan, Y., Yuan, R., Chai, Y., Zhuo, Y., Bai, L., Liao, Y., 2010. An electrochemical enzyme bioaffinity electrode based on biotin-streptavidin conjunction and bienzyme substrate recycling for amplification. Anal. Biochem. 405, 121–126.

Zhang, D.Y., Seelig, G., 2011. Dynamic DNA nanotechnology using strand-displacement reactions. Nat. Chem. 3, 103–113.

Zhang, H., Lai, M., Zuehlke, A., Peng, H., Li, X.F., Le, X.C., 2015. Binding-induced DNA nanomachines triggered by proteins and nucleic acids. Angew. Chem. Int. Ed. 54, 14326–14330.

Zhang, K., Wang, K., Zhu, X., Xie, M., Zhang, X., 2017. A new method for sensitive detection of microphthalmia-associated transcription factor based on "OFF-state" and "ON-state" equilibrium of a well-designed probe and duplex-specific nuclease signal amplification. Biosens. Bioelectron. 87, 299–304.

Zhu, X., Zhao, J., Wu, Y., Shen, Z., Li, G., 2011. Fabrication of a highly sensitive aptasensor for potassium with a nicking endonuclease-assisted signal amplification strategy. Anal. Chem. 83, 4085–4089.

Zhu, X., Feng, C., Zhang, B., Tong, H., Gao, T., Li, G., 2015. A netlike rolling circle nucleic acid amplification technique. Analyst 140, 74–78.

Index

Note: Page numbers followed by "*f*" refer to figures.

D

E

F

G

H

I

K

L

M

FSC
www.fsc.org
MIX
Papier | Fördert gute Waldnutzung
FSC® C083411

Kolibri 360 GmbH
Ferdinand-Jühlke-Straße 7
99095 Erfurt, Deutschland
produktsicherheit@kolibri360.de